TOBACCO IN HISTORY

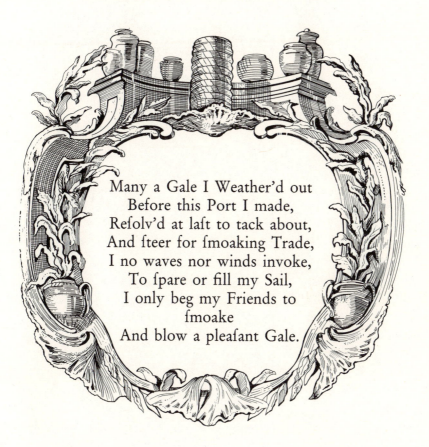

Many a Gale I Weather'd out
Before this Port I made,
Refolv'd at laft to tack about,
And fteer for fmoaking Trade,
I no waves nor winds invoke,
To fpare or fill my Sail,
I only beg my Friends to
fmoake
And blow a pleafant Gale.

TOBACCO IN HISTORY

The cultures of dependence

Jordan Goodman

London and New York

First published 1993
by Routledge
11 New Fetter Lane, London EC4P 4EE

Simultaneously published in the USA and Canada
by Routledge, Inc.
29 West 35th Street, New York, NY 10001

Phototypeset in 10 on 12 point Garamond by Intype, London
Printed in Great Britain by
T. J. Press (Padstow) Ltd, Padstow, Cornwall

Printed on acid free paper

British Library Cataloguing in Publication Data
A catalogue record for this book is available from the British Library

Library of Congress Cataloging in Publication Data
Goodman, Jordan.
Tobacco in history : the cultures of dependence / Jordan
Goodman.
p.   cm.
Includes bibliographical references and index.
1. Tobacco–History. 2. Tobacco–Social aspects.
3. Smoking–History. I. Title.
GT3020.G66 1993
394.1'4–dc20   92–43291

ISBN 0–415–04963–6

For my parents

CONTENTS

Part IV

Conclusion

FIGURE AND TABLES

FIGURE

TABLES

ACKNOWLEDGEMENTS

This book could not have been written without the kind assistance of librarians and archivists in a number of institutions. I would like to thank the staff of the following: the British Library, the Guildhall Library, the Wellcome Institute for the History of Medicine, the Museum of Mankind Library, the Library of the Royal Botanic Gardens, Kew, ASH (Action on Smoking and Health), the New York Public Library and the Arquivo Histórico Ultramarino in Lisbon. Thanks are also due to the Inter-Library Loan sections of the Albert Sloman Library, University of Essex and the Joule Library, University of Manchester Institute of Science and Technology for providing me with obscure books and journals.

Sections of this book have been given as individual papers at seminars and conferences. I would like to thank the participants at the University of Hull, the University of Humberside, the University of Manchester, the University of London and the University of Kent for the helpful comments and suggestions.

Several scholars provided me with much valuable unpublished material and bibliographical information: I would like to thank Ingrid Waldron, Woodruff Smith, Mac Marshall, Alexander von Gernet and Cathy Crawford for their help. The manuscript was read in full or in parts by Alexander von Gernet, Peter Earle, Nigel Bartlett and an anonymous reader at Routledge. To all of them I would like to extend my deepest thanks for their criticisms and suggestions and for their time. Nigel Bartlett was also responsible for collating the quotations and selecting the cover illustrations. The editorial staff at Routledge, especially Claire L'Enfant and Louise Snell, have been closely involved in this project from the beginning and I thank them warmly for their patience and perseverance.

But my final and most heart-felt thanks are reserved for Dallas Sealy who has seen me through this book, has read and re-read the whole manuscript and is the best critic anyone can have.

ABBREVIATIONS

AHU	Arquivo Histórico Ultramarino
BAT	British American Tobacco
BM Add. Mss	British Museum Additional Manuscripts
CSVP	*Calendar of State Papers Venetian*
FAO	Food and Agriculture Organization
RCP	Royal College of Physicians
RJR	R. J. Reynolds
UN	United Nations
UNCTAD	United Nations Conference on Trade and Development
USBC	United States Bureau of the Census
USDA	United States Department of Agriculture
USDC	United States Department of Commerce
USDHEW	United States Department of Health, Education and Welfare
USDHHS	United States Department of Health and Human Services

Introduction

It is now proved beyond doubt that smoking is one of the leading causes of statistics.

Fletcher Krebel Reader's Digest (December 1961)

Cigarettes just lie there in their packs
waiting until you call on one of them to help you relax
They aren't moody; they don't go in for sexual harassment and
    threats,
or worry about their performance as compared with that of other
    cigarettes,
nor do they keep you awake all night telling you of their life,
beginning with their mother and going on until morning about their
    first wife.

Fleur Adcock 'Smokers for celibacy', in Time-Zones (Oxford:
Oxford University Press, 1991), pp. 36–7

1

WHAT IS TOBACCO?
The botany, chemistry and economics of a strange plant

The origins of the tobacco plant are lost. Its history starts around eight thousand years ago, when two species of the plant, *Nicotiana rustica* and *Nicotiana tabacum*, were dispersed by Amerindians through both the southern as well as the northern American continent (Wilbert 1991: 179). Modern commercial tobacco is descended directly from the latter species. Until the very end of the fifteenth century no one outside the American continents had any knowledge of the cultivated varieties of this plant. Today it is grown in more than 120 countries, and its manufactured products are known to virtually everyone.

What is tobacco? An answer requires an analysis in several key areas. Tobacco exists in four principal dimensions: botany, chemistry and pharmacology, economics – production and consumption – and history. The last dimension is the main subject of this book, and the first three of this introductory chapter.

The tobacco plant is of the genus *Nicotiana*, one of the larger divisions of the family Solanaceae, otherwise known as nightshades. The nightshade family is one of the largest in the natural world and includes, among other plants, the potato, the pepper and, of course, the deadly nightshade (Heiser 1969). There are sixty species in the genus *Nicotiana* alone, 60 per cent of which are native to South America, 25 per cent to Australia and the South Pacific, and 15 per cent to North America (Goodspeed 1954: 8). According to Thomas Goodspeed, the origin of the genus lies in the South American continent from where it was dispersed to all other continents, Australia included. Most authorities in the field are in broad agreement with Goodspeed, though some dispute his interpretation of the inter-continental transfer of the genus (Feinhandler, Fleming and Monahon 1979). There is further agreement that of all the species in existence, only two, *tabacum* and *rustica*, have been cultivated, and it was these two that generally supplanted the wild species, in the Americas at least. By the time Europeans first sighted the New World, and long before then, *Nicotiana tabacum* was cultivated primarily in the tropical regions, while *Nicotiana rustica* could be found in many more areas, including the eastern woodlands,

Mexico, Brazil and at the extremes of agricultural activity in Chile and Canada (Wilbert 1987: 6).

Both tobacco species are annuals. *Tabacum* is a large plant between 1 and 3 metres high with large leaves; *rustica* is shrubby in comparison to *tabacum*, ranging in height from 0.5 to 1.5 metres, and produces small and fleshy leaves. *Rustica* is now the minor *subgenus*, being confined principally to only a few parts of the world – the former USSR, India, Pakistan and parts of North Africa (Akehurst 1981: 34).

Tobacco is grown from seed, microscopic in size – a one ounce sample may contain as many as three hundred thousand seeds (Akehurst 1981: 48). Wherever tobacco is cultivated, the crop needs to go through certain stages before it is ready for market. There are variations but the general pattern is as follows. Since the seed is minute and the seedlings produced very fragile, they need to be raised in seedbeds before being planted in the field. Once on their own, the growing plants are generally, though not always, topped and suckered – the flowers are removed as they appear, as are the suckers that grow subsequently. At maturity the plants are harvested either by priming the leaves from the stalk or by simply cutting the plant at the stalk. Curing is the next and most distinctive stage. The basic operation involves nothing more than drying the leaves or the entire stalk to reduce the moisture content and force the leaf chemistry to produce characteristic qualities and aroma. This can be done in one of several ways under different environmental conditions: in the open air and in shade, termed air-curing; in the open air but in full sunshine, termed sun-curing; in a barn with an open smoky fire, usually of wood, termed fire-curing; and, finally, in a barn with dry heat provided by flues running through the space, termed flue-curing (Akehurst 1981: 29–39).

Tobacco (except for Oriental tobacco) is now designated in two principal ways: it is classed as dark or light tobacco, according to its method of curing. Until the middle of the nineteenth century all of the world's cultivated tobacco was air-, sun- or fire-cured and dark. Light, flue-cured tobacco became of importance only around the turn of the twentieth century but it now accounts for the bulk of the world's output.

The tobacco plant has a general composition which can be found in most other plants. The chemistry of the leaf is straightforward: around 90 per cent is water and the rest is made up of mineral matter and organic compounds (Akehurst 1981: 522). Nitrogen is the most important element and the organic compounds the most important chemicals. The proportional representation of the chemical components of tobacco varies considerably according to the type and curing method used, as well as to the region where the tobacco is cultivated (Akehurst 1981: 578–604).

Nicotine is the most important nitrogenous compound in tobacco and in the smoke. It is an alkaloid, a plant substance of basic reaction, which produces physiological changes in the body. There are other alkaloids present in tobacco such as nornicotine and anabasine, but nicotine is the primary alkaloid in both commercial varieties of tobacco, *tabacum* and *rustica*: these two varieties, importantly, have higher concentrations of nicotine than do any of the wild species (Wilbert 1987: 134–6; Akehurst 1981: 543).

Tobacco smoke is chemically complex and is usually analysed in two parts, the particulate or solid and the gaseous phase. Some 4,720 separate compounds have already been identified in the smoke (Ginzel 1990: 430). The gaseous phase contains many chemicals that are well known: carbon monoxide (5 per cent), carbon dioxide, nitrogen oxide, ammonia, formaldehyde, benzene and hydrogen cyanide; the particulate phase includes nicotine, phenol, naphthalene and cadmium among other compounds (Davis 1987: 20). The compounds in the particulate phase, excepting nicotine, are collectively called tar. The higher the nicotine yield, the higher the tar yield and vice versa (Ashton and Stepney 1982: 29). In the particulate phase, the free nicotine is 'suspended on minute droplets of tar . . . less than one thousandth of a millimetre across . . .' (Ashton and Stepney 1982: 29). Tobacco smoke can also be further categorized into mainstream and sidestream smoke. About one half of the volume of the smoke is accounted for by each type. Mainstream smoke is drawn by the smoker down through the length of the cigarette and as it travels its temperature falls dramatically until it is comfortable to inhale. Sidestream smoke escapes as the cigarette burns and both the smoker and those present will inhale this smoke. As sidestream smoke is not diluted by passing through the cigarette or filter, the concentrations of chemicals in the smoke are much greater than in mainstream smoke, more than a hundred times for certain chemicals (Akehurst 1981: 642, 645–6; Ginzel 1990: 433). When tobacco is not smoked, nicotine is still present but in a water-soluble salt.

There are two facts about nicotine which are now irrefutable but which, until recently, were not confirmed. They are: first, that people consume tobacco in whatever form in order to administer nicotine to themselves; and second, that nicotine is highly addictive, in the sense that 'tobacco use is regular and compulsive, and a withdrawal syndrome usually accompanies tobacco abstinence' (West and Grunberg 1991: 486). Because cigarette smoke is acidic, the nicotine in cigarette smoke can be absorbed only by inhaling it into the lungs: the nicotine in both cigar and pipe tobacco smoke, being alkaline, can also be absorbed through the buccal mucosa, the membrane lining the mouth (Russell 1987: 29). Whether tobacco smoke is acidic or alkaline depends partly on curing methods and partly on the different strains of tobacco used (Akehurst 1981: 578–604, 647, 649).

Cigarette smokers who inhale absorb 92 per cent of the nicotine available in the smoke. What happens then is graphically described in the following account:

The modern cigarette is a highly effective device for getting nicotine into the brain. The smoke is mild enough to be inhaled deeply into the alveoli of the lungs from where it is rapidly absorbed. It takes about 7 seconds for nicotine absorbed through the lungs to reach the brain compared to the 14 seconds it takes for blood from arm to brain after an intravenous injection. Thus, after each inhaled puff, the smoker gets an intravenous-like 'shot' or bolus of blood containing a high concentration of nicotine which reaches the brain more rapidly than from an intravenous injection. The uptake of nicotine by the brain is also extremely rapid.

(Russell 1987: 26)

Within 15 or 20 seconds nicotine has reached every part of the body. Nicotine absorption by pipe and cigar smoking, without inhalation, is slower and less intense. Research has shown that confirmed pipe and cigar smokers are satisfied with this pattern of nicotine absorption, but when cigarette smokers switch to these alternative methods they invariably inhale the smoke in an attempt to replicate the pharmacological experience they had as cigarette smokers (Russell 1987: 29–30). Nasal or dry snuff, by contrast, offers the tobacco consumer as efficient an absorption of nicotine as cigarette smoke inhalation whereas the use of oral or wet snuff is akin to that of pipe and cigar smoking and chewing tobacco (Russell 1987: 31–2).

Nicotine is a powerful and complex drug. It reacts with excitable cells in many parts of the body and brain. One of the reasons for this is that nicotine is structurally similar to acetylcholine, a vital neurotransmitter, which acts to bridge the synaptic gap between nerve endings. Because it is structurally similar to acetylcholine, nicotine can unlock and combine with acetylcholine receptors throughout the body (Ashton and Stepney 1982: 37–8). The effect of nicotine is biphasic in that different dosage levels have differential impacts: a small dose produces a stimulant effect while a large dose acts as a depressant; an overdose blocks neurotransmission altogether leading to instant death (Ashton and Stepney 1982: 38–9). Besides its interaction and relationship with acetylcholine, nicotine has also been shown to release many other types of nerve transmitters, including norepinephrine, epinephrine, serotonin and dopamine, some of which have been shown to be related to hallucinogens (Martin 1987: 3; Wilbert 1991: 185). All of these chemical changes in the body result in physiological and psychological changes including changes in blood pressure and pulse rate; increasing and decreasing respiration; decreasing skin temperature; producing feelings of well-being, arousal, alertness and many others (Martin 1987: 2–3; USDHHS 1988). Nicotine seems to act in such a critical way in the body that there is more than a suspicion that it acts to release primary drives similar to hunger pangs (West and Grunberg 1991: 488).

6

Tobacco smoke also contains other, possibly mind-altering drugs (Janiger and Dobkin de Rios 1976; Siegel *et al.* 1977: 18). Unfortunately, it also includes many compounds that have been implicated as carcinogenic and disease-related. There are at least fifty such compounds, including cadmium, arsenic and formaldehyde (Davis 1987: 20; Ginzel 1990: 432–3).

In contrast to its chemical complexity, especially when burned, tobacco is comparatively simple to grow under differing climatic and soil conditions. The tobacco plant is prodigious in leaf growth at the same time as being economical on field space. Plant populations can range from 15,000 to 25,000 per hectare: a single plant can easily produce over 2 square metres of usable leaf (Akehurst 1981: 3). These characteristics alone suggest the vast economic potential of the tobacco plant.

In global terms tobacco is generally considered the most widely grown non-food crop though, in terms of overall area devoted to it, the tobacco crop is not that important, accounting only for about 0.3 per cent of cultivated land – this can usefully be compared to the proportion for grain, average 13 per cent; cotton, 2 per cent; and coffee, 0.7 per cent (FAO 1989: 1). In many countries, however, the proportion is much larger: in Malawi, for example, it is 4.3 per cent and in China it is over 1 per cent (FAO 1989: 1).

The tobacco plant is of enormous economic importance to many countries of the world, both developed and developing. Table 1.1 shows the distribution of the world's crop according to information available for 1990. Asia's enormous share of the world's tobacco crop is one of the most significant aspects of global tobacco cultivation. The grouping by regions in this fashion does, however, obscure the fact that production in individual countries of the developed world is considerable. This fact is revealed more clearly in Table 1.2, which shows the distribution of global production by the seven largest national producers. The position of the United States in the ranking of national producers is not surprising but that of China is important to note in the context of the historical discussion that follows in the succeeding chapters. One other important feature of

Table 1.1 World tobacco crop 1990 (000 metric tons)

	Production	*Share (%)*
North America	911	13
Europe	943	14
Asia	4,235	60
Africa	370	5
South America	575	8
Oceania	15	–
Other	6	–
World	7,055	

Source: *Tobacco Journal International*, May/June 1991: 61–3

Table 1.2 World tobacco crop 1990, percentage of world production, seven leading countries (000 metric tons)

	Production	*Share (%)*
China	2,692	38
United States	722	10
India	490	7
Brazil	435	6
Turkey	252	4
Former USSR	225	3
Italy	205	3

Source: *Tobacco Journal International*, May/June 1991: 61–3

contemporary tobacco cultivation is not revealed in Table 1.2. Most people do not associate tobacco growing with Europe but within the European Community it is extremely important. In 1990 the total production of the EC stood at 419,000 metric tons placing it in fifth position in the world's league table, but only marginally behind India and Brazil.

Most of the world's tobacco crop – estimated at 85 per cent of the total – ends up in cigarettes (FAO 1989: 6). To this end, therefore, most of the world's production of tobacco leaf is of the type suited for this purpose, that is, light air- and flue-cured tobacco (FAO 1989: 4; Chapman and Wong 1990: 30). This is a trend which has been in evidence for some time and is, according to most authorities on the subject, likely to continue into the future.

This prodigious output has many effects on the economy of each of the countries where tobacco is cultivated. One of the most obvious and most direct effects is on the demand for agricultural labour. It is difficult, and in some places almost impossible, to provide reliable information on labour, not only because of under-reporting but also because of the highly seasonal nature of the demand for labour and the fact that many farming families raise other crops in addition to tobacco. Nevertheless, while the degree of accuracy of the figures can be debated, the order of magnitude is clear enough. Recent figures on the numbers employed on the land in growing tobacco are shown in Table 1.3. China employs more people in tobacco cultivation than does any other country, about sixteen million people, according to recent estimates (FAO 1989: 6). In relative terms there is a great variation in the importance of tobacco growing in the demand for labour. In China, for example, where a large proportion of overall employment is on the land, tobacco growing occupies about 2 per cent of total agricultural employment: in Zimbabwe the comparable figure is 15 per cent, but in Greece and Italy it is 35 per cent and 17 per cent respectively (FAO 1989: 6–7; PIEDA 1992: 17). All of the figures are given in total numbers employed without regard for the nature of the work, whether full- or part-time, seasonal or annual. Using full-time equivalents as the

Table 1.3 Employment in tobacco growing 1987

	Number (000)
Africa	740
Asia	24,700
Latin America	1,500
European Community	800
United States	500
Former USSR	670

Sources: Chapman and Wong 1990: 50–1; FAO 1989: 7;
USDHHS 1992: 120

measure of labour force participation, the Greek figure, for example, would fall to around 10 per cent, which is still substantial enough (Joosens and Raw 1991: 1193).

In what was one of the most comprehensive analyses of its kind so far undertaken, two independent organizations in 1987 reported on the nature of the world's tobacco culture. According to this report, in 1983 the full-time labour demand for tobacco production, from growing to distribution, was 18.2 million people worldwide: adding in a proportion of labour from supply industries and relaxing the tight definition of labour demand, so that family members, part-time and seasonal workers are counted in full, the authors of the report estimated that tobacco was responsible for the livelihood of at least a hundred million people (Chapman and Wong 1990: 49; Warner 1990: 82).

In many countries of the world tobacco contributes significantly to agricultural incomes, being near the top of a league table in many places. Tobacco is particularly important in China, Zimbabwe, Malawi and Greece: available figures show that tobacco accounts for between 10 per cent and 25 per cent of total agricultural income in the last three countries (FAO 1989: 8). Even where the relative value is not as large as in these countries, tobacco still holds an important position in overall agricultural activities. In Japan tobacco ranks in fourth place of all crops; in Canada it is in fifth place; in the United States and Korea it is in eighth position (FAO 1989: 7–8).

There are many reasons why tobacco figures so importantly in the economy of so many countries, both developed and developing, but one of the most important and certainly most obvious reasons is that the return of tobacco per hectare of land is both absolutely and relatively high. In the mid-1980s, for example, the gross returns per hectare from tobacco in Zimbabwe were almost twice those of coffee, the next most profitable crop, and ten times more profitable than food crops: in Brazil, India and the United States tobacco is also the most profitable crop (FAO 1989: 8–9). The relative profitability of tobacco growing is largely accounted for by a series of factors including price supports, guaranteed prices, loans from governments and tobacco companies, provision of seed, fertilizer and other agricultural inputs as well as export subsidies (FAO 1989: 9).

As stated earlier, cigarettes account for as much as 85 per cent of the world's output of tobacco leaf. In 1988 over 5 trillion cigarettes were manufactured worldwide: Table 1.4 presents data on the global distribution of cigarette manufacturing for 1988 for the top six producers.

Table 1.4 World cigarette production (billion units)

	Number	Share (%)
China	1,525	29
United States	695	13
European Community	631	12
Former USSR	393	8
Japan	268	5
Brazil	158	3

Sources: Grise 1990: 22; FAO 1990: 15

While a national distribution for cigarette manufacturing activities, as presented in Table 1.4, is revealing, it is important to understand that the world's cigarette market is supplied predominantly by two main kinds of manufacturing enterprises: state monopolies and multinational tobacco companies. In 1988 eight multinationals accounted for 35 per cent of the world's cigarette output, and the state monopolies for 60 per cent (USDHHS 1992: 38). With few exceptions, the French state tobacco monopoly SEITA being the most important, state monopolies tend not to produce for export.

Less than 10 per cent of the world's output of cigarettes is exported, the United States having the largest share of this trade (Grise 1990: 22–3). It is the practice of multinationals to manufacture cigarettes for domestic consumption and, to this end, they have subsidiaries, affiliated manufacturing firms or licensing agreements throughout the world. There is not a single country outside the state monopoly system where a multinational tobacco company is not represented in some form or other. In recent years, these multinationals have made significant inroads into markets protected by state monopolies, either by exporting to them or, as in the case of China, by opening manufacturing facilities (Connolly 1990).

There are eight multinational tobacco companies, five of which are American and the remaining European. In terms of financial activity – sales and profits – as well as output, the largest multinationals by a wide margin over their competitors are Philip Morris and British American Tobacco (BAT). Most of the multinationals have a diversified base and the amount of sales and profits derived from tobacco products varies widely. In 1991, for example, Rothmans International derived almost 90 per cent of its sales from tobacco: BAT, by contrast, in 1990, earned 57 per cent of its sales from tobacco (Rothmans International 1991; BAT 1990). The profile of the six leading multinationals, giving their overall sales – tobacco and non-tobacco activities – and cigarette output in 1989 and 1988 respectively is portrayed in Table 1.5. Tobacco activities probably account for 60 per cent of overall sales (Connolly 1990: 143; RJR 1987). Based on 1988 figures, Philip Morris was the fourteenth largest company of any kind in the world and BAT the thirty-sixth (USDHHS 1992: 36).

Table 1.5 Multinational tobacco companies, economic activity and cigarette output

	Sales 1989 (US$ million)	Output 1988 (billion)
Philip Morris	39,069	555
BAT	23,529	575
R. J. Reynolds	15,224	285
Imperial Tobacco	9,900	43
American Brands	7,265	90
Rothmans International	2,210	220
Total	97,197	1,768

Source: USDHHS 1992: 36, 38

Cigarette manufacture is obviously very big business. So is central government revenue from taxation on tobacco products. The amount collected is in some cases extremely large. In 1983, for example, the British treasury collected in excess of $8 billion, the German government around $7 billion: tobacco tax revenue in the United States in 1986 amounted to $9.4 billion (FAO 1989: 11–12). For developing countries with a small tax base, tobacco tax revenues are, in relative terms, of critical importance. In many developing countries tobacco tax revenue accounts for at least 10 per cent of all central government tax revenue – in Haiti the figure for 1983 was 41 per cent and in Argentina 23 per cent (Chapman and Wong 1990: 53). But even in the developed world the proportional amount of tobacco taxation in overall taxation is quite large – 6 per cent in Britain, for example (FAO 1989: 12).

There is hardly any place in the world where tobacco is not consumed. The extent, degree and type of consumption does, however, vary widely. In 1985 per capita adult consumption of tobacco varied from under 0.5 kilos in parts of Africa to a maximum of 4.3 kilos in Cuba (FAO: 1990: 52–3). With some exceptions, per capita consumption in the developed world is substantially greater than in the developing world, though the trend in consumption now is generally up in the latter and down in the former.

The factors affecting the divergent experience of the developing and developed world are many, but of particular importance is the impact of anti-smoking activities and legislation in the latter, and a decisive shift towards the consumption of cigarettes as opposed to other forms of tobacco consumption in the former (Chapman and Wong 1990: 23). While few developing countries have experienced a decline in cigarette consumption per capita, it has been the norm in much of the developed world since the 1970s with some interesting exceptions: Japan, Greece, Spain, Iceland and Korea have all seen their consumption rise, in several cases by a substantial amount (Masironi 1990: 269). Table 1.6 presents data on per capita adult cigarette consumption for selected countries in the period 1985–8. The variations in per capita cigarette consumption are enormous. (Here and in similar statistics later, the figures are per head of total adult population, not per smoker.) The correlation is by no means perfect but there is a relationship between a country's wealth and its consumption of cigarettes. In rough terms, those countries with a high level of GNP per capita tend towards a high consumption of cigarettes, while the reverse is true for those countries with low levels of GNP per capita, but the actual picture is complicated by the involvement of many factors other than wealth in the determination of cigarette consumption (Chapman and Wong 1990: 24–5).

Table 1.6 Annual cigarette consumption per adult 1985–8

	Number
Cyprus	4,050
Cuba	3,920
Poland	3,300
Japan	3,270
United States	2,910
Canada	2,700
United Kingdom	2,120
Sweden	1,660
China	1,590
India	160

Source: Masironi 1990: 268

It is in Asia, in particular, that the modern commercial cigarette remains less important than in other parts of the world for reasons having little to do with strictly economic factors. In India, for example, the cigarette competes badly with traditional forms of tobacco consumption: in smoking, alternatives include the *bidi* (a cross between a small cigar and a cigarette made of locally produced dark tobacco, particularly popular in southern India), the cheroot and the hookah; and in chewing, there is tobacco alone or in a mixture with betel; snuff is also popular (Chapman and Wong 1990: 145). *Bidis*, in particular, outsell cigarettes: recent estimates suggest that the sales ratio between cigarettes and *bidis* is 1 to 7

(Chapman and Wong 1990). In Indonesia, despite a marked rise in the adult per capita consumption of cigarettes, the locally produced alternative, the *kretek*, commands the market for smoking. *Kreteks*, consisting of a mixture of tobacco and cloves, accounted for as much as 87 per cent of per capita consumption in the mid-1980s (Chapman and Wong 1990: 151).

Inasmuch as tobacco consumption varies considerably in type and extent, it also varies in degree, that is, in its prevalence within the population. Here again, there is a significant difference between the developed and the developing world. In general terms, a larger proportion of the male population consumes tobacco, especially through smoking, in developing countries. In the 1980s the developed countries of the world had an average prevalence for men of 40 per cent and for women of 27 per cent – the highest and lowest figure for men was 61 per cent in Greece and 27 per cent in Sweden: for women the corresponding figures were 42 per cent in Norway and 14 per cent in Portugal (Masironi 1990: 270). In the developing countries prevalence for men is much greater, and for women much less, than in the developed world. Scattered figures for the 1980s for selected countries give the following picture: Bangladesh, men 70 per cent, women 20 per cent; India, men 61 per cent, women, 7 per cent; Indonesia, men 61 per cent, women 5 per cent; Brazil, men 58 per cent, women 42 per cent; Argentina, men 68 per cent, women 36 per cent (Hendee and Kellie 1990: 874–5; Chapman and Wong 1990: 202, 206). Gender differences in some countries, such as Indonesia and other Islamic nations, are accounted for by proscriptions against women smoking for religious reasons, but, on a general level, it is argued that when constraints against female smoking are removed, prevalence rises near the point of convergence with male rates (Waldron *et al.* 1988).

Smoking prevalence varies not only by country and by gender but also by race, by locality (urban or rural) as well as by social class in general. Though the rule is not hard and fast, there is strong evidence that in developed countries, smoking prevalence is higher among lower social classes and among those with less formal education: in the United States, black Americans tend to have a higher smoking prevalence than do their white counterparts (USDHHS 1989: 269; Wald and Nicolaides-Bouman 1991: 66–7). In developing countries locality matters as well as social class (Chapman and Wong 1990: 87–231; USDHHS 1992).

The botanic, economic and, especially, medical literature on tobacco continues to grow rapidly and is focused primarily, but not exclusively, on the cigarette and its constituents. Authorities from other disciplines, including psychology, sociology, anthropology, politics and law have also become interested in the phenomenon of the cigarette. The main concern of this book is to explain how humankind became involved with the tobacco plant, and how the relationships between it and ourselves have changed over time. As the argument unfolds, it should become clear that

13

nothing about tobacco should be taken for granted and that, as in other matters, the history of tobacco is full of conflict, compromise, coercion and co-operation. It is through this historical process that tobacco has become a universal addiction for consumers, for growers and for governments.

Indeed, this is the overall theme of the book. Dependence unifies the history of tobacco whether seen from the vantage point of the consumer, of the producer or of the institutions concerned with its promulgation. In Amerindian cultures the sacredness of tobacco sustained its use at the same time as being sustained by it. Shamans depended on tobacco's unique pharmacological properties as they themselves became dependent upon it through its addictive powers. Under European control early colonial settlement became dependent upon tobacco and early settlers were addicted both as consumers and as producers to the culture of the plant. Governments, too, have become dependent on the tax revenue they derive by controlling its distribution. Those who grow tobacco describe their attachment to the plant in language more commonly used by those who consume it. This multi-faceted structure of dependence is what makes the history of tobacco fascinating while explaining why it has become so deeply entrenched throughout the world.

The organization of this book is simple and corresponds to a thematic division designed to develop the overall theme. Part I is concerned with two fundamental questions. First, what role did tobacco play in Amerindian cultures, and what meanings were given to, and derived from, the use of tobacco in the Americas before contact with Europeans? This is the subject of Chapter 2. The following chapter addresses the other question: how did Europeans react to Amerindian tobacco use, and how did they, and other cultures in their wake, extract tobacco from what was to them an incomprehensible cultural pattern, and incorporate it within their own?

Part II explores the consumption of tobacco in comparative perspective. Chapter 4 examines the structures and patterns of consumption before the emergence of the cigarette and generally before the isolation of nicotine, and its recognition as tobacco's primary pharmacological substance. Chapter 5 focuses primarily on the rise of the cigarette and its changing meanings and fortunes since the second half of the nineteenth century.

Tobacco was one among many exotic plants and substances that Europeans encountered in the New World, but none of the others was so successfully, and so rapidly and permanently, diffused cross-culturally. The worldwide spread of Amerindian tobacco generally preceded the appropriation of the plant, and its method of cultivation, by Europeans. Once, however, its value as a commodity was understood, tobacco rapidly became the plant of early colonization and, through its commercial circuits and cultures of consumption, acted to bind disparate economic regions in common purpose. The culture of tobacco cultivation and its relationship with colonialism is the general theme of Part III. Chapter 6 examines tobacco's transformation from a medicinal, uncommodified panacea into a commodity in the service of colonialism, and Chapter 7 pursues the meaning of tobacco to planters during the same period.

Part IV, the final section of the book, is concerned with two main themes. The first, covered in Chapter 8, is the antagonism in tobacco culture between the worlds of the small and the large producer that has underpinned the globalization of tobacco since the beginning of the nineteenth century, and has produced a class of dependent cultivators. This dependence is a historical process that also involved, and was shaped by, the transformation of tobacco from a commercial to an industrial product. The industrialization of tobacco was accompanied by the emergence and total domination of tobacco manufacture by huge companies and by the increasing involvement of the state in supporting both tobacco production and consumption. This is examined in Chapter 9.

The book ends, in Chapter 10, with an analysis of what many observers have termed a 'smoking epidemic'. Issues of health, ecology and the Third World, and the possible future of tobacco are discussed there.

This book should be read, in the first instance, as a history of tobacco from the past to the present. It is hoped, however, that it can also be read as an interpretation of certain grand themes in history and other disciplines, particularly, colonialism, cultural contact, consumption and its meanings, the growth of big business and dependence, using tobacco as a unifying concept. This is the principal objective of an historical approach which, in recent years, has been gaining in interest (Price 1984b). As an example of commodity history this study of tobacco shares the aspirations of other examples of this genre, such as Salaman on the potato, Mintz on sugar and, most recently, Adshead on salt and Foust on rhubarb, in believing that the study of the historical transformation of key commodities provides a rich perspective on the way we understand the world about us (Salaman 1949; Mintz 1985; Adshead 1992; Foust 1992).

The material upon which this study is based is drawn chiefly from published secondary studies primarily in the fields of history, anthropology, medicine and agriculture. Primary sources, both published and manuscript, have also been used. Though the secondary material provides an enormously rich and varied storehouse of information, it is not comprehensive. This study attempts a coherent and interpretative historical analysis of tobacco, but much fundamental research remains to be done.

Part I

Tobacco, divine, rare, superexcellent tobacco, which goes far beyond all panaceas, potable gold, and philosopher's stones, a sovereign remedy to all diseases . . . But, as it is commonly abused by most men, which take it as tinkers do ale, 'tis a plague, a mischief, a violent purger of goods, limb, health, hellish, devilish, out damned tobacco, the ruin and overthrow of body and soul.

Robert Burton (1577–1640) *Anatomy of Melancholy* II.4.2.2

2

FOOD OF THE SPIRITS
Shamanism, healing and tobacco in Amerindian cultures

Metsé inhaled deeply, and as he finished one cigarette an attending
shaman handed him another lighted one. Metsé inhaled all the smoke,
and soon began to evince considerable physical distress. After about
ten minutes his right leg began to tremble. Later his left arm began
to twitch. He swallowed smoke as well as inhaling it, and soon was
groaning in pain. His respiration became labored, and he groaned
with every exhalation. By this time the smoke in his stomach was
causing him to retch . . . The more he inhaled the more nervous he
became . . . He took another cigarette and continued to inhale until
he was near to collapse . . . Suddenly he 'died', flinging his arms
outward and straightening his legs stiffly . . . He remained in this
state of collapse nearly fifteen minutes . . . When Metsé had revived
himself two attendant shamans rubbed his arms. One of the shamans
drew on a cigarette and blew smoke gently on his chest and legs,
especially on places that he indicated by stroking himself.

(Dole 1964: 57–8)

The scene described above was witnessed among the Cuicuru Indians of
central Brazil in the twentieth century. It is a description of the 'death'
and restoration of a tobacco shaman. Metsé was experiencing what seems
to be an ordeal of self-inflicted pain and discomfort while being attended
by other shamans. That, however, is a modern reader's perception. For
the Cuicuru, Metsé is performing a central and ancient tradition, dating
back well before Europe discovered the New World. He is passing through
a hallucinatory experience accompanied by specific physical changes. Both
the experience and the physical changes are sought by the shaman, and
their meanings are clear to the Cuicuru. What is perplexing about the
scene, however, is that the hallucination is induced by tobacco. Why
tobacco? To answer that we need to begin by considering the meaning
and importance of hallucinogenic plants and altered states of consciousness
to Amerindian societies before contact with Europe.

The evidence on the Amerindian use of hallucinogenic plants and tobacco

19

on which the following account is based is drawn from a variety of sources, some historical, some ethnographic, some archaeological. The problems of marrying these sources and extracting from them an account that corresponds to Amerindian, as opposed to perceived European, reality is, of course, extremely difficult. These issues have been discussed by many in the field but there seems to be no perfectly satisfactory way of achieving the desired results (von Gernet 1988; 1992; Trigger 1991b). Dating Amerindian practices on the basis of the available evidence is also fraught with difficulties, but an attempt has been made to describe these practices as they might have existed on the eve of contact. This has required a degree of back projection, especially when using ethnographic information, but, with all its faults, it is a technique that achieves results, often confirmed by other sources (Wilbert 1987; Trigger 1991a).

Amerindian societies knew and employed as many as seven or eight times more narcotic plants than corresponding societies in the Old World (Schleiffer 1979: 1). According to Robert Schultes, a long-time student of hallucinogenic plants, the New World has as many as 130 separate plants that could be classified as hallucinogenic. The best endowed regions are to be found in South America and Mexico, though there is growing evidence that the United States and Canada were endowed with more hallucinogenic plants than earlier believed (Schultes and Hofmann 1979: 27, 29; von Gernet 1992: 8).

Our understanding of Amerindian hallucinogenic use remains incomplete, partly because many of the narcotic mixtures have not been thoroughly investigated botanically, and partly because the concoctions themselves are often complicated mixtures of various hallucinogenic and non-hallucinogenic plants. Nevertheless, it is possible to pick out major groups or families of plants that provide a considerable proportion of the hallucinogenic function (Schultes 1972). Although the age and importance of any particular hallucinogenic plant is still an open question, most authorities on the subject would agree to the following propositions: a very large number of New World hallucinogenic plants have been in continual use since the earliest peopling of the Americas: that the use of these plants has been so widespread that it is possible to speak about cultural networks of shared hallucinatory experiences: and that the hallucinatory experience itself was paramount in Amerindian life and played a critical role in its functioning (La Barre 1964; 1970; Dobkin de Rios 1984a; von Gernet 1992).

The nightshade family (Solanaceae) is possibly the main source of New World hallucinogenic plants. One of the most violent and certainly widespread hallucinatory experiences is derived from the datura plant. Most North American Indian tribes used datura to a greater or lesser extent (Schultes 1972: 46). Eastern woodland Indians often used datura as the base of a narcotic drink used in manhood initiation rites. These rites, during which adolescents passed through a prolonged crazed condition, were designed to transform boys into men (La Barre 1980: 76). The violent, mind-altering nature of the cultural rite and the powerful toxicity of the datura brew are, of course, related. To put it another way, datura was chosen as the drug for this rite because of the effects it had on the individual. Datura and its cultural manifestation were thus inseparable in both effect and meaning. Datura was also used in the North American south-west, California and in north-western Mexico (La Barre 1980: 76). In South America datura use was widely distributed throughout Colombia, Ecuador, Peru, Bolivia, Chile and in certain parts of the Amazon (La Barre 1980: 77; Dobkin de Rios 1984a: 38, 41).

The toxicity of datura was well understood by Amerindians. They used datura in specific cultural rites precisely because the violent hallucinations it produced were deemed necessary for the ritual, but they took great care in its use. There is no evidence that datura and other hallucinogenic plants were used casually (von Gernet 1992: 11–13). Where less violent experiences were required, Amerindians used less powerful hallucinogenic plants. Many non-dangerous narcotic snuffs in South America and the Antilles, for example, were prepared from plants belonging to the legume family (Schultes 1972: 24–8). Included in this family is the mescal or red bean used extensively as a narcotic in north-east Mexico as well as among some of the largest Indian tribes of the North American central and south-west, including the Apache, Comanche, Pawnee, Iowa and Wichita (La Barre 1980: 78). The antiquity and widespread nature of the use of the mescal bean have led some authorities to describe its consumption in cult terms and to identify, as in examples above, great social networks of shared hallucinatory experiences (La Barre 1957; Howard 1957). Other plant families that provide New World narcotics include the agaric family – the divine mushroom; the cactus family – peyote; the Ilex family and the family of Malpighiaceae (Schultes 1972: 7–17, 33–40; La Barre 1938; Dobkin de Rios 1984a; Hudson 1979; von Gernet 1992: 12–16).

That so many hallucinogenic plants were widely used in the New World provides evidence of the importance of narcotic substances and hallucinatory experiences in Amerindian cultures. Yet one can go further. It seems clear that, even though few Amerindian societies were without some narcotic, there were several important areas of the New World where narcotic use and the availability of hallucinogenic plants were particularly concentrated. Two areas, in particular, were well-endowed with hallucinogens – the Mexican culture area, circumscribed by Nahuatl speakers, and the Colombian area of Chibchan cultures (Schultes 1977; La Barre 1977). A full list would include hundreds of hallucinogenic drugs but very little is known about them. Many of the drugs mentioned in the herbals and chronicles of the Conquest period remain unidentified (Guerra 1967). Many others that are in use today are only now being investigated. And

21

many others have been lost in the destruction of the collective memory of Amerindian cultures.

The abundance and extent of New World hallucinogenic plants and their use has led one eminent anthropologist to speak of a New World narcotic complex. His argument is not, however, confined to plant geography, but embraces the cultural meanings attributed to the main types of hallucinogenic plants. These were regarded as sacred because they had physiological effects characteristic of supernatural powers (La Barre 1970: 77).

The classification of plants in this way derived partly from Amerindian religious ideology. Across the New World details of religious belief varied considerably, but there were certain features that most Amerindians shared. Reality consisted of a natural and a supernatural world. The natural world was continuous, expected and comprehensible; the supernatural was just the opposite (Hultkrantz 1979: 10). The Amerindian reality envisaged a social space in which the supernatural world impinged upon, and was visible within, the natural world. The space was inhabited by both human and supernatural beings. Spirits, inhabitants of the supernatural world, may have resided at the four cardinal points of the sky but they also resided in the natural world. Particularly potent and significant spirits found a home in hallucinogenic plants (Schleiffer 1979: 2). When someone consumed a hallucinogenic substance, they were introjecting the supernatural power within the plant into themselves. What they experienced, and what onlookers witnessed, was interpreted as a flight of the soul to the supernatural world. In other words, not only was narcotic use a method of altering the state of consciousness but, more importantly, it was only in an altered state of consciousness that communication with the spirits of the supernatural world was possible.

The fact that hallucinogenic plants were sacred, and that the hallucination was a spiritual communication, meant that their consumption was strictly regulated. The responsibility of experiencing, and employing, an altered state of consciousness fell to the shaman, the most spiritually gifted vision seeker in Amerindian societies (Hultkrantz 1979: 74–80). The vision quest was fundamental to Amerindian religious experience. Often these visions were sought en masse; sometimes in special groupings, such as medicine societies; and sometimes by individuals on their own. The shaman, however, stood above all other vision seekers. Though he or she (women were frequently shamans) did not monopolize religious experiences (as did the shaman in Siberia), shamans nevertheless dominated religious life (Eliade 1989: 297–336). Being more spiritually adept than common visionaries, the shaman not only travelled extensively through the spirit world but also had access to many more spirits, particularly those helpful to mankind, than anyone else.

While vision seeking was not unique to the shaman, as a healer he/she had no rivals. It is this vision, or, as some scholars have put it, ecstatic performance, that distinguished the shaman from what has been called the medicine-man. This distinction is not simply one of semantics: rather it is cultural and practical. The shaman was the one who mediated between the natural and spiritual world with the aid of ecstatic devices; the medicine-man typically without these (Hultkrantz 1985). Yet in most cases it was to the shaman that patients went (the exception was in the case when the disease could not be connected directly to some accident or misfortune – a broken bone or a wound, for example) (Hultkrantz 1989: 334–5). The reason for this lay principally in the Amerindian belief that illness was caused by supernatural forces. These forces acted on the body through disease to make the subject ill. Generally there were two main causes of disease. The first was intrusion, that is, when the illness was caused by the presence in the body of a foreign spirit or object. In some societies the intrusion was considered as a literal but magical object, injected by a malevolent spirit; in other societies the intrusion was not so much the physical object as its essence (Hultkrantz 1979: 88–9; Silver 1978: 209; Lamphere 1983). The second cause of disease was soul loss. The sufferer's soul was believed to be drawn away, and/or to have wandered off into reaches of the supernatural world, often into the land of the dead. Regardless of its precise cause, illness was the result of an intervention by the supernatural into the natural world. To cure illness, and heal the victim, was to restore the patient to this world. Healing required a deep understanding of the ways of the supernatural.

Naturally enough, because of his visionary experiences and the fact that he was so spiritually acute, only the shaman could be expected to heal. The shaman was required to travel through the contours of the supernatural, locate and extract the magical object or its essence, in the case of disease by intrusion, or retrieve the runaway soul. The shaman's function was positive but not without danger; if, for example, the sufferer's soul had crossed into the land of the dead, the shaman's soul itself might be caught by the inhabitants of that land (Hultkrantz 1979: 89; Eliade 1989: 327–8).

Access to the supernatural world was through an altered state of consciousness, perceived by onlookers as a trance. These trances were induced primarily by ingesting hallucinogenic plants, and, though some writers associate the use of hallucinogenic plants with settled agriculture, there is some evidence that hunter-gatherers also followed the practice (La Barre 1980; Wilbert 1987: 149–50; von Gernet and Timmins 1987: 41–2).

Shamanistic trances, healing and narcotics formed the complex of Amerindian medicine. The entire system was very carefully and precisely handled. In particular, it appears that certain hallucinogenic plants were used for certain purposes, depending on the extent and nature of the shaman's flight as well as on the nature of the cure (Dobkin de Rios 1984a). In much the same way as hallucinogenic plants were differentiated by use in cultural rites previously described in this chapter, they had specific uses in medicine. Datura and ayahuasca, for example, produced very different hallucinatory experiences and contributed to different kinds of healing programmes: the former as a diagnostic tool for prescribing remedies and the latter for identifying ill-doers (Dobkin de Rios 1984b: 38–42). It is also true that shamans made use of whatever hallucinogenic plants were at hand, a shaman in the Brazilian rain forest having access to different plants than his/her counterpart in northern Canada.

However, when one looks more carefully at what plants shamans actually used, one discovers a most remarkable phenomenon: regardless of location, the one plant used more than any other was tobacco. Virtually every Amerindian society knew tobacco. In the pre-Columbian period tobacco consumption was certainly common from Canada's eastern woodlands to southern Argentina; from the Atlantic to the Pacific and stretching up the north-west coast towards the Aleutian Islands (Brooks 1937: 18; Wilbert 1987: 9–132). Wherever it could be grown, it was, and its cultivation was often isolated from that of other crops, especially foodstuffs, in specially designed gardens (Herndon 1967: 296–7; Hurt 1987: 31–3, 47; Russell 1980, 160–4; Linton 1924: 4–6; Heidenreich 1978: 381, 385; Trigger 1986: 159–60). Even among Amerindians who practised no other form of agriculture, tobacco was planted and cared for (Bean and Vane 1978: 667; Linton 1924: 7); the Haida Indians of the Queen Charlotte Islands, off the coast of British Columbia, and the Tlingit Indians further to the north in Alaska, both typified as hunter-gatherers, nevertheless reserved some of their time and precious land for tobacco cultivation (Turner and Taylor 1972: 249). The same was true of the Siuslawans, Coosans and Takelma Indians of the Oregon coast, all hunters and fishers (Zenk 1990: 573; Kendall 1990: 590); and of the Plains Indians, notably the Blackfoot, Crow and Sarci Indians (Lowie 1919; Haberman 1984: 270). The more agricultural societies from Mexico southward all grew tobacco, to a greater or lesser extent, with the exception of the north-eastern coastal region of South America where, under Inca rule, coca cultivation and consumption prevailed (Wilbert 1987: 4, 21, 30–41, 51, 65; Cooper 1963: 525–8).

Most narcotics and stimulants used by Amerindian societies in the New World had very specific, though limited, uses. Datura, for example, was used mainly in magico-religious functions, specifically on shamanistic occasions, and only rarely in other instances (Cooper 1963: 555). Ayahuasca use among the Jivaro of eastern Ecuador, and the Mestizos of the Peruvian Amazon, was concentrated on its hallucinogenic qualities in achieving trance-like states and visions (Dobkin de Rios 1984b). Peyote was employed in much the same way (Stewart 1987). What made tobacco unique among New World plants was that its effects were largely predictable, relatively short-lived and not life-threatening (as datura could be) and thus had a vast functional repertoire. Its uses ranged from the purely symbolic to the medicinal; from its role as a hallucinogen in shamanistic practice and ritual to ceremonial and formal social functions; from profane to religious use; from its identification with myth and the supernatural to the formal ritualism of social experiences. As we shall see, none of the uses was mutually exclusive and, though there is a temptation in some scholarly circles to distinguish tobacco use north and south of the Mexican border, the distinctions are not as clear as some argue (Cooper 1963: 535–6; von Gernet 1992: 17). Tobacco use formed a complex continuum.

Tobacco's main function was to induce hallucinations in shamanistic rituals. It may seem surprising to find tobacco in this role, but it is important to recognize that there are big differences between the way tobacco was used then and now. First of all, it is certain that the species of tobacco used were *Nicotiana rustica* and *Nicotiana tabacum* or varieties of them. *Nicotiana tabacum* was generally used south of Mexico and *Nicotiana rustica* north of that country (Goodspeed 1954). Whatever the species, there is little doubt that the nicotine content was many times greater than that of present-day commercial species and varieties, and that it was capable, by itself, of inducing hallucinations (Haberman 1984; Adams 1990; Wilbert 1987: 134–6; von Gernet 1992: 20–1). There is also some evidence that alkaloids other than nicotine are present in non-commercial varieties. These may be hallucinogenic in their own right, and possibly even more so in combination with high concentrations of nicotine (Janiger and Dobkin de Rios 1976). Finally, there is also some evidence that tobacco was often mixed with other more potent substances (Siegel *et al.* 1977; Wilbert 1987: 27–8, 100–1). Growing evidence leaves little doubt that at the time of contact tobacco was valued primarily for its psychoactive powers, especially since they were mild when compared to other substances (La Barre 1980; Dobkin de Rios 1984b: 37–51; von Gernet 1992).

Tobacco was regarded as having supernatural origins as well as supernatural powers. Myths of tobacco's origins that have been documented make this clear. This is how the Winnebago Indians of southern Lake Michigan explain tobacco in a father's advice to his child:

Earthmaker created the spirits who live above the earth, those who live on the earth, those who live under the earth, and those who live in the water; all these he created and placed in charge of some powers . . . In this fashion he created them and only afterwards did he create us. For that reason we were not put in control of any of these blessings. However, Earthmaker did create a weed and put it in our charge, and he told us that none of the spirits he had created would have the power to take this away from us without giving us something in exchange. Thus said Earthmaker. Even he, Earthmaker, would not have the power of taking this from us without giving up something in return. He told us if we offered him a pipeful of tobacco, if this we poured out for him, he would grant us whatever we asked of him. Now all the spirits come to long for this tobacco as intensely as they longed for anything in creation, and for that reason, if at any time we make our cry to the spirits with tobacco, they will take pity on us and bestow on us the blessings of which Earthmaker placed them in charge. Indeed so it shall be, for thus Earthmaker created it.

(Tooker 1979: 74–5)

It is remarkable how often versions of the Winnebago origin myth crop up throughout the New World. In the mythology of the Pilaga Indians of the Gran Chaco, in Paraguay, for example, tobacco first appeared out of the ashes of a cannibal-woman killed by the culture hero (Wilbert 1987: 151). The Fox Indians, on the western side of Lake Michigan, inherited tobacco from the Great Manitou (Wilbert 1987: 182). As the Manitous were addicted to the plant, and as they could not grow it themselves, they entered into a contract of mutual benefit with humans, tobacco in return for care and protection (Callender 1978: 643). Among the Chippewa of Lake Superior, tobacco was held in a similar supernatural esteem (Ritzenthaler 1978: 754). The Yecuana of Venezuela believe that women were created from clay over which tobacco smoke was blown (Wilbert 1987: 154); among the Yaqui, on the other hand, tobacco came into existence through the metamorphosis of a woman (Moisés et al. 1971: 95).

Offering tobacco to the spirits in exchange for their care and good works was clearly an important way to reinforce and maintain the mutual dependence between humans and supernaturals. Gifts to the spirits often took the form of smoke from tobacco thrown on a fire or leaves left at some sacred spot. Among the north-east woodland Indians, for example, the guardian spirits and patrons were considered to be both the source and the means of material wealth; medicine bundles frequently contained expressions of this material wealth, such as glass beads, together with tobacco (Hamell 1987: 77). Many times upon first contact with Europeans these Amerindians offered tobacco by casting it at the feet of the newcomers; or even, in the extraordinary case of the chief of the Menomini, upon encountering his first white man rubbed the sacred plant into the stranger's forehead (Hamell 1987: 88). For the tobacco shaman, however, the offering took the form of ingested tobacco which, in the first instance, allowed contact to be made with the supernatural world. This happened, of course, through the shaman's own hallucinatory experience as he/she, disembodied, travelled to the spirits. Tobacco smoke symbolized this contact (Hugh-Jones 1979: 231). One such tobacco trip has been recorded among the Tapirapé of the Central Brazilian rain forest.

I smoked much and then I smoked again . . . I travelled singing as I

26

walked. I spent three days walking. I climbed a large mountain on the other side of Araguaya. There it is that the sun comes up ... There were many ... souls of shamans. I did not talk but came back.

(Wagley 1977: 209)

In addition to facilitating the trip, the ingested tobacco was the food of the supernaturals. When the shaman consumed tobacco, he was feeding the spirits within him. The craving for tobacco that the shaman experienced through nicotine addiction was the hunger pangs of the spirits who crave tobacco (Wilbert 1987: 173, 177; von Gernet and Timmins 1987: 40).

The significance of tobacco for the magico-religious reciprocity between the shaman and the spirits lay in two areas. First, the pharmacological properties of nicotine and its manifest symbolic expression were structurally related. Both of these were clearly perceived and exploited by shamans. From the pharmacological viewpoint nicotine's biphasic effects reinforced the shamanistic act. The rapid pharmacological impact of nicotine, manifested in the shaman's physical and mental changes, symbolically translated into the shamanistic trance and flight. The rapid metabolism of nicotine likewise translated itself into the shaman's return to this world both physically and mentally. The actual time between flight and return was shorter for tobacco than for any other narcotic substance available in the New World (Wilbert 1987: 157). Second, ingested tobacco fed the spirits within. Tobacco smoke was the symbol of life-giving energy and carried the supernatural food upwards, to appease the spiritual craving (Wilbert 1979: 29–32).

The extent of tobacco shamanism in the New World is not entirely clear, but the recent and exhaustive study by Johannes Wilbert certainly convinces one of a wide distribution on the South American continent. The Warao of Venezuela, for example, practise tobacco shamanism today as they have done for centuries. It is their only form of shamanism. This is all the more remarkable since the Warao do not themselves grow tobacco but are dependent upon others for their supply of this spiritual food (Wilbert 1972). The extent of tobacco shamanism in North America is less well documented, though recent work suggests a comparatively wide distribution, from the west and south-west to the eastern woodlands (Kroeber 1941: 20; Bean and Vane 1978: 667; Mathews 1976; von Gernet 1992).

As a medical expert the shaman mediated between health and illness. Tobacco was universally upheld as a medicine of unrivalled application and efficacy, and used in all stages of treatment, from diagnosis to remedy. As a diagnostic tool tobacco was particularly revered when intrusion was suspected of causing illness. The shaman would blow smoke over the patient's body in order to locate the bodily dysfunction. The Yuman and Piman shamans of the American south-west used tobacco in this way to 'see' evil substances (Lamphere 1983: 760, 762).

Blowing smoke over the sick body also symbolized the power of the shaman, as this was frequently associated with breath (Métraux 1949: 592; von Gernet 1992: 23). In a more practical sense, tobacco smoke prepared the body for the shaman's surgery:

> The shaman takes deep puffs until about a quarter of the cigar is consumed and then starts to blow large mouthfuls of smoke over the afflicted part of the patient's body. He places his lips close to the skin of his patient and lets the smoke roll out from his mouth so that the smoke will hover over the diseased area for some time. After the smoke has rolled away, the shaman repeats the process until he believes that the patient's skin has been 'softened' sufficiently . . . The time needed to soften the skin of the patient varies greatly according to the disease . . .
>
> (Wilbert 1987: 187–8)

Once the body, or afflicted area, was sufficiently softened, and the precise location of the intrusion ascertained, the shaman would begin to extract the foreign body by sucking it out through a straw, or directly from the patient's skin (Hultkrantz 1979: 88–9). The object, such as a stone, was displayed and then disappeared – the patient recovered (Silver 1978: 209; Smith 1978: 441). Tobacco smoke was also employed in a more direct way. Blown by the shaman, the smoke would penetrate the patient's skin and, depending on the precise nature of the illness, would either drive out the evil essences or appease those spirits in the body who had a particular liking for tobacco (Wilbert 1987: 189–90). Whether the patient experienced physiological changes during such practices is unclear. Girolamo Benzoni, one of the first Europeans to witness tobacco therapy in the New World, certainly thought so. In his *History of the New World*, Benzoni made the following observation:

> In La Española and the other islands, when their doctors wanted to cure a sick man, they went to the place where they were to administer the smoke, and when he was thoroughly intoxicated by it, the cure was mostly effected. On returning to his senses he told a thousand stories, of his having been at the council of the gods and other high visions.
>
> (Benzoni 1857: 82)

This practice would appear to have had wide circulation in the New World with the possible exception of the Aztec, Maya and Inca. Though their medical philosophy clearly allowed for a supernatural cause of illness, their practice did not include sucking at the intrusion (Hultkrantz 1979: 184–285; Guerra 1964; Guerra 1966a, b; Ortiz de Montellano 1989; 1990). Yet their medical philosophy included magical incantations in which tobacco was particularly esteemed. In the *Treatise* of Ruiz de Alarcón, written in 1629, which describes Indian superstitions and practice in Mexico, tobacco played a central role in conjuring and fortune-telling (Andrews and Hassig 1984). Among the Quiché Maya of Highland Guatemala tobacco was used in divination in medical diagnosis (Orellana 1987: 57).

Shamans would also cure patients by directly appealing to the spirits for help by offering them tobacco. In an example of a Winnebago curing ritual the shaman offers tobacco consecutively to the spirits of the earth, almost all of whom inhabit animal bodies, and each of whom has bestowed on the shaman during his trip to their world certain powers of healing. He asks for these powers to aid him in his patient's cure, in return for tobacco offerings. The offering is also extended to the Sun, the Moon, the Earth and to the one called Disease-Giver (Tooker 1979: 96–8).

Once the diagnosis was completed, tobacco was often prescribed as a remedy. As an analgesic tobacco was used widely in Amerindian medicine. Several methods of application were practised. One way to reduce pain was for the shaman to massage the afflicted part of the patient's body with tobacco spit (Wilbert 1987: 190; Barbachano 1982: 38–9). Other methods involved applying wet tobacco leaves, snuff plasters and tobacco juice to the body (Wilbert 1987: 143–4). Toothache was a common source of pain, and here the use of tobacco was particularly important. Among the Quiché Maya the painful tooth was first washed with tobacco juice, after which a wad of tobacco was applied directly on to the tooth (Orellana 1987: 84); most Mayan texts refer to tobacco as a treatment for toothache (Robicsek 1978: 42–3). The Indians of central Mexico used pounded green tobacco leaves together with a few drops of copal (Andrews and Hassig 1984: 172–3). The Iroquois of New York followed a similar course, placing tobacco leaves inside tooth cavities to relieve pain (Vogel 1970: 247). The Tunebo of Colombia and the Campa of Peru both employed tobacco preparations as a remedy against toothache (Wilbert 1987: 189). Earache, like toothache, could be treated with tobacco. Among the Indians of central Mexico, tobacco not only acted as an analgesic in the cure of earache but also performed a magical function. In the incantation for earache, tobacco was used to search for the location of the pain; chest pains received similar treatment and tobacco was ordered by the healer to pursue the pain (Andrews and Hassig 1984: 289, 292). The Cherokee also used tobacco as a pain killer (Tooker 1979: 286). In the treatment of open wounds, snake and insect bites tobacco was used for its alleged antiseptic characteristics (Wilbert 1987: 189). Among the Maya tobacco was widely used in this way (Thompson 1970: 118–19). The Guatemalan Indians squeezed tobacco juice on to the open wound and then placed tobacco leaves which they had chewed up on to the bite (Orellana 1987: 81–2). Aztec medicine also made use of tobacco for bites (Elferink 1984: 55–6).

These were tobacco's most common medicinal uses but there was hardly an ailment for which tobacco was not prescribed somewhere in the New World – asthma, rheumatism, chills, fevers, convulsions, eye sores, intestinal disorders, worms, childbirth pains, headaches, boils, cysts, coughs, catarrh and so on. Few complaints did not respond to tobacco therapy, according to Amerindian beliefs. Tobacco also found use as a preventative (Barbachano 1982: 37–8). The Mazatecs, a Mesoamerican nation on the Pacific coast, believed that a paste mixture of tobacco and lime protected pregnant women from witchcraft (Robicsek 1978: 30). The Aztecs believed that tobacco protected the unborn child, as it protected adults, from poisonous snakes, insects and evil spirits (Robicsek 1978: 30). Tobacco was widely held to alleviate the pains of hunger and thirst, to vitalize and to strengthen (Wilbert 1987: 154, 172–3; Andrews and Hassig 1984: 84–6).

Tobacco as a medicine in Amerindian societies cut across areas of pragmatic, spiritual and magical experiences. There is nothing contradictory in any of this since, as we have seen, medicine, health and illness were deeply embedded within a cosmology including supernatural and natural phenomena. The shaman, medicine-man and high priest were often one and the same.

As the next chapter will show, what impressed Europeans most about tobacco was its use in healing. It was this function that Europeans understood though, of course, in their own terms. Nevertheless, it will be argued, this fact alone meant that tobacco could be incorporated easily into European medical philosophy. For Amerindians, however, the medical role of tobacco could not be separated from any of its other functions since their approach to tobacco was holistic. Not surprisingly, tobacco played a central role in a large body of ceremony and ritual in which the plant's sacredness was displayed and confirmed – as a spiritual offering, it had no equal (Kroeber 1941: 19). To describe all of the various themes and their variations in the Amerindian ceremonial use of tobacco would fill volumes. Rather than attempting an in-depth coverage, what follows should be taken as representing the deep ideational and practical significance of tobacco in Amerindian belief systems.

The ceremonial use of tobacco was based firmly on the idea of the reciprocal gift. Among the south-western Chippewa, for example, the spirits of the supernatural world who inhabited living and non-living things and resided in the earth, sky and water were soothed and honoured by propitiatory offerings of tobacco. All religious and ceremonial occasions began with the ritual smoking of tobacco. The smoke ascended to the spirits who were comforted, as their demands for tobacco were satisfied, and were made aware that the Chippewa were mindful of their presence (Ritzenthaler 1978: 754). The Lacandon Maya offered the first harvest of tobacco to the gods in cigars:

Each is lighted in the new fire or with the aid of a crystal to concentrate the sun's rays. It is momentarily held in front of the mouth of a sacred jar, and then is leaned against the mouth of the god whose head is in relief on the side of the incense burner and who is the recipient of the offering.

(Thompson 1970: 112)

Among the Aztecs tobacco was regarded as an incarnation of the body of the goddess Cihuacoahuatl, and also offered to the war god Huitzilopochtli and to lesser gods (Robicsek 1978: 28). The Tlaxcalans, another Mexican tribe, also offered tobacco to their war god (Robicsek 1978: 29).

Tobacco and divination were inseparable in both North and South America (Robicsek 1978: 30). The Chippewa, for example, would leave a pinch of tobacco on a rock to alert the spirit to ward off a bad storm (Ritzenthaler 1978: 754). The Highland Guatemalans consumed tobacco to learn of future events and 'to consult on the requests and petitions of others with which they had been entrusted' (Orellana 1987: 57). Other uses of tobacco included smoke offerings in rain-making ceremonies but there were many others as well (Mason 1924: 8; Robicsek 1978: 30; Bolton 1987: 151; Kroeber 1941; Springer 1981: 219).

The Maya deified tobacco – many of the deities represented in stone monuments, ceramics and in paintings are depicted as smokers, either heavy or occasional. Mythological animals, too, appear in Mayan depictions as smoking tobacco (Robicsek 1978: 31, 59). Tobacco gourds and pouches were described as symbols of divinity, and were the insignia of the Aztec priesthood, as well as that of women doctors and midwives (Thompson 1970: 111–12, 119). Tobacco's association with sacredness thus permeated gods, spirits and people. The Tzotzil Maya's tobacco world-view shows the interconnections between the supernatural, the environment and mankind:

Tobacco was an anhel, a term used to describe the rain and mountain deity and protector of mankind, because it takes care of our bodies ... There are people who chew moi (ground tobacco) ... every day from the time they get up in the morning ... On hearing thunder, people will bring out their moi and keep it in their cheek and in that way it will not thunder too loudly. When we die the moi defends us.

(Thompson 1970: 116)

As a social offering tobacco also played an important role. Tobacco smoking was already well established among the Aztec upper classes as an after-dinner activity before Europeans first witnessed it. In an early description of the court of Montezuma III the after-dinner scene is remarkably modern:

very handsome women served Montezuma when he was at table ... They also placed on the table three tubes, much painted and gilded,

31

in which they put liquid amber mixed with some herbs which are called tobacco. When Montezuma finished his dinner, the singing and dancing was over and the cloths had been removed, he would inhale the smoke from one of the tubes. He took very little of it and then fell asleep.

(Robicsek 1978: 4)

Fray Bernardino Sahagún, who witnessed the social lives of merchants and nobility, frequently commented on the use of tobacco in ceremonial, as well as social, occasions (Sahagún 1950–69). Among the Chippewa tobacco usually accompanied an invitation to a feast, and when it was offered to a shaman he was obliged to undertake the request of his client (Ritzenthaler 1978: 754).

Tobacco was ceremoniously offered in the harvesting and gathering of food crops (Ritzenthaler 1978: 754). The Hasinais of eastern Texas offered the first cutting of tobacco to bless the harvest and at the same time they blew tobacco smoke to the four winds (Bolton 1987: 161). This link between tobacco and the fertility of the land had parallels in the use of tobacco for enhancing the fertility of game, and of women, a practice common among South American Indians (Wilbert 1987: 153–5). The Plains Indians acknowledged the remarkable power of tobacco and went even further. Neither the Blackfoot nor the Crow cultivated any crops except tobacco. For both tribes tobacco cultivation was attended by an elaborate ceremonial and cultural experience. Among the Crow tobacco culture formed the basis of an institution that not only underlined the centrality of tobacco but also provided Crow society with a structure. Tobacco cultivation was a privilege. The responsibility for cultivation rested in a tobacco society, a substructure of Crow social life. The tobacco society was itself organized into chapters. These chapters were founded on individual revelations, and chapter members shared in the essential truth of these revelations. The revelations conferred the right to cultivate tobacco; each of them was different enough in detail to produce different ceremonies and rituals, as well as different procedures in cultivation. In one detail, however, all the revelations agreed: that tobacco was identified with the stars. Each chapter alleged that tobacco came to the Crow in mythic time when a star was planted. Tobacco was therefore the botanic fusion of heaven and earth; not surprisingly, it was upheld as the distinctive medicine of the Crow (Lowie 1919).

In communicating between the natural and the supernatural world, whether in healing, in divination or in offering, tobacco was critical to the Amerindian concept of the relationship between the individual and the spiritual world. This is how the Iroquois understood tobacco and its magical power in forging this relationship. It can be considered typical of Amerindian tobacco belief:

32

The Iroquois believed that tobacco was given to them as the means of communicating with the spiritual world. By burning tobacco they could send up their petitions with its ascending incense, to the Great Spirit, and render their acknowledgements acceptably for his blessings. Without this instrumentality, the ear of Ha-wen-ne-yu could not be gained. In like manner they returned their thanks at each recurring festival to the Invisible Aids, for their friendly offices, and protecting care.

(Springer 1981: 219)

The power of tobacco was not, however, limited to the individual. Special social functions exploited tobacco's magical qualities. Any agreement or obligation sealed in the presence of tobacco, typically by passing the pipe, made it binding. Amerindian societies, especially of eastern North America, developed this function of tobacco into an elaborate ritual known as the calumet, one of the most important formalized uses of the pipe.

The calumet ceremony consisted primarily in an exchange of political obligations or goods (Springer 1981: 221–7). The ceremony was often accompanied by a dance and singing, but at its core was the ritual pipe or calumet shared among participants. Anthropologists and historians have long been fascinated by the calumet ceremony, and many attempts have been made to explain its origin and diffusion (Turnbaugh 1975; Blakeslee 1981; Brown 1989). Most agree that the calumet was not a unique expression of the power of tobacco but part of a larger smoking complex which diffused throughout eastern North America (von Gernet 1988: 291–302). This complex also included, according to a recent commentator, the Eagle Dance of the Iroquois and the medicine bundles of the Plains Indians. Though the purpose of each ritual was different, what was common to all was the presence of tobacco and, to a lesser extent, the pipe (Springer 1981: 228–9).

What is so striking about tobacco in the New World is the extent of its penetration through both continents, and the way it was so deeply integrated into so many diverse cultures, to the extent that even societies who practised no form of agriculture nevertheless cultivated some tobacco. Despite variations across Amerindian societies it is remarkable how widely diffused were the spiritual meanings and the functions attributed to its power.

By the time Europeans made contact, Amerindians had experimented with every conceivable method of consuming tobacco and had developed the technology necessary for its use. Amerindians practised five principal methods of tobacco consumption: smoking, chewing, drinking, snuff and enemas. Smoking, without any doubt, headed the list, an observation which is perhaps not surprising given the symbolic value of smoke and the fact that smoking is the most efficient and potent way of absorbing nicotine.

It is probably safe to say that, south of the Mexican border, the most common way of smoking was in cigars. Smoking was not a simple act. It could consist either of inhaling or, less commonly, smoke-blowing. Lionel Wafer, a surgeon on an expedition to Panama in the 1680s, left a particularly vivid account of one of these smoke-blowing sessions:

> These Indians have Tobacco among them . . . When 'tis dried and cured they strip it from the Stalks: and laying two or three Leaves upon one another, they roll up all together side-ways into a long Roll, yet leaving a little hollow. Round this they roll other Leaves one after another, in the same manner but close and hard, till the Roll be as big as ones Wrist, and two or three feet in length. Their way of Smoaking when they are in Company together is thus: a Boy lights one end of a Roll and burns it to a Coal, wetting the part next to it to keep it from wasting too fast. The End so lighted he puts into his Mouth, and blows the Smoak through the whole length of the Roll into the Face of every one of the Company or the Council, tho' there be 2 or 300 of them. Then they, sitting in their usual posture upon Forms, make, with their Hands held hollow together, a kind of Funnel round their Mouths and Noses. Into this they receive the Smoak as 'tis blown upon them, snuffing it up greedily and strongly as long as ever they are able to hold their Breath, and seeming to bless themselves, as it were, with the Refreshment it gives them.

> (Wafer 1934: 63)

In South America the cigar also predominated. The length of cigars varied considerably – they could be up to a foot in length and an inch in diameter (Wilbert 1987: 64–121). Pipe smoking was less common in South America, being confined principally to the Gran Chaco in Paraguay, and the Peruvian Amazon (Wilbert 1987: 66, 121–3). The Maya and Aztecs smoked cigars, though pipes were not unknown (Mason 1924: 6–8). Mexican Indians used tubular pipes, and there are examples of elbow pipes from Nicaragua and Costa Rica. Some pipes had two stems, and were used for nasal inhalation (Robicsek 1978: 9). Small cigars were often smoked in the form of crushed tobacco leaves in a reed cover (Mason 1924: 6).

North American Indians, by contrast, were generally pipe smokers (Linton 1924: 8–20). The smoking complexes described above clearly point to the enormous symbolic value of the pipe to North American Indian cultures. The mythology surrounding the pipe was no less cosmic than that of tobacco itself (Paper 1988). The pipe was a work of art. Its features, shapes and, above all, the highly sculpted bowls carried enormous symbolic meaning (Turnbaugh 1980: 16; Mathews 1976; von Gernet and Timmins 1987). The manufacture of a pipe was no less sacred than the tobacco itself (Furst 1976: 30). The pipe became an elaborate article, especially in the calumet ceremony, when it would symbolically represent peace or war (Linton 1924). The enormous symbolic value of the pipe derived partly from the close association between its stem and the shaman's device for sucking illnesses from patients (Hultkrantz 1979: 80; von Gernet 1992: 23).

In South America tobacco chewing, snuff taking and drinking were probably equally common. The latter two were common among Peruvian and Guianese Indians (Wilbert 1987: 30, 51). Tobacco chewing – or, more accurately, mastication – by contrast appears to have been much less geographically concentrated. Frequently the tobacco was mixed with lime or with ash in order to release nicotine more efficiently. How tobacco was chewed is not known. The Yanomamö of present-day Brazil, who are inveterate tobacco chewers, masticate tobacco by inserting a wad of pre-pared tobacco between the lower lip and teeth (Chagnon 1983: 65). Whether this method was followed by other Amerindians is not certain. The ancient Maya also chewed tobacco together with lime (Robicsek 1978: 21–2). In North America tobacco chewing was far less common, and concentrated entirely on the western side of the continent. The Haida and Tlingit Indians of the north-west coast chewed their tobacco mixed together with lime prepared from burning shells, as did the Indians of coastal central California, though further inland, among the Shoshone of Death Valley, tobacco eating with lime was more common (Turner and Taylor 1972: 251; Kroeber 1941: 17–19). Interestingly, it was not until the late eighteenth century, when Russian fur traders introduced them to the pipe, that the Tlingit started to smoke (de Laguna 1990: 212).

Finally, some Amerindian societies adopted tobacco in rectal adminis-trations, using enema syringes and clysters, both ritually and medicinally. How common this was is not known but it had a following (Furst 1976: 27–9). Ritual enemas were certainly practised by the Maya, Incas and Aztecs though what substances were used is debatable (de Smet 1983: 150–2). The famous Aztec herbal, known as the Badianus Codex and written in 1552, recommended the tobacco enema, mixed with other herbs and flavourings, for abdominal rumblings:

> For one whose bowels are murmuring because of diarrhea, make a potion, let him take it with an early clyster, of the leaves of the herb tlatlanquaye, the bark of quetzalaylin, the leaves of yztac ocoxochitl and these herbs . . . , the tree tlanextia quahuitl ground in bitter water and ashes, a little honey, salt, pepper and alectorium, and finally picietl (Tobacco).
>
> (Robicsek 1978: 38)

Other examples of tobacco enemas have been noted for both North and South America (de Smet 1983: 142–3: Wilbert 1987: 46–8).

There was very little about tobacco that Amerindians did not know.

Everything about it, from the shape of the leaf to the effect of nicotine, was incorporated within a rich cosmology imbued with enormous symbolic significance. There were thus no contradictions about the plant, no asymmetry between it and religious and medical ideology. The stupefaction of Metsé by tobacco, the description of which introduced this chapter, is perfectly understandable within an Amerindian world-view. To modern readers it is counter-cultural: to Amerindians it was culture itself. The last word can be left with the Winnebago who, in their Night Spirits Society ceremony, capture the very essence of tobacco as the interface between the world of spirits and the world of people. In his greeting one of the guests turns to tobacco:

> This instrument for asking life is the foremost thing we possess, so the old people said. We are thankful for it. We know that Earthmaker did not put us in charge of anything, and for that reason the tobacco we received is our greatest and foremost thing . . . Those whom we call Nights have been offered tobacco, and the same has been offered to the four cardinal points, and to all the life-giving plants. To this many tobacco has been offered. It will strengthen us. This is what we call imitating the spirits, and this is why we are doing it. Children of the night-blessed ones who are seated here, I greet you all. The song we will now start is a pipe-lighting song.
>
> (Tooker 1979: 134–5)

3

WHY TOBACCO?
Europeans, forbidden fruits and the panacea gospel

On 15 October 1492 Columbus was offered a bunch of dried leaves as a present and, one month later, two of his crew members, returning from a trip into the interior of Cuba, reported seeing Indians smoking leaves (Columbus 1990: 39, 73). A bunch of dried leaves would have made very little impression on the Admiral – not only was he interested in something with a little more glitter, but he would have entirely missed the point of being offered these dried leaves (Morison 1974). We do not know how the sailors reacted to the sight of smoking. This episode in the contact between Amerindians and Europeans was not publicized for many decades but it was the beginning of a long series of encounters between the two cultures in which tobacco was exchanged.

As the previous chapter showed, Amerindian societies located tobacco within a specific cosmology of which the art of healing was a critical component. The Amerindian cosmology was certainly incomprehensible to Europeans, and even those who attempted to understand it – in particular people such as the historian and champion of Amerindian rights, Bartolomé de las Casas, and the inveterate chronicler of the Aztec way of life, Bernardino de Sahagún – could not accept it for what it was. The meaning of the bunch of tobacco leaves presented to Columbus was incomprehensible to him. Yet within no more than fifty years of Columbus's first voyage tobacco made a formal appearance in Europe, at the Portuguese court in Lisbon. By 1570 the plant was growing in Belgium, Spain, Italy, Switzerland and England, though on a very small scale (Brooks 1937; Dickson 1954; von Gernet 1988: 60–1). By the turn of the century tobacco was also growing in the Philippines, India, Java, Japan, West Africa and China. Chinese merchants introduced the plant into Mongolia, Tibet and eastern Siberia so that, only one century after Columbus's voyage, tobacco was either grown or consumed in most of the known world. Magellan's circumnavigation of the globe was a remarkable achievement; tobacco's was no less so.

How was it then, that a substance embedded within an incomprehensible cosmology, with meanings that were bewildering to outsiders, could find a place not only within European culture but within so many diverse cultures throughout the world, in such a relatively short time?

The story, naturally enough, begins in the New World. There Europeans encountered a world of flora (and fauna and indigenous peoples) that was utterly puzzling (Elliott 1972; Ryan 1981; Greenblatt 1991; Sauer 1976: 815). In the first half century of exploration few learned men crossed the ocean to report on the plant world; their understanding and perceptions were drawn largely from accounts offered to them by explorers, administrators, ships' captains and sailors (Talbot 1976: 834–5, 838). In vain early botanists searched through the classical literature hoping to match the received descriptions with the written word (Sauer 1976: 823).

The process of cataloguing and classifying New World flora was not, however, inspired simply by an intellectual challenge to assimilate New World phenomena within European cultural traditions, though this was important. Precious metals notwithstanding, Spanish interests in the New World (as well as those of later colonizers) recognized plants as economic assets. Of particular significance were plants with medicinal value – edible plants, it should be noted, were not given much attention (Hamilton 1976; Davies 1974: 141–96). Philip II, the architect of Spanish colonialism in the New World, clearly understood the value of medicinal plants, but found the indirect method of gathering information less than satisfactory for his purposes. In 1570 he took personal charge of gathering information about his New World possessions. He issued a royal edict appointing Francisco Hernandez as a special protomedico (or royal physician) for New Spain (Risse 1987: 31). The written instructions to Hernandez made the purpose of his visit clear: 'to gather facts from all physicians, surgeons, Spanish and native herbalists, and other inquisitive persons with such abilities who can possibly know something, and, in general, obtain an account of all medical herbs, trees, plants, and seeds that exist in a given place' – the place was originally within a radius of fifteen miles of Mexico City, but in the event his travels took him farther afield (Risse 1987: 31, 43). According to one recent authority, Hernandez's task was monumental – the medicinal flora of Mexico at the time has been estimated at 5,000 plants (Risse 1984: 35). Nevertheless Hernandez, over a period of seven years, collected and described over 3,000 plants – he even ran clinical trials, on patients in the Royal Hospital for Indians in Mexico City, to establish indications for their use (Risse 1984: 36).

Economic considerations prompted Philip's remarkable project. One important consideration was the extent to which a colony could supply commodities that were being imported from elsewhere. Medicines were perhaps the single most important commodity that Europe imported from the East, and there was every justification for Spain in the sixteenth century (as well as for England, France and Holland in the seventeenth century) to reduce their dependence upon the Venetian and Portuguese middlemen

(Wake 1979; Lane 1940; Steensgaard 1985; Roberts 1965). To find substitutes for such expensive substances as Chinese rhubarb; to grow in the colonies medicinal plants imported from abroad; and to be able to export prepared medicines from Spain to the New World – these were the economic motives lying behind Hernandez's mission (Dermigny 1964: 373–87). Some of these objectives were already satisfied before 1570: jalap, from Mexico, was already ousting rhubarb from the Spanish pharmacopoeia; ginger was introduced into Mexico in 1530 leading to the abandonment of ginger imports from China and India; cassia fistula was introduced into Santo Domingo from Ethiopia as early as 1514; and finally, simples and compounds prepared in Seville were already bound for the New World (Guerra 1966b: 38; Fernández-Carrión and Valverde 1988: 29). Seville imported sarsaparilla from the New World, in part because the alternative cure for syphilis, guaiacum, was handled by the Fugger family whose monopoly controlled the prices for this so-called wonder drug (Munger 1949; Guerra 1966b: 38; Lorenzo Sanz 1979: 604–5). Notwithstanding these achievements in reducing Spain's import bill for medicines, much remained to be done.

Hernandez's long sojourn in the New World has been acknowledged rightly as a pioneering botanical investigation and one which would be replicated in later centuries of European colonialism throughout the world. Philip's desire to know his assets was not, however, confined to this single act. Lesser officials were enjoined to gather 'information on the herbs, aromatic plants the Indians use to cure themselves, and their medicinal virtue or poisonous characters they have' (Guerra 1966b: 49). The problems encountered in collecting this information must not be discounted, but they were probably not as great as those encountered in interpreting the evidence. Sixteenth-century Europeans were poorly equipped to make sense of the New World. Not only did their own prejudices stand in the way, but even when they attempted to interpret with an open mind, their own lack of knowledge proved a constraining force. The botanist was unfortunately severely hampered in interpreting floral evidence because, although he relied upon one of the greatest herbals of antiquity, the materia medica by Dioscorides, this compendium of some 600 plants naturally featured only those of Mediterranean origin (Stannard 1966). Once the stage of cataloguing and classification was completed then the task of evaluating and classifying the plant's medicinal properties began.

Medicinal remedies, including spices, were all interpreted through the Galenic or humoral system (Teigen 1987). The main characteristics of this medical philosophy can be summarized as follows. All matter had an essence formed from the binary combination of four opposing qualities – hot and cold, moist and dry. The human body, at the core of Galenic medicine, had four humours – blood, phlegm, black bile and yellow bile. Each of these humours, because they were matter, had an essence. Phlegm, for example, was cold and moist; blood was hot and moist. A healthy body was defined as a body in humoral equilibrium: an unhealthy body was in humoral disequilibrium, caused by a disease, the symptom of which was the ailment. A physician's agenda was to diagnose the nature of the humoral disturbance and restore equilibrium by drawing off the excess humour through bloodletting, purging, vomiting and sweating. This could be accomplished either directly, by using leeches, or by prescribing remedies that effected the fluid depletion. The medicinal therapy conformed to the theory of opposites: a hot remedy for a cold ailment, and a cold remedy for a hot one. It is important to understand that humoral disequilibrium defined not only a physiologically unhealthy state but also what we would now call psychological illnesses including mood changes. This medical system also extended to diet, and the use of spices therein: physicians and cooks were encouraged to co-operate in order to regulate health and temperament (Sass 1981).

Despite Philip II's patronage and direction, Hernandez's work in Mexico was a failure on the whole. Very few New World medicinal remedies managed to cross the Atlantic and find their way into the European pharmacopoeia. The gulf between the aims and the results is a problem that has attracted the historian's attention. Charles Talbot, a student of the European drug trade, has attributed the lack of European assimilation of New World medicines to a combination of ignorance, conservatism and fear, especially of the unknown. European physicians, he argues, were not willing to commit themselves to prescribing exotic simples and compounds when known and trusted substances were available (Talbot 1976). Additionally, even if physicians had been less sceptical, by the time the medicinal plants arrived in Europe they were frequently in such a poor physical state as to render their medicinal properties ineffective (Talbot 1976: 837–8). It is not easy to dispute these claims for conservative attitudes, and practical problems in shipping certainly existed. Yet these same circumstances did not affect the import of exotica from the East which, since the early Middle Ages, had defined European foreign trade. The incorporation of Chinese and Indian medicinal remedies must have been at least as difficult as those from the New World but the former succeeded whereas the latter did not. Talbot may, however, be too pessimistic about the extent of transmission. If one widens the definition of medicinal remedies to include spices, as would have obtained in the sixteenth century, then the failure to accept New World products does not appear as great as Talbot argues (Lorenzo Sanz 1979; Roberts 1965).

Other historians of New World medicine, notably Guenter Risse, contend that the Galenic system was the filter through which New World medicines passed on their way to Europe: those that could not be assimilated within the Galenic system were rejected; those that could be incorporated stood a much better chance of being transferred (Risse 1984). This assertion has led to the suggestion that native informants 'whose views on disease actually resembled those of classical humoralism . . . had arrived at a set of physiological notions remarkably similar to those held by their European conquerors' (Risse 1984: 37). Medical anthropologists have been debating this point for decades: that is, whether South and Central American Indians followed a humoral or some other medical philosophy; the debate turns on the question of whether the resemblance between Spanish medicine and native medicine was genuine or depended too much on the interpretations of Spanish physicians (Foster 1987; 1988; Messer 1987; Logan 1977).

Risse is certainly correct in pointing to the Galenic system of late Renaissance Europe as the cultural framework through which European physicians and herbalists confronted New World medicines. To argue, however, that medicines stood or fell on the Galenic system itself would be to attribute to it a degree of canonical force it did not possess. Not only did the Galenic system incorporate a high level of subtlety, but it allowed for a fair degree of empirical investigation and differing opinions (Teigen 1987). Moreover, given the fact that diseases far outnumbered medicinal therapies, one might expect physicians to use the Galenic system to assimilate rather than reject exotic substances.

The Galenic system did not really act as a filter because exotic substances were often transferred before they were placed under the lens of the learned physician. Its main use was to ascribe properties to medicines after they had crossed the cultural divide. The prevailing medical philosophy legitimized rather than determined choice. It still remains, therefore, to explain the choice in the first place. An argument put forth recently by the anthropologist Marshall Sahlins will help to clarify the issues involved (Sahlins 1988). Sahlins was interested in exploring a process he terms commodity indigenization; in other words, the way in which native cultures responded to, and absorbed, commodities brought to them by Europeans (Sahlins 1988: 5). Drawing on the experiences of the Chinese in the eighteenth century, the Hawaiians in the nineteenth century and the Kwakiutl Indians of British Columbia both in the nineteenth and twentieth centuries, he argues that native cultures were not passive in their contact with European commodities but employed them actively in their own social, political and cultural modes of behaviour. Native cultures, he points out, accepted European commodities actively by providing them with meanings derived from their own belief systems (Sahlins 1988: 6–9). One can elaborate this argument into a possible model of cultural contact; the success with which a commodity crosses from one culture to another depends on whether this new object can be given a meaning within the host culture.

This notion of commodity indigenization can be applied to the European case, in reverse, so to speak, and help to explain how tobacco became

41

incorporated in the European materia medica. As previously argued, Europeans encountered a New World containing an enormous range of exotica. It was not the first, nor would it be the last, time that Europeans encountered exotic commodities. In the Middle Ages, Europeans saw and tasted the products of India and China, none of which were indigenous to Europe itself. Aside from raw silk, the eastern commodities imported into Europe were primarily what we now call spices. To Europeans of the sixteenth century spices were medicines. Pepper, ginger, nutmeg, anise and cinnamon, all imported from the East before the sixteenth century, were the stock in trade of urban medicine and cookery, both of which were intimately related. What the New World might offer in the area of medical therapy was not insignificant to sixteenth-century Europeans.

The success of tobacco in crossing cultures is one of the most intriguing aspects of the sixteenth-century European encounter with the Amerindian world. Analytically, what occurred largely supports Sahlins' model. To understand why tobacco was so successful we need to explore the European cultural context of the sixteenth century as well as the paths of cultural transmission.

One part of this cultural context involved the use of narcotics. According to historians such as Piero Camporesi and Carlo Ginzburg, the urban and rural poor lived in a world where what we would now call hallucinatory or ecstatic experiences were common (Camporesi 1989; Ginzburg 1990). These experiences can be conveniently divided into two kinds: those which were induced by the regimen of poverty and those principally associated with witchcraft. Hallucinations were, as Camporesi points out, a by-product of a world of subsistence. The worst off in society did not actively search out hallucinatory experiences, but became victims of them through the lack of adequate and frequent nourishment; through eating either contaminated food or bread made with grains spoiled by fungi that were themselves hallucinatory – ergot, for example; or by mixing together various plants as food substitutes (Camporesi 1989: 26–39, 120–50). Camporesi is persuasive on this point. Certainly, prescriptions for how to deal with famine were not uncommon. Hugh Platt wrote one such pamphlet in 1596 urging his readers to consider the following possibilities when faced with nothing edible to eat: try, he wrote, eating fresh turf or a clod of earth, sucking one's blood, drinking one's urine, or eating wheat-straw bread (Platt 1684: 163–5). The importance of suppressing hunger was one of tobacco's main attributes and was repeated frequently in the medical literature of the sixteenth and seventeenth centuries. Witchcraft, and the witch's sabbath, were undoubtedly associated with hallucinatory plants. Ergot, according to Ginzburg, is one possibility, but other writers have suggested plants of the solanaceous family – that is deadly nightshade, henbane, mandrake and especially datura or thorn apple – and cannabis (Ginzburg 1990: 303–5; Harner 1973; Abel 1980: 106–8). Despite the paucity of the evidence, the impression of European culture as being punctuated by hallucinatory experiences can be sustained, but only if it is understood as being a phenomenon of a particular social class, or of folk culture, and not institutionalized. One hypothesis hinted at by Ginzburg, and worth following up, is that the matrix of mind-altering substances in Europe belonged to what Ginzburg calls female medicine, or to what Camporesi refers to as the medicine of the poor (Ginzburg 1990: 304; Camporesi 1989: 108–14).

If one accepts the hunger/mild-hallucination pairing as representing certain aspects of European society and culture, then it can be argued that the possibilities existed here for the incorporation or acceptance of some non-food substance that suppressed hunger, without inducing violent mind-altering effects. The four solanaceous plants would not have served this purpose since, whatever their influence on appetite, their hallucinations were violent and thus counter-productive – hence their use in witchcraft (Harner 1973).

It is likely, therefore, that there was a cultural wisdom about psycho-tropic plants, and that this wisdom was concentrated in the folklore of urban and rural populations. Much more research is needed in this area, but enough has already been done to portray sixteenth-century Europeans as inhabiting a very complex cultural space where mind-altering substances and their experiences played a significant role.

Those Europeans who set out to chart the resources of the New World were certainly not searching for hallucinogenic plants and experiences. Yet it would not be stretching the imagination too far to suggest that those they employed, sailors in particular, had some acquaintance with hallucinogenic preparations. We know, by what was said of them, that sailors did return to Europe with tobacco and it is not inconceivable that their first few puffs produced an experience that was not entirely unfamiliar and one that was entirely pleasing (Dickson 1954: 44–5; von Gernet 1988: 23).

Plant investigators, Hernandez among them, by contrast, were primarily interested in plants that would feature in a healing therapy. In the sixteenth century, there were relatively few curative agents which were capable of treating a large array of diseases (Slack 1979; Peter 1967). Specific remedies were of limited use, and there was a widely held belief in the existence of a universal panacea, among both Galenists and Paracelsians (Teigen 1987; Pagel 1982; Akernecht 1964: 8). The roots of this belief extend far back into European history but were rekindled by the renaissance of classical medicine in the fifteenth and sixteenth centuries. Galenists held firmly that panaceas were organic, since herbal substances were considered more natural than inorganic types. Followers of Paracelsus, however, contended that the most efficacious remedies were inorganic, indeed mineral – in the treatment of syphilis, for example, Galenists prescribed guaiacum or sassa-fras, while Paracelsians advocated mercury (Risse 1987: 24; Webster 1979;

Multhauf 1954). For Galenists, at least, the cornucopia of New World flora promised to yield a panacea of greater efficacy than so far uncovered in the Old World.

It was into this context that the plants of the New World vied for assimilation, and it was this bundle of assigned meanings that determined the fate of the exotic substances. One of the first to get attention was tobacco, a plant of the solanaceous family which includes the psychotropic plants mentioned above as well as the potato and the chilli pepper. Following the accounts of tobacco that were published, or available in manuscript form, in Europe between Amerigo Vespucci's description of tobacco-chewing witnessed in 1499 (but not published until 1505), and the 1570s, it is clear that tobacco was becoming accepted as a herbal therapy capable of curing an increasingly large number of ailments. Under the scrutiny of a host of European botanists, physicians, churchmen and bureaucrats, who either grew tobacco in their gardens, tried it themselves, read about it in the Spanish accounts of the New World or spread the gospel about it, tobacco picked up one accolade after another (Dickson 1954: 31–56).

Until 1571, however, in the world of printed books, tobacco was frequently mentioned alongside other plants, though a reading through this literature suggests that, in relative terms, the space devoted to tobacco was growing, and the terms in which it was being discussed were becoming more expansive. One of the most important landmarks in the history of the formal recognition of tobacco in European literature was the publication, in 1571, of the celebrated and widely read second part of Nicolas Monardes' history of medicinal plants of the New World. Monardes himself had not been to the New World but, as the leading physician of Seville, he did grow the plant in his own garden and was keenly aware of what was being said and written about tobacco as well as being involved in commerce between Seville and the New World (Pike 1972: 83–4, 89–91; Guerra 1961: 24–6, 79–82). Monardes provided all the justification needed for locating tobacco at the heart of the European materia medica. Not only did he establish its humoral essence – hot and dry in the second degree – but he listed more than twenty specific ailments which tobacco cured, from toothache to cancer. He also emphasised that tobacco alleviated hunger and thirst (Monardes 1925: 75–91). Writing several years later in Mexico, Juan de Cardenas, a Spanish physician, echoed Monardes and in many ways showed just how great an influence Monardes had on nicotian thinking (Dickson 1954: 95–7). He wrote thus:

> To seek to tell the virtues and greatness of this holy herb, the ailments which can be cured by it, and have been, the evils from which it has saved thousands would be to go on to infinity . . . this precious herb is so general a human need not only for the sick but for the healthy.
>
> (Dickson 1954: 95)

Besides listing and extolling its curative qualities, Monardes had commented favourably on tobacco's mind-altering effects, a not inconsequential fact. In a section in which he described the use that Indians made of tobacco, he referred to its role in inducing visions and quickly legitimized the experience therein by reminding his readers that no less an authority than Dioscorides wrote about ancient herbs that induced similar states (Monardes 1925: 86–7).

Monardes made it clear that there were three main methods of using tobacco: green wet leaves, usually warmed, were prescribed in topical applications mostly for pain, as well as for sores, cuts, wounds, etc.; leaves mixed with lime could be chewed to alleviate both hunger and thirst; finally, dried leaves could be smoked, to counteract weariness and induce relaxation. Writers on tobacco, from Monardes on until at least the turn of the nineteenth century, often simply reproduced his arguments and indeed his prose. Much of the medical literature debated points of humoral argument – such as pinpointing who should and should not take tobacco; or attempted to refine Monardes on particular aspects. For example, Monardes suggested in his treatise that it was the juices of the tobacco wad which alleviated hunger and thirst directly (Monardes 1925: 90); Edme Baillard, a Parisian physician who advocated the use of snuff, argued instead that tobacco silenced those membranes and fibres that gave the soul of the body the idea that it was hungry or thirsty – this was written in 1668 (Baillard 1668: 93–4). The *Counterblaste* on tobacco written by James I and published in 1604 constructed its argument against the use and abuse of tobacco by debating humoral points and the alleged existence of a panacea (James I 1982: 87–99).

Monardes' discussion of nicotian therapy included smoking, though it is not given great prominence in his discussion. Undoubtedly he was drawing on observations by André Thevet, whose *Les singularitez de la France antartique* had been published in 1557 and contained a description of Amerindian smoking practices, reproduced below:

Another curiosity is a plant, which they call in their language *petun*, which they generally carry with them, because they believe it to be wonderfully useful for several things. It resembles our ox-tongue. They carefully gather this herb, and dry it in the shade in their little cabins. When it is dry they enclose a quantity of it in a palm leaf, which is rather large, and roll it up about the length of a candle. They light it at one end and take in the smoke by the nose and mouth. They say it is very good to drive forth and consume the superfluous moisture in the head. Besides, when taken in this way, it makes it possible to endure hunger and thirst for some time. Therefore they use it often, even when they are taking counsel they inhale this smoke and then speak; this they do ordinarily one after the other in war, where it is very useful. Women do not use it at all. It is true that if they take too much of this smoke or perfume it will go to their head and make them drunk like the smell of strong wine. The Christians here today have become very attached to this plant and perfume . . .

(Dickson 1954: 119)

Monardes does not appear to have drawn on information from Benzoni's history of the Americas published in 1565 – quoted in the previous chapter – and indeed it was by avoiding his account of tobacco shamanistic practices that Monardes could assimilate smoking into the humoral paradigm (Dickson 1954: 121; Benzoni 1857). In the English edition of Monardes translated by John Frampton and published in 1577, an addition was made to Monardes' original description but was not written by him; rather, it was translated from Charles Estienne and Jean Liébault's manual on agricultural techniques first published in 1567 (Dickson 1954: 36, 76). The addition contains a succinct, if not entirely clear, description of smoking, moving beyond tobacco's ability to alleviate hunger and thirst to its humoral powers in expelling excess moisture from the body, and, significantly, evoking the image of sucking the smoke up through a pipe: in the original Spanish version the discussion about smoking is set within a description of Amerindian ways and not offered, directly, as a method for European consumption:

The leafe of this hearbe being dried in the shadowe . . . being caste on a Chaffying dishe of Coales to bee burned takying the smoke thereof at your mouth through a tonnell or cane, your hed being well covered, causeth to avoyde at the mouthe great quantitie of slimy and flekmaticke water, whereby the body will be extenuated and weakened, as though one had long fasted, thereby it is thought by some, that the dropsie not having taken roote, will bee healed by this Perfume.

(Monardes 1925: 97)

Monardes was not alone in singing the praises of tobacco. More than any of his contemporaries, however, this physician from Seville added a voice of authority to advocates of nicotian therapy and, thereby, legitimized its use among physicians and herbalists and put it squarely within a European cultural framework. Monardes' compendium and commentary on the medicinal plants of the New World was translated into the major European languages and appeared in a number of editions and, for at least two centuries, little of substance was added to his account of tobacco by all those who followed him (Dickson 1954: 83; Talbot 1976: 836; Stewart 1967).

The critical position of Monardes in the history of tobacco and of nicotian therapy – a medical practice that continued well into the nineteenth century in official circles and later on the fringe of orthodox medicine – has been generally acknowledged by historians of tobacco (Brooks 1937; Dickson 1954; MacKenzie 1984; Kell 1965). It is, however, important to remember that tobacco was being consumed in Europe, to some extent, before Monardes published his work. The manufacture of pipes in London, for example, is believed to have begun around 1570 (Ayton 1984: 4). What happened in the 1570s was that two historical trajectories fused. The first consisted of the exchange of tobacco between Amerindians and European explorers, sailors and settlers along the eastern seaboard of the Americas. Matthias de l'Obel, in his celebrated herbal of 1570, gave a clear account of how the intrepid Europeans appeared after being initiated into nicotian rites:

> For you may see many sailors, all of whom have returned from there carrying small tubes . . . [which] they light with fire, and, opening their mouths wide and breathing in, they suck in as much smoke as they can . . . in this way they say that their hunger and thirst are allayed, their strength is restored and their spirits are refreshed; they asseverate that their brains are lulled by a joyous intoxication.
>
> (Dickson 1954: 44–5)

The second was the intellectual assimilation of the New World 'herbe', which Monardes participated in and developed. Once the fusion occurred, the process of the exchange of tobacco across two cultures was completed.

It is clear that European herbalists and physicians learned about tobacco from several sources: the written accounts of Amerindian practices; empirical investigation of the plant itself; and, perhaps most importantly, oral reports of returning sailors. What also seems to be clear is that tobacco permeated all European social classes at about the same time, in contrast to other exotic substances of the period such as sugar, chocolate, coffee and tea, all of which appear to have entered at the top and percolated downwards (Mintz 1985: 74–150; Braudel 1981: 249–60; McCracken 1988). Even as herbalists and physicians were raising its therapeutic profile, tobacco was given a seal of approval higher up the social scale. Again this came from several directions. In France Jean Nicot, who was the French ambassador to the court in Lisbon from 1559, was perhaps more important than anyone else in promoting tobacco not only as a superb remedy but as one that was perfectly suited to life at court. Nicot had obtained a specimen of the species *Nicotiana tabacum* (probably originating in Brazil) while in Lisbon from Damião de Goes, the Portuguese humanist and keeper of the Crown's manuscripts (Dickson 1954: 68; von Gernet 1988: 32). Having witnessed its miraculous curative powers, Nicot sent seeds and plants to the French court, in particular to the queen mother, Catherine de Medici (Baudry 1988: Falgairolle 1897: 50; Laufer 1924b: 49–50). Jacques Gohory, in his treatise on tobacco published in 1572, argued that, because the queen mother herself was responsible for its cultivation in France, it should be named 'Medicée' (Bowen 1938: 356). Nicot was the source for Liébault's work that presented tobacco as a herbal panacea and, in return, Liébault popularized the story about Nicot, as well as suggesting that, in his honour, the plant be called nicotiane (Dickson 1954: 72–5). The wonder plant was introduced to the papal court in Rome by Prospero di Santa Croce, the papal nuncio in Lisbon in 1561; and in Tuscany credit is given to Bishop Niccoló Tornabuoni who, between 1560 and 1564, was the Grand Duke of Tuscany's ambassador to the French court (Dickson 1954: 151–2). The royal and noble associations of tobacco on the Continent were clearly important in providing a meaning for the plant and legitimating its use. In England, it was not noble approval so much as distinguished approval which proved salient in tobacco's acceptance. Thomas Hariot, the highly respected scientist who accompanied the first English colonists to Roanoke Island, in 1585, reported on tobacco use among the Indians, but also described his own experiences:

> We our selves during the time we were there, used to sucke it after their maner, as also since our returne, and have found many rare and woonderfull experiments of the vertues thereof: of which the relation would require a volume by itselfe: the use of it by so many of late men and women of great calling, as els, and some learned Physicians also, is sufficient witnesse.
>
> (Quinn 1979: 146)

The allusion above to 'Physicians' included a reference to Monardes, a copy of whose book Hariot took on his voyage, in order to help him identify the flora present (Quinn 1979: 146).

In accounting for the incorporation of tobacco into European culture several factors have been emphasized. First, during the sixteenth century, Europeans were being introduced to tobacco and smoking both directly, in their encounter with Amerindians in the Americas, and indirectly, through those returning from transatlantic voyages, including those Amerindians who appeared in Europe from time to time. Most of those with such first-hand knowledge of tobacco did not record their impressions but we can learn a great deal from those very few who did. Thomas Hariot's account is one such revealing testament. That Hariot became addicted to tobacco, and introduced the practice to others, is well established, but his experiences could not have been unique – Hariot, it should be added, died of cancer of the nose, caused, no doubt, by the practice of exhaling smoke along the nasal passage (Shirley 1983: 432–4). Second, the language of tobacco, employing terms such as panacea, holy herb, sacred weed, did much to attract medical and popular attention to the plant's wondrous curative powers – 'holy herb' first appeared as a description of tobacco in

48

Damião de Goes's chronicle of the reign of King Manoel (Dickson 1954: 65). Matthias de l'Obel, as one of the strong advocates of nicotian therapy, placed tobacco as a panacea beyond question when he wrote, 'it should be preferred to any panacea, even the most celebrated' (Dickson 1954: 44). Finally, one must take account of the fact that tobacco was being hailed as a miracle plant in the seats of European power, where publicists such as Jacques Gohory suggested nomenclatures appropriate to its status: thus, Nicotiane, *l'herbe du Grand Prieur*, Medicée and Tornabona (Dickson 1954: 151–2). Other factors are more difficult to support with evidence, but it is not too far-fetched to argue that the novelty of smoking must have been an attraction as were the undoubted psychological effects, from hallucinations to feelings of comfort and ease, that were reported (von Gernet 1988: 26).

Tobacco was grafted on to European culture by several different agents operating contemporaneously. Which agents were the most important is impossible to judge but there is no doubt that each reinforced the other. Monardes, for example, did not address his remarks to a specific audience: the information in herbals rapidly diffused through both literate and non-literate societies. These are the factors that operated on the European side of the exchange, but there were also factors present in the Americas that helped tobacco's passage. The important ones have already been discussed in the previous chapter but it is well to repeat them briefly. First, tobacco had a very wide geographic distribution. In terms of contact, this meant that no one set of European explorers had a unique encounter with tobacco. Portuguese, Spanish, French and English had experiences broadly similar. Second, tobacco had the status of a panacea in Amerindian medicine and, therefore, the work of identifying which plant might acquire this standing had already been done. Aside from its special use in shamanistic practices, tobacco was not used by Amerindians to promote any special political or social authority: the divisions between sacred and profane, between popular and reserved, were blurred. Tobacco was widely consumed for a great many reasons (von Gernet 1982). Tobacco's sacredness was there to be shared, not monopolized. Offering tobacco to the newcomers from across the waters was a reflection of the significance of the plant in engaging social contact. This social role of tobacco would prove to be highly mobile across the cultural chasm, as the next chapter will argue (von Gernet 1988).

To complete this analysis of the cultural transfer of tobacco it is useful to contrast tobacco's success with coca's failure. Coca was not 'discovered' by Europeans until after the conquest of the Incas, that is until after 1531; though Ramon Pané, a missionary on Hispaniola, did refer to something like it, in a manuscript written around 1497 (Gagliano 1979: 39). Coca thus entered European perceptions of the New World some three decades after tobacco. Coca was used by highland Indians for reasons that were not unlike those of tobacco-using Indians (Martin 1975). That is, the medicinal use prevailed, with certain uses being reserved for shamanism and divination. Unlike tobacco, however, which was widely consumed throughout the Americas, coca use concentrated in the Inca kingdom, principally, of course, in Peru. Coca was also central to many Inca religious rites and was one of the privileges of the royal family and priests. There is some evidence, in fact, that before the conquest coca was not used in peasant communities and that its use after the mid-sixteenth century occurred in the wake of the destruction of Inca rule (Parkerson 1983). Nevertheless, as far as Europeans were made aware, the early accounts of the use of coca virtually ignored its significance as a medicinal plant. Indeed coca became the subject of a very intense debate among missionaries, administrators and merchants, but mostly among missionaries (Gagliano 1963). Basically the debate centred on the question as to which of two aspects of coca consumption would hold sway. One group of missionaries was convinced that coca use thwarted the missionaries' attempt to christianize the Indians since coca not only linked the Indians to the Inca past but also was used in what were seen as heathenish practices. These prohibitionists appealed to the crown to destroy the coca plantations. Other missionaries took the opposite line: that coca served as a nutritive and stimulant substance for the undernourished Indians and, more to the point, without their ration of coca they would refuse to work in the silver mines.

The fact that a substance was being debated on points that had nothing to do with its medicinal efficacy made it very difficult to prescribe for use by Europeans. Indeed Pedro de Cieza de Léon, one of the early chroniclers of Peru, went so far as to decry coca chewing, as he put it, as a habit fit only for Indians (Gagliano 1979: 40). The coca controversy raged on for more than a century. In the seventeenth century in every accusation of witchcraft that came before the judges in Peru, coca was implicated (Gagliano 1979: 43). Monardes wrote about coca in the third part of his history of New World medicines, but aside from one sentence pointing out that coca chewing alleviated hunger and thirst the entire description of coca (only two pages) concerns the plant's botany while tobacco is given twenty-four pages of text (Monardes 1925: 31–2). Avid readers of Monardes, and there were many among physicians and scientists, would find little here to kindle an interest in coca. Even when towards the end of the sixteenth century scattered reports of coca's wide medicinal applications were circulating, coca was still not being taken seriously by European physicians, and the reason here had nothing to do with the coca controversy. The reports spoke about coca's power in healing diseases among the sierra and altiplano Indians (Gagliano 1979: 39–44; Saignes 1988). This alone would have precluded its use in Europe, since it was commonly believed that diseases and their cures occurred in the same location. Thus the highland Indian diseases were believed not to be the same as those of Europeans since coca did not grow in Europe. It might be argued that this problem applied also to tobacco since it was found growing in the New World and not Europe. European herbalists of the first half of the sixteenth century, however, thought tobacco was the third variety of henbane, the first two of which grew in Europe. One of the first to do this was Pier Andrea Mattioli, the personal physician to Archduke Maximilian and the author of a celebrated commentary on Dioscorides, but other herbalists, notably Rembert Dodoens, agreed with this classification (Anderson 1977: 163–80; Dickson 1954: 33–9).

Coca remained outside of the European materia medica. It was not until the very end of the eighteenth and early part of the nineteenth century that Spanish officials urged the Crown to consider marketing coca in Europe as a substitute for coffee and tea (Gagliano 1965: 166–7; Gagliano 1979: 46–7). The suggestion was turned down and it was left to others, Sigmund Freud included, later on that century to extol the virtues not of coca but of its principal alkaloid, cocaine.

The Europeanization of tobacco was fundamental in the plant's subsequent history for it was Europeans who were active in its initial diffusion beyond the Americas. While this is beyond doubt, the precise timing of tobacco's appearance in other cultures remains obscure as do the routes of transmission. But despite these problems and reservations, it is remarkable how quickly tobacco was absorbed into very different cultural systems. On the other hand, the reason why tobacco passed through the cultural divide separating Europe from Asia and Africa was strikingly similar to the transmission from the Americas to Europe.

It is generally accepted that tobacco travelled next to Asia after its initial appearance in Europe, but this did not happen until the European meanings so succinctly described by Monardes were established. Tobacco first made its appearance in Asia around 1575 when the Spanish brought the plant to the Philippines from Mexico, as part of the galleon trade that the Spanish had established in 1571 when they first arrived in Manila (Reid 1985: 535). In the Philippines the tobacco plant established itself quickly as a satisfactory cash crop, as both people and land were well suited to its cultivation (de Jesus 1980: 2–3). Chinese, particularly Fukienese, sailors and merchants were responsible for its introduction into the Fukien province of southeastern China around the turn of the seventeenth century, although a more precise dating is not possible (Goodrich 1938: 648–9; Spence 1975: 146–7). The Spanish themselves were not responsible for the diffusion of tobacco from their base in the Far East. It was the Portuguese who from their base in Macao carried the tobacco plant to other parts of the region. Their role in tobacco's movements has been documented for India, possibly as early as 1595 (Gokhale 1974: 485); Java in 1600 (Reid 1985: 535; Höllmann 1988: 35); Japan some time around 1605 (Laufer 1924b: 2; Satow 1878). These areas became the staging post for tobacco's journey to other parts of the Far East. From Japan tobacco spread to Korea (Laufer 1924b: 10);

from India to Ceylon around 1610 (Laufer 1924b: 14); and from China to eastern Siberia, Mongolia, Turkestan and Tibet (Laufer 1924b: 15). By the 1630s, tobacco was being cultivated in Indochina as well as in Taiwan (Höllmann 1988: 24, 28, 50–1).

Tobacco seems to have reached the Near East at about the same time as it appeared in the Far East. By the turn of the seventeenth century its use was known in Persia having been introduced, according to one authority, by the Portuguese (Comes 1900; Laufer 1924b: 15). It was introduced into the Ottoman Empire at about the same time by the English (Birnbaum 1957: 24–6).

As for Africa, the whole issue of tobacco's introduction is steeped in controversy. Yet it is clear that by 1630, at the very latest, tobacco had penetrated West Africa (Ozanne 1969: 24). On the eastern side of the continent tobacco was being cultivated in Madagascar as early as 1638 (Laufer 1930: 11); and in the Maghreb definitely by the turn of the seventeenth century (Ozanne 1969: 35). Given the nature of the historical contacts between different regions of Africa and other parts of the world, there is growing evidence that the introduction of tobacco into Africa was made at different points by different people: Morocco by the English (Ozanne 1969: 36–7); West Africa possibly by the French (Philips 1983: 317–18); and East Africa by a combination of Portuguese and Arabic intermediaries (Laufer 1930: 10).

We can therefore safely argue that by the third decade of the seventeenth century tobacco had completed its circumnavigation not only as a substance but also as a cash crop. How can we explain this remarkable penetration?

In the same way as Europeans were attracted to tobacco because of its avowed efficacy as a herbal panacea, so too, to the extent that it is possible to be precise, were the Chinese, Javanese, Japanese and Indians. One of the first Chinese to write about tobacco underscored its beneficial medicinal properties. Chang Chieh-pin, a physician from Shan-yin in Chekiang province, in tones that are strongly reminiscent of contemporary European medical authority, particularly that of Monardes, recommended tobacco for expelling colds, for reducing swellings in cases of dropsy, for malaria and cholera (Laufer 1924b: 3; Goodrich 1938: 648). Its efficacy in counteracting malaria was, according to Chang, the main reason for its success. In his words:

> Inquiring for the beginnings of tobacco-smoking we find that it is connected with the subjugation of Yünnan Province. When our forces entered this malaria-infested region, almost every one was infected by this disease with the exception of a single battalion. To the question why they had kept well, these men replied that they all indulged in tobacco. For this reason it was diffused into all parts of the

country. Every one in the south-west, old and young without exception, is at present addicted to smoking by day and night.

(Laufer 1924b: 3)

It is interesting that Chang, as well as his contemporary the poet and
essayist Yao Lü who also wrote about tobacco, refer only to smoking and
not to other methods of tobacco consumption. Yao Lü is precise in his
description of tobacco:

There is a plant called tan-pa-ku produced in Luzon . . . You take
fire and light one end and put the other in your mouth. The smoke
goes down your throat through the pipe. It can make one tipsy but
it can (likewise) keep one clear of malaria . . . It is commonly called
gold-silk-smoke.

(Goodrich 1938: 649)

Other accounts stress tobacco's role in alleviating diseases brought on by
'extreme cold' (Goodrich 1938: 651). The fullest account of tobacco as a
panacea appeared in a Chinese herbal written some time between 1644 and
1661.

Tobacco has an irritating flavor and warm effect and contains poison.
It cures troubles due to cold and moisture, removes congestion of
the thorax, loosens the phlegm of the diaphragm, and also increases
the activity of the circulation. The human alimentary and muscle
systems are aided in their smooth operation as the smoke goes directly
from the mouth to the stomach and passes from within to outside,
circulating around the four limbs and the hundred bones of the
body . . . it gives man satisfaction whenever he is hungry . . . it makes
man hungry when he is sated. If a person smokes when he is hungry,
he feels as though he has taken plentiful food; and when he smokes
after eating sufficiently, it affords good digestion in a most satisfactory manner. For this reason many people use it as a substitute for
wine and tea, and never get tired of it, even when smoking all day
long.

(Laufer 1924b: 8–9)

The first introduction of tobacco into India was made under the guise
of a remedy. Asad Beg, a chronicler of India, wrote in 1605 that the
Mughal Emperor Akbar was introduced to tobacco as a medicine which
European doctors had praised in their writings (Elliot and Dowson 1875:
166; Sangar 1981: 207–11). In south-east Asia tobacco also assumed a
medicinal role (Höllmann 1988: 101–3). While the Chinese yang–yin (hot–
cool) medical system easily absorbed and defined tobacco use, in south-
east Asia a similarly easy incorporation of tobacco was effected by the

striking similarity of medical systems in this region (a combination of Indian Ayurvedic and Greek and Arabic theories) to the humoral system prevailing in contemporary Europe (Reid 1988: 52–7). While tobacco was smoked, in both pipes and cheroots, it also became incorporated in the ritual associated with betel chewing, though precisely when tobacco was so used remains unclear (Reid 1985: 535–8). As a supplement to betel tobacco was also appreciated for its narcotic properties and was assigned ritualistic meanings derived from betel consumption, namely as a polite offering to a guest, as a relaxant and as a substitute for food (Reid 1985: 530–2).

Moreover, in the same manner that tobacco entered systems of healing in Europe as a herbal panacea, south-east Asian medicine employed herbal remedies to great effect. Betel in particular was employed to prevent tooth decay, aid digestion and prevent dysentery (Reid 1988: 54). No wonder then that tobacco, with its avowed properties, could not only take its place alongside betel as a medicine but could be consumed together with it.

In the Near East tobacco seems to have had the same meanings as it did in south-east Asia. Simon Contarini, the Venetian ambassador to the court in Constantinople, in his report to the Doge in 1610 referred to tobacco as 'a certain herb which comes as a medicine from England' (CSPV 1607–1610: 505–6). İbrahim Peçevî, in his history of the Ottoman Empire written in the sixteenth century, confirms that 'about the year [1600/1601] the English infidels brought it [tobacco] and sold it as a cure for "wet" diseases' (Birnbaum 1957: 24). Later in the same century, John Fryer, an official of the East India Company, on his travels through Persia was impressed by tobacco's role in social relationships. 'Tobacco is a general companion', he wrote, 'and to give them their due, they are conversable Good-Fellows, sparing no one his Bowl in their turn . . .' (Fryer 1912: 210). In contrast to the use of tobacco as a supplement to betel in India and south-east Asia, in the Near East tobacco was clearly associated with coffee and with the coffeehouse, itself a Near Eastern social invention (Fryer 1899: 234; 1915: 34; Hattox 1988).

Whether medicine was the vector for tobacco's incorporation into African cultures is less certain. Available sources make no mention of tobacco as a remedy. On the contrary, what is stressed in these sources is the juxtaposition of tobacco with other narcotic substances, principally kola and cannabis (Philips 1983: Ozanne 1969: Laufer 1930). Until further research is carried out, the relationship between tobacco and medicine in Africa must remain unclear.

The Europeanization of tobacco in the sixteenth century involved offering tobacco as a herbal panacea rooted firmly within European medical tradition. It thus entered European culture as a popular remedy though there were strong objections to its use. Nevertheless, Europeans offered tobacco to the rest of the world in their terms. From the Near East to

China medical systems embraced the notion of herbal remedies, as well as understanding their actions in terms broadly similar to those prevalent in Europe. Just as in the European case, a suitable niche for tobacco was already present in these diverse cultures, and incorporation was virtually automatic.

Part II

Custom, in this small article I find
What strong ascendance thou hast o'er the mind.
My friend's advice the first inducements were
'Take it', said she, 'it will your spirits cheer.'
All resolute the offered drug to take
But in the trial sickened with my hate.
By repetitions I was brought to bear,
Then rather liked, now love it too, too dear.
Be careful, oh, my soul! how thou let'st in
The baneful poison of repeated sin;
Never be intimate with my crime,
Lest Custom makes it amiable in time.

Elizabeth Teft (1747), from Roger Lonsdale, *Eighteenth-
century Women Poets: an Oxford Anthology*
(Oxford: Oxford University Press, 1988)

Peart and chipper and sassy,
    Always ready to swear and fight,–
And I'd larnt him ter char terbacker,
    Jest to keep his milk-teeth white

John Hay, from 'Little Breeches' (1871)

4

RITUALS, FASHIONS AND A MEDICAL DISCOURSE
Tobacco consumption before the cigarette

It was probably not until the 1570s that tobacco began to enter into consumption patterns outside of North and South America. Its role as a medicine has been outlined in the previous chapter where it was also argued that it was tobacco's avowed and advertised therapeutic value that made it acceptable in many diverse cultures. An obvious question that presents itself is: when did the underlying reason for tobacco consumption change from therapy to recreation? This is a difficult question to answer with any degree of confidence but, before attempting to do so, it is necessary to outline consumption patterns in more detail.

There is little doubt that within a relatively short period of time after its introduction into European culture the demand for tobacco grew at a bewildering rate. Statistics on the importation of tobacco into England during the seventeenth century provide the clearest evidence of this phenomenon. In 1603, the first year for which there are satisfactory data, 25,000 pounds of tobacco, all of it from Spanish America, was imported into England; in 1700 the corresponding figure was almost 38 million pounds (Gray and Wyckoff 1940: 18, 24).

The pace of increase of imports was greatest in the second half of the century but, according to a recent interpretation of the tobacco phenomenon in England, tobacco was a commodity of mass consumption certainly by c. 1670 and possibly several decades before: that is, enough tobacco was available for at least 25 per cent of the adult population to have a pipeful at least once daily (Shammas 1990: 78). If one accepts this definition, then tobacco was the first of the range of non-European exotica to establish itself permanently as a European commodity, well before chocolate, coffee, tea and even sugar. In England the latter was not a commodity of mass consumption until the end of the seventeenth century and tea did not have a similar status until the 1720s (Shammas 1990: 81, 84). Evidence from other parts of Europe suggests a rapid penetration by tobacco, though, with the possible exception of Holland, neither the pace nor the extent was as great as in England. Using the above definition of a mass-consumption commodity, tobacco was not in this class in France until the middle of the eighteenth century (Price 1973: 8).

The progress of tobacco consumption in England until at least the eighteenth century is much easier to chart than elsewhere, largely because of the relative abundance of information. The availability of fairly reliable trade and demographic statistics allows us to witness the rise of tobacco consumption on a per capita basis. Table 4.1 outlines the trend as best as can be done though it is far from perfect. With the exception of the final entry around the turn of the century, the figures refer only to legal consumption and therefore exclude smuggled or contraband tobacco. Though there is more than enough evidence that smuggling occurred it has not been possible to quantify it, but most authorities seem to agree that smuggling did not become a very serious problem until towards the end of the seventeenth century when customs duties rose considerably in a very short time (Rive 1929: 554–8; Ramsay 1952; Nash 1982: 357; see also Chapter 7).

Table 4.1 Tobacco consumption, England and
Wales 1620–1702

Years	Annual consumption (lb per capita)
1620–9	0.01
1630–1	0.02
1669	0.93
1672	1.10
1682, 1686–8	1.64
1693–9	2.21
1698–1702	2.30

Source: Shammas 1990: 79

Much less is known about tobacco consumption in other European countries. Historians give a strong impression of the Dutch as avid tobacco consumers, and, although precise data are lacking, an amount of judicious manipulation of figures and assumptions does provide general support for this assertion (Schama 1987: 193–201). Around 1670 the Dutch consumed about 3 million pounds of tobacco, or 1.5 pounds per inhabitant (Roessingh 1978: 42; Price 1961: 90; Price 1973: 186). In other words, tobacco was being mass-consumed in Holland and possibly even earlier than in England. Certainly, there is sufficient literary and pictorial evidence that attests to the widespread use of tobacco by all social classes from early in the seventeenth century; there is also evidence that the Dutch paid less for their tobacco than did the English, and, given their highly efficient and extensive inland water transport, there is every likelihood that imported and domestic tobacco was more easily available in Holland than in England

(Schama 1987: 193–201; Price 1973: 852; de Vries 1978). French consumption, as already stated, was relatively low during the seventeenth century, probably no more than 0.33 pounds per person towards the end of the century (Price 1973: 10). This level increased considerably in the eighteenth century but it was probably not before the middle of that century that tobacco became mass-consumed (Price 1973: 377–8). Portugal is the only other country for which an estimate of per capita consumption can be made, in the absence of precise data. At the turn of the eighteenth century Portugal consumed about 1 million pounds of tobacco, equivalent to a per capita figure of 0.5 pounds, but by mid-century tobacco was being mass-consumed (Lugar 1977: 36, 47–9; Hanson 1982: 156–7). Other evidence also points to increasing per capita consumption. The output of snuff from the royal manufactory in Lisbon increased by 30 per cent between 1700 and 1750 (Nardi 1986: 19). It is possible to give an indication of the extent of tobacco consumption in Austria though this is for the end of the eighteenth century. The Austrian State Tobacco Monopoly, with its manufacturing facilities in Vienna, began operations in 1784 to supply tobacco consumers in the Austrian state with manufactured tobacco, snuff and smoking tobacco. In its first year of operation the monopoly produced just over 14 million pounds of tobacco, which translates into a per capita consumption of more than 1 pound, suggesting that Austrians were mass consumers of tobacco by the mid-eighteenth century (Hitz and Huber 1975: 195). Information for other parts of Europe is not available but it would be reasonable to argue, in general, that tobacco emerged as a European mass-consumed commodity in the eighteenth century, probably by 1750.

While it is possible to get some idea of the amount consumed, it is quite another matter to find out who was consuming tobacco. Some commentators on tobacco and its medical uses and abuses were clear in their prescription as to who would benefit from nicotian therapy. James Hart, writing in 1633, agreed with the prevailing wisdom, as derived from Monardes, that tobacco expelled phlegm and generally warmed and dried the body (Hart 1633: 317). Children, he advised, should not consume tobacco, for they tended to be hot-brained, as he put it; and the same proscription was extended to pregnant women. Hart went on to add that tobacco was best for older men, 'where the brain is cold and moist', and generally for all those who live in 'moist, fenny, waterish . . . places; as in Holland, in Lincolnshire' (Hart 1633: 320). Towards the end of the century another famous tobacco polemicist argued, in opposition to Hart, that pregnant women should smoke, as they cannot properly nourish the foetus if the stomach is not working properly: tobacco smoke was understood to stimulate gastric functions (van Peima 1690). Interestingly, except for the references to pregnant women, there is little sense in seventeenth-century tobacco literature of any prohibition of tobacco use by gender. Women were constituted, according to humoral theory, as cold and moist, and could theoretically withstand or benefit from tobacco consumption (Maclean 1980). Dutch genre painting of the seventeenth century often portrayed women smoking, though it is not clear whether the depiction of smoking women is a form of opprobrium (Schama 1987: 203–13).

It is doubtful whether we will ever know if there was any gender or age prohibition about smoking before the nineteenth century. Odd references do appear that give some indication of who was consuming tobacco but no conclusive statement is possible. Both Egon Corti and G. L. Apperson, two early writers who were particularly interested in the social history of tobacco, could not be definitive about the social profile of smokers (Corti 1931; Apperson 1914). Until other evidence is forthcoming it is best to leave the issue as it stands, by stating that no conclusive proof exists that points to any proscription against anyone of whatever age, gender or social class consuming tobacco. Though we may find the image of a pipe-smoking woman uncomfortable because of our own gender assumptions and constructions, there is little evidence of this in the seventeenth century. In a typical example of a seventeenth-century allegory on the abuse of tobacco and its dire effects, both physical and moral, *Bacchus's Wonder Wercken* by the Dutch engraver Gillis van Schenyndel, published in Amsterdam in 1628, both men and women are depicted smoking pipes and succumbing to the temptation of the weed (Schama 1987: 214–15). David Teniers, the seventeenth-century Flemish painter, who often chose tobacco as a subject for his canvases, depicted a woman smoking a pipe in one of his paintings. In it a woman is sitting in an inn or tavern with a man who has just filled her pipe with tobacco. The woman, smartly dressed, is at the point of lighting her pipe in a distinctly elegant way. The scene is extremely serene, even touching, an atmosphere created not only by the lack of commotion (so typical of many other tavern scenes of the period) but also, and more importantly, by the warmth implied by the gaze of the woman's companion as well as by another, older woman, looking in from a nearby window (Brongers 1964: 195). The entire feeling of the painting is of refined dignity, an atmosphere which is replicated in another, delicate, painting entitled *The Sleeping Beauty* by the seventeenth-century Dutch master Gabriël Metsu. Refinement is also the theme in an eighteenth-century engraving by N. Arnoult of a smoking tavern in France. Entitled *La Charmante Tabagie*, the engraving shows a group of three elegantly dressed women sitting around a table. Two of the women are smoking long-stemmed clay pipes at the table while the third woman is standing up cutting the tobacco roll in preparation for smoking (Vigié 1989: 79). But perhaps the most famous painting of a woman smoking a pipe is a portrait of the painter Madame Vigée-Lebrun in the second half of the eighteenth century. In this picture the painter is depicted elegantly dressed smoking a long-stemmed clay pipe at a table on which are two or three other pipes as well as a few devices for lighting pipes (Corti 1931: 206). Other literary references to women smoking are scattered but include the famous botanic compendium of Paul de Reneaulme published in Paris early in the seventeenth century where, in a passage devoted to the therapeutic value of tobacco, he singles out women for particular mention: 'How many women have I seen, almost lifeless from headache or toothache or catarrh in the lungs or elsewhere, restored to their former health by the use of this plant?' (Dickson 1960: 155–6). In the English-American colonies both visitors and travellers, when commenting on the habits of the colonists, stressed the near-universal use of tobacco. One such traveller reported in 1686 his observations on a backwoods settlement:

> Everyone smokes while working and idling. I sometimes went to hear the sermon; their churches are in the woods and when everyone has arrived the minister and all the others smoke before going in. The preaching over, they do the same thing before parting. They have seats for that purpose. It was here I saw that everybody smokes, men, women, girls and boys from the age of seven.
>
> (Robert 1952: 99)

Peter Kalm, the Swedish botanist who was dispatched by the Swedish Academy in search of plant seeds, left vivid descriptions of tobacco consumption in Canada and Pennsylvania around the middle of the eighteenth century, both of which refer to the ubiquity of the habit. In referring to the people of French Canada he wrote:

> It is necessary that one should plant tobacco, because it is so universally smoked by the common people. Boys of ten or twelve years of age, as well as the old people, run about with a pipe in their mouth . . . People of both sexes and of all ranks, use snuff very much.
>
> (Kalm 1966: 510–11)

On his way to Philadelphia Kalm made the following observation about the inhabitants of the region:

> The English chewed tobacco a great deal, especially if they had been sailors. Not an hour passed when they did not take as much cut tobacco as they could hold in the fingers of the right hand and stuff it in the mouth. Young fellows of from fifteen to eighteen years of age were often as bad as the older men.
>
> (Kalm 1966: 637)

Despite the fragmentary nature of the evidence it is reasonable to argue for a widespread diffusion of the tobacco habit. If anything constrained the pace of diffusion, it was probably only economic factors of price and availability: specific social proscriptions, if they existed at all, were unimportant. At the beginning of the seventeenth century tobacco was a

63

luxury product, but this situation did not last long. Fragmentary retail prices suggest that Spanish-American tobacco averaged £1. 10s. per pound and this price remained stable until the middle of the second decade (Lorimer 1973: 270). At the time a labourer could expect a wage of 8d per day (Clarkson 1971: 222). As supplies of tobacco increased, especially after the Virginia colony started to concentrate entirely on this crop, the price of tobacco fell precipitously. The farm price of Virginia tobacco, which in 1618 stood at an average of 40d. per pound, collapsed to 3d. per pound in the 1630s and to 2d. in 1660 (Menard 1976: 404–8). For the rest of the century farm prices rarely reached one penny per pound (Menard 1980: 158–60). Wholesale prices in 1683 stood at 5½d. pence per pound and farm prices at ⅞d. per pound (Nash 1982: 369; Menard 1980: 159). Meanwhile, a labourer's wage had risen to 12d. per day (Clarkson 1971: 222).

The sharp drop in price certainly affected the demand for the product and it is reasonable to argue that as tobacco became cheaper it was consumed more widely. Yet this would be a hasty conclusion to draw. During the first half of the seventeenth century prices came down much faster than per capita consumption rose. Farm prices in 1640 had fallen to a tenth of their 1620 level; the wholesale price in London fell from 8s. to 1s. per pound and prices for Virginia tobacco in Amsterdam show a similar fall (Menard 1980: 150). Retail prices are not available for the period but it seems reasonable that they should fall also though perhaps not as quickly as wholesale prices and certainly not as steeply as farm prices. Yet, over the same period, as Table 4.1 shows, per capita consumption barely changed. In 1640 one can estimate per capita consumption at a shade under 0.02 pounds (Menard 1980: 158; Wrigley and Schofield 1981: 532).

One likely reason for the lack of growth of per capita consumption was the problem of availability. For the first half of the seventeenth century London dominated the tobacco trade. In the 1620s and 1630s London's share of tobacco imports fluctuated between 80 per cent and 90 per cent (Gray and Wyckoff 1940: 18–19; Pagan 1979: 256–62; Williams 1957: 418–20). The remainder was imported through south coast ports such as Plymouth and Southampton as well as Bristol. Shipments of tobacco inland or coastwise must have raised the final price of tobacco, offsetting, to some extent, the fall in wholesale price at the port of entry.

Direct evidence pertaining to the spatial diffusion of English tobacco consumption is lacking, but remains of clay pipes (the chief way of consuming tobacco in this period) confirm the predominant role of London. Clay pipe manufacture began in London towards the end of the sixteenth century, probably in the 1570s, and the capital continued to be the centre of this industrial activity until the middle of the following century (Walker 1983: 2; Atkinson and Oswald 1969; Oswald 1960: 42). Pipes found in sites outside London dating from before the middle of the century tended to be London-made (Lawrence 1979: 68). The provincial pipe making industry did not get off the ground until well into the seventeenth century. The available evidence suggests that the industry diffused from London in accordance with the primary trade routes linking the capital with the rest of the country. It is therefore not surprising to find that the clay pipe production outside London before the 1640s was mostly confined to Bristol, Newcastle and Gateshead which between them had seven pipe makers established in the 1630s (Davey 1988: 4; Oswald 1960: 43–4; Walker 1971; Walker 1983: 3). Pipe making was established in Northampton, Hull, Liverpool, Chester and York by the 1640s and Norwich by 1660 (Watkins 1979: 85; Wells 1979: 123; Karshner 1979: 295). Once established in the provincial centres, pipe making continued to grow for the rest of the century and into the early decades of the eighteenth century. Table 4.2 presents a partial outline of the growth of pipe making in England by assembling available data for Hull, Chester, Newcastle and Gateshead.

Table 4.2 Pipe makers in England 1630–1700

Years	*Number*
1630–9	7
1640–9	18
1650–9	30
1660–9	37
1670–9	46
1680–9	61
1690–9	66

Sources: Watkins 1979: 104; Rutter and Davey 1980: 49; Davey 1988: 4

Though it is not complete, in that it includes only a few centres, the message from the data displayed in Table 4.2 confirms what the import data showed: namely, that after 1640 or 1650 there was a substantial increase in the availability of tobacco in England. That smoking spread after 1640 and possibly diffused throughout the social structure in rural England is also confirmed by remains from Cheshire which show that pipes from the gentry-owned priory dated from 1600, whereas those from the village dated from the mid-1650s (Davey 1985: 164).

The art of pipe making spread from England to Germany, Scandinavia, possibly to France, but most especially to Holland (Laufer 1924b: 57–8). Dutch pipe makers were to be found in Holland's major cities, including Amsterdam, Haarlem and Groningen, but their greatest concentration was in Gouda. There were in the seventeenth century as many as fifteen thousand workers in this industrial activity, about half of the city's labour force (Schama 1987: 195). The main period of activity of the Gouda pipe industry was after mid-century but Gouda pipes were widely exported to France, Germany and Scandinavia (Israel 1989: 267). Even as late as the middle of the eighteenth century there were over 350 pipe makers in Gouda (Kellen-benz 1977: 507).

The pipe was the symbol of tobacco smoking in most of northern and north western Europe. Horatio Busino, a visitor to London in the second decade of the seventeenth century, recorded his impressions of the capital's smoking public. His description of the ritual of the pipe, as he witnessed it in 1618, is particularly striking:

> One of the most notable things I see in this kingdom and which strikes me as really marvellous is the use of the queen's weed, prop-erly called tobacco, whose dried leaves come from the Indies, packed like so much rope. It is cut and pounded and subsequently placed in a hollow instrument a span long, called a pipe. The powder is lighted at the largest part of the bowl, and they absorb the smoke with great enjoyment. They say it clears the head, dries up humours and greatly sharpens the appetite. It is in such frequent use that not only at every hour of the day but even at night they keep the pipe and steel at their pillows and gratify their longings. Amongst themselves, they are in the habit of circulating toasts, passing the pipe from one to the other with much grace, just as they here do with good wine, but more often with beer. Gentlewomen moreover and virtuous women accustom themselves to take it as medicine, but in secret. The others do it at pleasure . . .
>
> (*CSPV* 1617–1619: 101–2)

Pipes were generally made of clay, but more elaborate constructions existed, including silver pipes, and, as ceramic techniques were perfected, sumptuous, sculpted and decorated pipes began to appear (Laufer 1924b: 35: Brongers 1964). It was not, however, the only accoutrement of the smoker. Tobacco boxes were an indispensable accompaniment and these contained not only the smoking tobacco but all of the devices necessary to produce the smoke, including flint and steel and ember-tongs, in addition to the pipe itself (Laufer 1924b: 38). These boxes could be simple and made of domestic wood; or they could be quite extravagant, con-structed out of scarce woods such as nutwood, expensive metals such as silver and copper, or exotic materials such as ivory, tortoise-shell and mother-of-pearl (Laufer 1924b: 38). It was the box, rather than the pipe, that was coveted: tobacco boxes were frequently exchanged as gifts and they also appeared in long-distance commerce – the Portuguese, for example, imported lacquered and gilded tobacco boxes into Japan in 1637 (Laufer 1924b: 38; Schama 1987: 195; Boxer 1959: 196).

Sources pertaining to the culture of pipe smoking present a mixed image of the pipe smoker. On the one hand, pipe smoking was part of a social, public occasion. Smoking clubs and smoking schools appeared in England and France in the late sixteenth and early seventeenth century (Mackenzie

1984; 90–2). The pipe at the mouth and the glass in the hand were a common image portrayed in seventeenth-century French, Dutch and Flemish paintings, as well as in personal accounts (Schama 1987: 193–201; Brongers 1984: 137–56; Vigié 1989: 56–61). The French referred to these smoking taverns as *tabagies* and in descriptions of them the parallel with Amerindian smoking sessions was often drawn (von Gernet 1988: 113–19). On the other hand there is little doubt that tobacco was being consumed in a private manner, as many still-life paintings evoking a tranquil domestic scene suggest. Perhaps pipe smokers puffed in the confines of their own home for medicinal reasons while choosing the public space of the tavern for recreational consumption.

For the most part Europeans, whether in Europe or the New World, consumed tobacco by smoking it: pipes in northern Europe, and the corresponding colonies across the Atlantic, and the cigar in Spain, possibly Portugal and its possessions. Whether Europeans smoked pipes or cigars seemed to depend on which form of consumption they had encountered in the New World. The smoking of cigars, that is prepared tobacco leaves wrapped either in tobacco leaves or some other vegetable matter, was the preferred Amerindian form of smoking in Central and South America while the pipe was common elsewhere (Wilbert 1987: 64–121; von Gernet 1988). Unlike the pipe and its manufacture which was diffused widely across Europe (and to other parts of the world), the cigar, in Europe, at least, remained confined to the Iberian peninsula until the end of the eighteenth century. The royal tobacco manufactories in both Seville and Cadiz were producing both large and small cigars by the end of the seventeenth century (Perez Vidal 1959: 90–5). But for the rest of Europe, the sight of a cigar was still unusual. John Cockburn, an English traveller in Costa Rica in 1735, left this account of cigar-smoking, a description that confirms the novelty of the practice to his eyes at least:

> These gentlemen [he had just encountered three friars] gave us some seegars which they supposed would be very acceptable. These are leaves of tobacco rolled up in such a manner that they serve both for a pipe and tobacco itself. These the ladies, as well as gentlemen, are very fond of smoking; but indeed, they know no other way here, for there is no such thing as a tobacco-pipe throughout New Spain, but poor awkward tools used by the Negroes and Indians.
>
> (Mackenzie 1984: 225)

Chewing tobacco, another form of consumption that Europeans witnessed among Amerindian tobacco consumers, remained distinctly marginal in European culture. It is difficult, if not impossible, to quantify the amount of tobacco that went into chewing in Europe. The records of the French tobacco monopoly, which give some of the best indications of the distribution of forms of consumption, yield little concrete information. In

1708, for example, the Dieppe manufactory's specifically designated chewing tobacco represented less than 1 per cent of the year's output: more than two-thirds of the output, however, was of the standard rolled form – called an *andouille* – that could be cut up by the consumer and either smoked, snuffed or chewed (Price 1973: 193). On the other hand, in the same year, the tobacco monopoly purchased 500,000 pounds of Brazilian tobacco – one-sixth of the year's total – particularly renowned in Europe as a chewing tobacco (Price 1973: 187). This figure should not, however, be taken as anything more than an indication of the potential use of Brazilian leaf in chewing tobacco. André Antonil, the author of an important description of Brazil near the beginning of the eighteenth century, makes it clear in his account of tobacco growing around Bahia that the best leaf was used for smoking, chewing and snuffing (Antonil 1965: 315). Partly because of the vogue for snuff which was sweeping Europe at the time, Antonil included in his account a detailed description of how to grind and perfume tobacco, the results of which he praises: he was far less impressed by chewing, not because of any lack of therapeutic efficacy, but because if used continually, he noted, 'the [beneficial] effects decrease, it alters taste, taints the breath, blackens the teeth and fouls the lips' (Antonil 1965: 315–17, 321).

The fact that Antonil chose to criticize chewing tobacco on the grounds of the appearance of the consumer, rather than the virtues (or not) of the product, has more to do with the issue of respectability than of the use of chewing. Père Labat, writing about a slightly earlier period, while acknowledging that the tell-tale effects of chewing tobacco were unsightly – as he said, 'the bad odour of the breath . . . cannot be corrected even after bathing the mouth with a quantity of eau-de-vie' – nevertheless recommended it for its therapeutic effects, including the well publicized influence on hunger and thirst (Labat 1742: 280). Edme Baillard, who wrote a very influential treatise on tobacco in the second half of the seventeenth century, was, however, impressed by chewing tobacco, highlighting its efficacy in standard humoral terms as well as praising its power in suppressing hunger (Baillard 1668: 93–4). Yet even such a distinguished physician had little impact on consumption patterns and it must be assumed that chewing tobacco in the seventeenth century was confined to sailors and those who worked outdoors or where the danger of fire was great. It should be pointed out that while Europeans sought to imitate, or absorb, Amerindian habits of consuming tobacco, they were least successful when it came to chewing. When Amerindians chewed tobacco they did it with the addition of an alkalizing substance, usually ash, but also burnt seashells. The effect of mixing tobacco with an alkaline catalyst is to accelerate and intensify the action of nicotine, though nicotine is liberated naturally by mastication (Wilbert 1987: 138). Why Europeans did not follow this example, as they did in smoking and snuffing, is uncertain, but it may be that the inefficiency in nicotine administration through chewing when compared to smoking led Europeans to prefer the latter over the former.

If smoking and, to a lesser extent, chewing were un-European practices, then snuffing must have appeared particularly bizarre. One of the first accounts of tobacco consumption among Amerindians to circulate in Europe was that of Friar Ramon Pané, a Catalan priest who remained in Hispaniola on Columbus's second voyage and whose writings were reproduced in Peter Martyr's history of the New World which appeared in 1511. The passage is as follows: 'wherefore, when their chiefs consult the zemes (magical objects) about the outcome of a war, about the harvest or about their health they go into the house dedicated to the zeme, and there, having snuffed cohoba into their nostrils, they say that the house is turned upside down' (Dickson 1954: 25). This description was a paraphrase of Pané's more detailed original.

> The cogoiba (cohoba) is a certain powder which they take to purge themselves, and for the other effects of which you will hear of later. They take it with a cane about a foot long and put one end in the nose and the other in the powder, and in this manner they draw it into themselves through the nose and this purges them thoroughly.
>
> (Dickson 1954: 25)

Even if Pané's original did not enjoy the wide circulation of other writers in the sixteenth century, including Columbus himself and Oviedo y Valdes, the description of snuffing must have entered into the European imagination at the time. Imagination, however, is not the same thing as practice. Despite the story that Jean Nicot delivered tobacco to Catherine de Medici in the form of snuff – of which there is no corroboratory evidence – snuffing tobacco was probably not practised until the second or third decade of the seventeenth century, at least to any noticeable level (Dickson 1954: 92). Certainly snuffing tobacco must have been practised by Spanish priests in Peru as early as 1588. In that year the rules about priestly behaviour included the following statement:

> It is forbidden under penalty of eternal damnation for priests, about to administer the sacraments, either to take the smoke of sayri, or tobacco, into the mouth, or the powder of tobacco into the nose, even under the guise of medicine, before the service of the mass.
>
> (Dickson 1954: 150)

This charge against Spanish priests in the New World was repeated in other parts of Spanish America and in Spain itself. A Papal Bull of 1642 expressly forbad clerics to 'take tobacco in leaf, in powder, in smoke, by mouth or nostrils in any of the churches of Seville, nor throughout the archbishopric' (Dickson 1954: 154). The practice of snuffing seems to have become particularly popular in Spain, not only by clerics, who clearly enjoyed tobacco in this way, but by less endowed individuals (Perez Vidal 1959: 73–4). Snuff was being produced by the Spanish royal monopoly by the second half of the seventeenth century (Rogoziński 1990: 68).

As a European form of tobacco consumption, snuff, therefore, had its origin in Spain. The Portuguese followed suit very quickly; in one of the first years of its operations, the snuff manufactory of the royal tobacco monopoly in Lisbon produced nearly one-third of a million pounds of snuff (Nardi 1986: 19). The significance of this should not be overlooked. Snuff in both Lisbon and Seville was a manufactured product and can be considered as the first example of manufactured, as opposed to processed tobacco. Indeed, wherever tobacco manufacture and distribution were the responsibility of a state monopoly as in Spain, France, Portugal and Austria, an important, if not the most important, aspect of the monopoly was to provide centrally produced snuff. The manufacture of snuff entailed not only the washing and grinding of the tobacco leaves but the addition of both colours and perfumes (Brooks 1937: 158). Snuff was thus a tobacco product distinctly different from tobacco for chewing and smoking in which manufacturing consisted only of cutting or shredding.

Spain and Portugal were the leading producers of snuff in seventeenth-century Europe but by the end of the century snuff was being manufactured in many other parts of Europe. The French tobacco monopoly was one of the first to manufacture snuff centrally, providing either prepared snuff, or *tabac ficelé*, which could be ground into snuff by the consumer (Price 1973: 174–5, 187–8, 465–76). On the eve of the Revolution snuff, in both of these forms, represented over 80 per cent of the 15 million pounds of manufactured tobacco (Price 1973: 426).

Other tobacco monopolies followed the lead into snuff. The Portuguese royal tobacco monopoly, as previously stated, began producing snuff at its Lisbon manufactory in 1675 (Nardi 1986: 15, 19–20). At the beginning production focused on the finest snuffs, but over the following century there was a shift towards cheaper types, reflecting the appeal of snuff to a wider market (Nardi 1986: 19–20). Italy, too, seems to have been gripped by the snuff fashion. Comparative data are unavailable but the reign of snuff is confirmed by the fact that in 1821, for example, more than half of the tobacco sales in Lombardy and Venetia was snuff (Rogoziński 1990: 56). Evidence from Austria and Scandinavia shows a similar trend towards snuff production during the eighteenth century (Hitz and Huber 1975: 195; Price 1961: 97–8; Rogoziński 1990: 103–4).

Elsewhere in Europe, where tobacco monopolies did not exist, snuff production also expanded. In Holland, the making of French-style *carottes* (pressed and shaped tobacco) dates from the 1740s, for example (Roessingh 1976: 401). Snuff mills were located in the vicinity of the Zaan, which was also the location of the shipbuilding industry (Roessingh 1976: 403–4). Dutch tobacco farmers benefited particularly from the upsurge in the

European demand for snuff even if they themselves continued to smoke pipes. Snuff prices on the Amsterdam bourse began to rise steeply after 1740 (Roessingh 1976: 456). In response Dutch tobacco farmers began to produce a heavier leaf well suited to snuff manufacture: heavy manuring was the key to the cultivation of this leaf (Roessingh 1978: 31). What would have been totally unsuitable for smoking tobacco soon found a ready market outside Holland. As the differential between Havana leaf and Dutch domestic leaf widened after 1740, Dutch snuff manufacturers began to export their product to southern Europe and it was only as manufactories were established, principally in Italy, that the Dutch switched from exporting snuff to exporting leaf (Roessingh 1978: 46).

British consumers also turned to snuff. When, precisely, is hard to ascertain, but certainly the habit was not uncommon around the middle of the seventeenth century, more in Scotland and Ireland than in England (Brooks 1937: 160; Mackenzie 1984: 115, 164). There is little doubt that the practice of snuffing grew over the course of the eighteenth century (Mackenzie 1984: 165–72). Direct evidence is lacking, but it is suggestive that the number of pipe makers in selected English towns shows an unmistakable decline, especially after the middle of the eighteenth century. In Chester, for example, pipe making as an industrial activity peaked in the 1730s, in Bristol in the 1740s and in Newcastle and Gateshead and Hull possibly two or three decades earlier (Rutter and Davey 1980: 49; Davey 1988: 4; Watkins 1979: 104; Walker 1983: 3). According to one authority on the subject, the English pipe making industry was in decline during the second half of the eighteenth century (Oswald 1960: 45).

This evidence certainly supports other, anecdotal, evidence pointing to the growing popularity of snuff particularly during the second half of the eighteenth century. Dr Johnson in 1773 proclaimed that 'smoking has gone out' and the number of snuff manufacturers and the range of snuffs were growing (Kiernan 1991: 27; Mackenzie 1984: 170). One tobacco manufacturer, Lilly and Wills, the forerunner of the more famous firm W. D. and H. O. Wills in Bristol, produced both smoking tobacco and snuff, in roughly equal proportions (Alford 1973: 27).

One of the problems that has puzzled historians of tobacco consumption in the eighteenth century is the apparent slowdown in per capita consumption in England over the century. Alfred Rive in 1926 summarized the most reliable statistics of the amount of tobacco kept for home consumption and these showed an unmistakable change during the eighteenth century. In per capita terms, he argued, consumption fell from just under 2 pounds per annum in 1700 to under half that amount in 1786 (Rive 1926: 63). With a fuller data set and better population statistics Jacob Price was able to refine Rive's figures but essentially arrived at the same conclusion: that per capita consumption, despite fluctuations, fell through the eighteenth century but mostly during the first half (Price 1954b: 89–90; Nash 1982:

356; Shammas 1990: 79). Recently, however, the entire exercise has been thrown into doubt because the arguments were based entirely on statistics of legal imports and, even though all participants to the discussion acknowledged that smuggling was a problem that affected the confidence of the results, none was able to offer a quantitative assessment of the extent of the contraband trade, until Robert Nash who constructed a new series of per capita consumption for the period 1698 to 1752 reproduced in Table 4.3. What these figures show clearly is that consumption did not fall during the first half of the eighteenth century but remained stable at around 2 pounds per capita per annum. Figures for the second half of the eighteenth century have not been adjusted for the contraband element. These show a small decline in per capita consumption to around 1.5 pounds – during the years of the American Revolution and the wars at the end of the century the figures fell below this level (Shammas 1990: 79).

Table 4.3 Tobacco consumption, England and Wales 1698–1752 (*lb per capita annual average*)

1698–1702	2.30
1703–7	1.56
1708–12	2.23
1713–17	1.80
1718–22	2.62
1723–7	2.13
1728–32	2.23
1733–7	2.00
1738–42	1.65
1743–7	1.56
1748–52	1.94

Source: Nash 1982: 367

Attempts have been made to explain either the decline, as it was, or the stability, as it is perceived now, in consumption. These have included the suggestion that the growth of spirit consumption especially between 1720 and 1760 might, in some way, have dampened the demand for tobacco (Price 1954b: 90–2); and more recently that, when in the later eighteenth century the alehouse began to fall into decline as a social institution, tobacco consumption suffered since publicans were a cheap and ready source of both tobacco and pipes (Shammas 1990: 81). Neither explanation, however, has much merit: the first begs the question – why should increased spirit consumption affect tobacco consumption? Alcohol and nicotine are not substitutes for each other. One could in fact just as easily make the argument that increasing alcohol consumption leads to an increase in tobacco consumption. Besides which, the social history of drinking in the eighteenth century is much more complicated than a simple statement of the correlation between alcohol and tobacco consumption would allow

(Clark 1988; Brennan 1991). As for the role of the alehouse, this again begs a question about the comparative social history of drinking and smoking, about which we know very little. Moreover by the eighteenth century the alehouse was only one of many retail outlets for tobacco (Earle 1989: 47).

A more likely explanation is the one that Alfred Rive suggested earlier on this century: that of an increase in the consumption of tobacco in the form of snuff rather than smoke (Rive 1926: 63–4). The evidence for the increasing preference for snuff over smoking tobacco in eighteenth-century Europe is very strong and there is more than a suggestion that snuff taking may have resulted in less pressure on per capita consumption of tobacco leaf. A given amount of tobacco leaf was stretched further in the manufacture of snuff than in that of smoking tobacco: not only did snuff contain many additives absent in smoking tobacco but there was much less waste in the former (Rive 1926: 63; Price 1961: 98; Price 1973: 423–5, 788). The fact that the English consumer was becoming increasingly attracted to snuff also suggests that the problem of explaining the stability in English consumption needs to be viewed within a European context. Where data are available, the slowdown in consumption levels was apparent in other parts of Europe. In Portugal, for example, the production of snuff at the Lisbon and Oporto manufactories fluctuated considerably over the century, though in the long run output kept pace with the rise in the country's population (Nardi 1986: 19–20). In Norway tobacco consumption appears to have levelled off after 1760 from a maximum of 2.5 pounds per inhabitant and then fell off to a level of 1.5 pounds per inhabitant around 1815: the shift to snuff has been implicated in this decline (Hodne 1978: 118–20). If we take a total European perspective, the trend of consumption registered for individual countries is confirmed. In 1710 one can estimate that western Europe consumed about 70 million pounds of tobacco and towards the end of the century this figure had likely risen to 120 million pounds (Price 1973: 732; Lugar 1977: 48; Roessingh 1978: 42–7). In per capita terms, however, the rise in consumption was far less, from about 1 pound to 1.2 pounds.

Snuff taking differed from tobacco smoking in many ways. In the first place, the taking of snuff became highly ritualized. For those who prepared their own snuff, a practice that became increasingly fashionable during the eighteenth century, the principal devices were the snuff box and the rasp. The former had its origins in the tobacco box of the seventeenth century but, unlike its predecessor, it contained only the tobacco product and therefore could be quite small. Snuff-box making became an art in the eighteenth century and the range of design, materials and size was bewildering. Not only was every known metal used but so were natural materials such as ivory and shell as well as the increasingly popular fine porcelain. In late seventeenth- and eighteenth-century France the manufacture of snuff boxes became the nexus of creative forces involving artists and artisans; some of France's leading artists were involved in designing snuff boxes for an aristocratic clientèle (Le Corbeiller 1966; Vigié 1989: 68–9). Giving a snuff box as a present became a sign of exalted gift-giving: Marie-Antoinette had fifty-two gold snuff boxes in her wedding basket (Vigié 1989: 70). While this may seem extravagant, it should be remembered that in the eighteenth century the snuff box was the equivalent of jewellery and not only did the snuff box change with artistic fashion but anyone who was anyone needed to have a variety of these boxes. As Louis Sebastian Mercier noted in his description of Paris in the second half of the eighteenth century: 'One has boxes for each season. That for winter is heavy; that for summer light. It's by this characteristic feature that one recognizes a man of taste. One is excused for not having a library or a cabinet of natural history when one has 300 snuff boxes . . .' (Vigié 1989: 71). Not only was the passion for snuff boxes enjoyed by the consumer but snuff boxes became a highly demanded collector's item – the Prince of Conti at his death had a collection of more than 800 snuff boxes (Vigié 1989: 73). Snuff rasps were also elaborate, though because they were used to prepare the snuff the material used was limited to ivory or metal (Brongers 1964: 160).

The artistry manifested in the snuff box was matched by the ingenuity in preparing snuff. Edme Baillard offered one of the first descriptions of how snuff should be prepared and the precise purpose of each ingredient. He recommended a formula of six parts Virginia to three parts St Christophe tobacco as the foundation for the mixture and then he goes on:

> the virtue of melitiot purges it of its sulphurous narcotic and softens it. The spirit of flowers of orange moderates its bitterness. The sandalwood takes the edge off its heat. The dye of Indies wood or alkanet gives it its colour. Angel water and its flowers takes away its strong and piquant scent leaving the former in its place.
>
> (Baillard 1668: 87–8)

The whole process required an elaborate operation of moistening with the liquors containing the flavourings, colours and essences and consecutive drying processes taking at least three weeks before it was ready (Brongers 1964: 86; Vigié 1989: 63–4). The results of preparing the snuff were made available under a variety of names each of which carried the message of the essence of the product. Thus in France *tabac d'Espagne* was reddish, perfumed with civet, musk and cloves and *tabac de Pongibon* was yellow, perfumed with civet and sweetened with sugar, orange flowers and jasmine. Other ingredients included bergamot and mint, for a refined snuff, or cumin and mustard for a stronger concoction (Vigié 1989: 64; Brooks 1937: 159). Snuffs also carried the name of their makers: in eighteenth-century England there were at least 200 kinds of snuff commercially available (Brooks 1937: 159). The precise recipes were jealously guarded either by the individual or more likely by snuff manufacturers themselves. One Bristol snuff manufacturer presenting evidence to a parliamentary committee of 1789 estimated that his recipes were worth more than £1000 – this at a time when the capital costs of a typical tobacco manufacturer were less than a fifth of this (Alford 1973: 12, 20).

Once the devices and the snuff were in place, then there were well followed rules governing the technique of taking snuff. One Dutch poem of the period captures the essentials of taking snuff

> Bend with great pump
> Two feet backward in space
> Gracefully forward again
> Now bend without disgrace
> According to the manner of the French.
> Now take from the pocket of your camisole –
> As though you did at the command
> Taught you at the French School –
> Your snuff-box big, small or middling,
> Tap it well-mannered,
> Sniff with a noble grace,
> Open it quickly,
> Offer it to your friend
>
> (Brongers 1964: 157–8)

Outside the privileged circles, however, dignified and ostentatious modes of snuff behaviour were of far less appeal. Most of the centrally manufactured snuff destined for the mass market lacked the exotic flavourings and colours and was no more than ground tobacco.

It is perhaps difficult to understand the appeal of snuff. Certainly today, in the West at least, it is rare, but the change to snuff from smoke was a remarkable occurrence in the social history of tobacco and one that had important consequences for the future history of the plant. The reasons for the shift to snuff can be found in two areas: the progress and results of the medical debate on tobacco consumption, and changing perceptions of respectable consumption.

In 1571, as the previous chapter related, Nicolas Monardes drew together all of the information, botanic and medicinal, then current about tobacco in the second part of his famous natural history of New World plants. According to Monardes there were three principal ways of taking tobacco, each having specific purposes. Green, fresh tobacco, often mixed with its own juice, was to be employed topically against pains, sores, wounds etc.; the same could be chewed to alleviate hunger and thirst; while dried tobacco that was burned was to be taken internally as smoke to dry the cold humours and expel excess phlegm. What he never discussed was the snuffing of powdered tobacco into the nostrils, though he does mention drawing smoke that way.

Monardes' treatise was both popular, in that it went through several editions and was translated widely throughout Europe, and very fashionable. It also, however, inspired a controversy that lasted for at least a century during which time many learned scholars, doctors, moralists, writers and others took sides in the great nicotian debate. Though this part of tobacco history has often been sited in England, in truth the debate was Europe-wide.

The debate was about many things but in almost all cases it was carried on within a medical discourse informed by the prevailing humoral theory. Once tobacco had been ascribed with its humoral properties – hot and dry in the second degree – the door was left open for debate and disagreement about tobacco's administration and efficacy. The reason for this was quite simple. Humoral theory provided a dynamic understanding of the human body: not only did the balance of humours change as a result of illness or environmental conditions but it also changed in the long term as the body grew older. Ageing was understood humorally as the process of drying up and, therefore, old people were frequently advised not to smoke. Disagreements about the use of tobacco focused very much on its humoral effects and the main participants to the debate fell naturally into two camps: those who were overwhelmingly committed to nicotian therapy because of the wide range of disorders it could cure; and those, less committed or even hostile, who felt that the therapy was either at best, ineffective or dangerous. Those of the Monardes school insisted on the therapeutic value of expelling moist humours while extending the list of diseases and ailments that tobacco alleviated. To a large extent Monardes' followers from all parts of Europe simply repeated his assertions or went a step further. Johann Magnen, for example, published in 1648 a treatise on tobacco that not only summed up the prevailing medical insights but provided a lengthier defence of tobacco that included many more nicotian cures. Cornelis Bontekoe, the Amsterdam doctor who was to tea what Monardes was to tobacco, stated that the discovery of tobacco was one of humanity's greatest achievements; he advocated universal smoking not only to cure illnesses but also to satisfy hunger and stimulate the brain (Brooks 1938: 492; Schama 1987: 172). In England one of tobacco's staunchest advocates was William Kemp who, in 1665, wrote a treatise on the plague. In it the author makes the interesting recommendation that smoking tobacco not only fumigates the air – pestilence and bad air were inextricably linked – but also draws out the unwanted humours. Nicotian critics, by contrast, attacked tobacco more specifically. The 1602 anonymously penned pamphlet *Work for Chimney-Sweepers: or, A Warning for Tobacconists*, argued in an attack on nicotian therapy that was echoed in many other publications afterwards, that the humoral qualities were dangerous, and not efficacious as its promoters maintained. Smoking, the author warned, 'withereth our unctious and radical moisture'; the sperm and seed of man was greatly altered and decayed; it had a stupefying effect; it increased melancholy and 'wasteth the liquid and thin part of the blood' (Philaretes 1602). Other commentators offered similar interpretations and, in the case of Simon Paulli, who published an attack on tobacco in 1665, thoroughly disapproved of the taking of tobacco to any extent, preferring to rename it *Herba Rixosa* or *Herba Insana*, and advocating its total destruction (Stewart 1967: 262; Brooks 1938: 378–80).

Both sides of the debate, and the themes they raised, continued to be discussed into the eighteenth century and, for the most part, nothing very new was added until the end of the eighteenth century. In addition to medical and proto-medical authorities, a serious debate about tobacco broke out among theologians and clerics. There were several issues that emerged in the debate. One of the first, and one that was hinted at by Monardes himself, was the paradox about the origin of tobacco. The paradox was real and serious. That tobacco was a New World plant and different from European henbane had become accepted among European botanists by the time Monardes was writing. This meant, however, that while tobacco was being hailed as a holy herb, in European discourse its pagan origins among heathenish worshippers labelled it as the food of the devil. Philaretes's attack on tobacco concerned not only its purely medicinal features but also its origins: 'this hearbe seemed to bee first found out and invented by the divell', he wrote, 'and first used and practised by the divels priests, and therefore not to be used of us Christians' (Philaretes 1602: F4). Other similar and sometimes more vicious remarks appear in some of the theological literature alluding to tobacco practices (von Gernet 1988: 322–3; Dickson 1954: 139–62). There were, however, answers to the charge of the devilish origin of tobacco. Leonardo Fioravanti, a physician, follower of Monardes and therefore an early advocate of nicotian therapy, described in his publication of 1582 the wealth of cures effected by tobacco and urged that 'everyone should make use of tobacco, since it is the plant which has been revealed in this century for human health through the goodness of God' (Dickson 1954: 91; Dickson 1959: 78). Those who argued that tobacco was divine, even though it was discovered in the New World, finally held the day.

Once the debate about the origin of tobacco waned, an issue concerning the use of tobacco by clerics during the service emerged. Part of the problem had to do with the unclean state of the churches caused by spitting, sniffing and a disagreeable odour (Dickson 1954: 154). Several Papal Bulls, of which the most important was that of 1642 (previously referred to in this chapter), were proclaimed to counteract this annoyance. But the problem was more serious than one of uncleanliness. It also turned on whether the Blessed Sacrament might be expelled upon sneezing or spitting after taking tobacco (Corti 1931: 132). Another issue that burned the ecclesiastical mind was whether tobacco was a food and, therefore, when taken, broke the fast during Lent, a problem handled in great depth by Antonio de Léon Pinelo in a book published in 1636 and echoed by others after him (León Pinelo 1636; Tedeschi 1987; von Gernet 1988: 331). As to the alleged aphrodisiac qualities of tobacco – an observation made in some of the early narratives on the New World – this seems also to have been turned on its head and ecclesiastical authorities actually urged their clergy to take tobacco to combat lust (von Gernet 1988: 332–3). Benedetto Stella, who wrote an influential nicotian treatise in 1669, in which he debated all of the issues about tobacco, also pontificated on the connection between tobacco and lust and declared:

> I say . . . that the use of tobacco, taken moderately, not only is useful, but even necessary for the priests, monks, friars and other religious who must and desire to lead a chaste life, and repress those sensual urges that sometimes assail them. The natural cause of lust is heat and humidity. When this is dried out through the use of tobacco, these libidinous surges are not felt so powerfully . . .
>
> (Tedeschi 1987: 112–13)

In the same way as the medical debate settled down considerably in the eighteenth century, so too did this ecclesiastical controversy. Many popes were addicted to tobacco, and the final statement of the inevitable surrender to tobacco came in an ordinance of Pope Benedict XIII of 1725 in which he permitted the use of snuff in St Peter's (Tedeschi 1987: 112). Not to be outdone by secular initiatives, the Papacy opened a tobacco factory of its own in 1779 (Camporesi 1990: 167).

The trend of debate during the seventeenth and eighteenth centuries was towards an acceptance of what was an actual situation; that tobacco consumption was increasing both geographically and socially. But the debate was more than a backdrop. It was an essential part of the public discourse about tobacco. Consumption and discussion, albeit through publications, were interrelated. And as long as the therapeutic value of tobacco was being debated, the use of the plant for medicinal purposes could be invoked. Benedict XIII's proclamation contained just such a statement reflecting the inexorable link between medicine and consumption. This is how the argument for permitting tobacco in the Vatican was made:

> To provide for the needs of everybody's conscience, and especially for the good order of the basilica, which is seriously compromised by the frequent walking out of those who can't abstain from the use of tobacco which is so widespread today, partly due to the opinion of the physicians who recommend it as a remedy against many

infirmities, especially for those people who are obliged to frequent cold and humid places in the early morning hours.

(Tedeschi 1987: 112)

The nicotian debate was at its height during the first half of the seventeenth century and it is more than interesting that snuff does not appear anywhere in it. Though Johann Magnen mentioned snuff in his 1648 publication, it escaped a full treatment until Louis Ferrant published his treatise on the subject. Ferrant was the first to discuss snuff explicitly in print focusing on the medicinal features of the product as well as offering tips on how to prepare it. It was also Ferrant who apparently began circulating the story that Jean Nicot, on his return from France, offered Catherine de Medici a box of powdered tobacco to be used for the relief of her headaches (Arents 1938: 70; Laufer 1924b: 50). Whether there is any truth behind this story or not, there is little doubt that an association between snuff and the French royal family could only raise its profile among tobacco consumers (Dickson 1954: 92).

Coming as it did towards the end of the most intense part of the nicotian debate, Ferrant's book did not, like others before his, receive any rebuke from the anti-nicotian quarter. The next publications on snuff, by Baillard in 1668 and Brunet in 1700, both of whom were, interestingly, French, pushed the cause of snuff even further. Baillard, in particular, argued that snuff evacuated unwanted humours, particularly phlegm, simply through the action of sneezing; snuff did not pass into the brain as some argued was the case with smoke (Baillard 1668: 35–49). Brunet, by contrast, dealt entirely with the preparation of snuff and did not comment on its medical efficacy. That the panacea gospel, popularized by Monardes and others and promoted through smoking and topical applications of tobacco, had become transposed to snuff can be seen clearly in the eighteenth century in texts other than those of a more medical nature. Père Labat, the intrepid naturalist and traveller in the Antilles, who reproached French colonists for abandoning tobacco in favour of sugar, had only the highest praise for snuff. Drawing on the words of others before him, Labat nevertheless put the case cogently. Snuff, he wrote:

heals colds, inflammation of the eyes, involuntary tears, headaches, migraines, dropsy, paralysis, and generally all those misfortunes caused by the pungency of the humours, their too great amount and their dissipation from their normal conduits. Nothing is better to increase the fluidity of blood, to regulate its flow and circulation. It is an unfailing sternutory to revive those with apoplexy or those in a death trance. It is a powerful relief for women having the pains of childbirth; a certain remedy for hysterical passions, dizziness, restlessness, black melancholy, mental derangement. Those who use it having nothing to fear from bad and corrupted air; the plague, syphilis, purpura, one does not have to guard against approaching those with popular illnesses that are easily communicated. It strengthens memory, it stimulates the imagination. Scholars are never afraid to tackle very abstract and difficult problems with their nose full of tobacco.

(Labat 1742: 278–9)

The anonymous author of an early eighteenth-century pamphlet praising tobacco also advocated snuff in terms similar to those of Labat. Contrasting it with smoking and chewing, the author put the use of snuff in the following terms:

By its gently pricking and stimulating the membranes, it causes Sneezing or Contractions, whereby the Glands like so many squeezed Sponges, dismiss their Serosities and Filth. And it serves for a drain to excessive Moisture in the Eyes or Head; so when a sufficient Moisture is wanting, its quick and noble Spirit opens the Vessels that afford Supplies thereto; and pure Snuff put into the Corners of the Eyes is found to alter and destroy the sharp Humours that occasion Bloodshot etc. So beneficial is this Powder for the Preservation of that most dear and valuable of all our senses, the Sight.

(Anon. 1712)

Even Simon Paulli, the inveterate critic of nicotian therapy, had to admit that not only did he take tobacco as a 'salutary medicine' but snuffing was not as dangerous as smoking (Paulli 1746: 21, 25). The alleged efficacy of tobacco in a powdered form that made one sneeze was a winning combination. Sneezing was generally considered beneficial in many ways, not least of all in its ability to expel 'corrupt humours'. Since classical times sneezing had been viewed positively. Aristotle thought that sneezing was sacred, as it emanated from the head, a holy place; and Socrates attached enormous significance to sneezing in making decisions (Kanner 1931: 553, 556). Pliny the Younger, in his natural history, related remedies derived from sneezing, including heaviness of the head (Kanner 1931: 556). The association between sneezing and other human passions was also widespread; sneezing and love, sneezing and luck, sneezing and happiness. Girolamo Baruffaldi, a cleric and poet in eighteenth-century Ferrara, praised the use of snuff not least because of the fact that it made one sneeze. One formula, in which tobacco was mixed with pulverized root of the white Hellebore, made one sneeze frequently. This to Baruffaldi was all to the better, for the sneeze was what mattered:

> The sneeze is a good omen
> (as long as it isn't a false sneeze
> like that of a cold)
> Welcome

Honoured
and worshipped as God
(Baroni 1970: 5)

There were a few who voiced complaints against snuffing – Johann Cohausen in 1716 being the most important – but, generally speaking, the tobacco discourse of the second half of the seventeenth and eighteenth century is positive about snuff. Perhaps one should not be surprised. After all, snuff, as opposed to pipe smoking, had its alleged beginnings in royal circles and was promulgated by the clergy who, judging by literary accounts, were the first to use snuff habitually across Europe. The association between snuff and respectability was extremely important in the dissemination of the practice.

The association was perhaps most evident in France which, if not the first country in Europe to take to snuff, certainly promoted it with the greatest energy. Once the French state tobacco monopoly took over the sale of tobacco, retail sales were licensed. At the beginning of the eighteenth century there were between 1,100 and 1,200 tobacconists in Paris, but the most frequented by the sophisticated snuffers of the city was in the Place de Palais-Royal (Vigié 1989: 61). The signs on the shops had since the 1670s changed from representing pipes to representing aspects of snuff – the *carotte* itself, or the civet, from which animal the precious scent was extracted that went into the finest snuffs (Vigié 1989: 61).

French mores, fashions, language and culture, in general, were much sought after in the eighteenth century and these criss-crossed Europe (Jones 1973: 207–10). Underlying this cultural movement was respectability. This, perhaps more than the medical and moral disputations, was the key to the success of snuff and its victory over smoking.

It is difficult to provide a completely satisfactory definition of respectability, but several elements of its construction and action are clear. Of crucial importance to respectability was that it was

> an assertion of a person's moral worth *as an individual,* as demonstrated primarily by behaviour . . . anyone could achieve or lose respectable status by his or her behaviour . . . the essence of the newer idea of respectability was behaviour – something over which an individual had control . . . the display of respectable behaviour constituted a demand (based in part on demonstrated moral worth) for deference from inferiors, acceptance by social peers and respect from superiors – at any social level.
>
> (Smith 1991: 24)

As the claim of status shifted from gentility, based on descent, to one of individual behaviour, so too did the rituals of consumption.

The nature of this new consumption has attracted a great deal of interest from historians and others but most of the literature stresses the fact that consumption became more individual and more private (McCracken 1988). Certainly there is ample evidence that the consumption of the new beverages of the period, tea, coffee and chocolate, became structured within new rituals. The tea ritual as it developed mostly in Holland and England in the late seventeenth and eighteenth centuries displayed the definition of respectable behaviour. It was not ostentatious but refined; it was domestic and private; and it engendered moderation (Smith 1991: 24–5). The ritual of coffee consumption also changed in the same period. In Paris, until the last decade of the seventeenth century, coffee was available either from street vendors or from shops frequented by the aristocracy and run by Armenians: 'the retreats where coffee was sold, were, however, merely shops reeking of tobacco smoke' (Leclant 1979: 89). The café that opened in 1686 was totally different from the normal coffee-drinking milieu. It was elegant but spacious, luxurious but not over-bearing. In short, 'the café became a worthy meeting place for respectable people . . .' (Leclant 1979: 90). Coffee drinking, like tea drinking, involved the use of a special set of utensils, not only for preparation but also for consumption. Both drinks had to be sipped and an air of respectability and security pervaded the ritual (Bizière 1979; Bödecker 1990). In eighteenth-century Germany the coffeehouse, as a social institution drawing upon and affirming these crucial values, also struck out across the gender divide, in the form of the *Kaffeekränzchen*, a daily or weekly social meeting of women (Albrecht 1988).

Into this new sensibility snuff accommodated itself perfectly in a complex of soft drugs including the new beverages of tea, coffee and chocolate, as well as alcohol. Snuff proclaimed the individual. The range of concoctions was enormous; even those who were forced to purchase from a monopoly or from small retailers could choose specific brands or doctor the standard package to make the product more individual (Alford 1973: 10). A probable scene described by Brooks perfectly reflects the individuality within the snuffing ritual:

Each leader of society and the coteries which attached themselves had their favorite snuff, and, in consequence, an extraordinary mingling of scents pervaded each court or ballroom where the well-bred met. In a room where the conversation was punctuated by discreet sneezes, the lady who adored *Jassamena* (made especially precious to her because of the exquisite box from which she took it) would condescend to take a pinch from the proffered box of the dandy who preferred *Orangery*. This she would do in the approved manner, whereby a delicate, bejewelled wrist and a well-turned arm would be displayed to advantage, while her companion, on his part, was in perfect position to indicate the handsome rings he wore, without

82

apparent ostentation. This exquisite technique for the correct means of taking snuff was developed by the French mentors of etiquette, to which native touches were given when the habit invaded London, Rome and elsewhere.

(Brooks 1937: 159)

This ritual was acceptable in other ways that reaffirmed notions of respectability. Although it is difficult for us to appreciate it readily, smoking in the seventeenth century was anything but dainty. Whether it was the tobacco itself that did it, or for some other, unknown reason, pipe smoking was accompanied by a considerable amount of expectoration (Brongers 1964). A spittoon or cuspidor was an essential accompaniment to a refined smoking scene, the floor for one that was not. The spittoon seems to have been a Dutch invention, and certainly in many Dutch engravings and paintings of the seventeenth century depicting pipe smoking the spittoon is usually visible somewhere in the scene (Brongers 1964: 163–70; Schama 1987: 196–217). It is very likely, as has been suggested by several authorities on the subject, that in an age of respectability the discharging of one's glutinous liquids into a vessel either on the table or at the foot became a sign of disgust (Douglas 1979: 21; Elias 1978: 143–60). Jean-Baptiste-Louis Chomel, at the end of the eighteenth century, echoing a debate that existed a century earlier about the link between hunger and tobacco consumption, clearly drew a connection between ingesting tobacco and expectoration and, interestingly, also the disease of hypochondria, the male equivalent of the eighteenth-century hysteria (Mullan 1988). As he wrote:

In a sense, one may say that Tobacco appeases one's hunger and thirst. The first thing, it does, by diminishing the sensation or feeling, the other, by tickling the salivary glands in the mouth – thus promoting the flow of saliva thither. From this it follows, that one spoils one's appetite by smoking a short time before a meal, and one upsets one's digestion if one smokes directly after a meal, especially if one spits a lot, – as the saliva one ejects whilst smoking, should have been a great help in the digestion of the food. This is more serious than one generally imagines, for a slow weakening and upsetting of the digestion is one of the causes of that unpleasant and obstinate disease, called hypochondria ... Even Hippocrates knew that people who spit a lot, become melancholy; and even strong youths rapidly tend to get thin, they eat little and become melancholy if they chew things that promote the flow of saliva or suddenly become heavy smokers.

(quoted in Brongers 1964: 164–5)

As early as 1669 Benedetto Stella, in his treatise on tobacco, advocated snuff principally for those who suffered from chronic spitting (Stella 1669: 288). Snuff was not implicated in the production of spittle, and, as we have seen, the sneeze was considered beneficial. Within the snuff ritual, in respectable circles, even the sneeze became part of the scene, and the trappings of the tobacco consumer now included a delicate handkerchief (Vigié 1989: 71–2). The daintiness of the sneeze, as part of the snuff ritual, was emphasized in a telling way by André Antonil, in his eighteenth-century work on Brazil. In relating the various methods of taking tobacco and their medicinal effects, Antonil referred to a manner of making small pellets of tobacco that were placed in the nostrils. These pellets, kept in the nostrils either overnight or during the day, were believed to draw moisture from the nasal cavities, clear the head, prevent catarrh and promote breathing (Antonil 1965: 321). Though he did not think the method very efficacious, Antonil's main criticism of taking tobacco this way was that it was an unsightly activity. As he put it, 'one only recommends it to those who, in using it, can avoid the indecency that appears when the pellets, being discharged from the nostrils and the drop of snot that is always suspended, soils the chin and nauseates the person with whom one is speaking' (Antonil 1965: 321).

Snuff was an improvement upon smoking in a practical sense too. Before the nineteenth century and the invention of the safety match, pipe smokers needed to carry with them a panoply of tools in order to set a light. Steel and flint and tinder box, not to mention the tobacco box, holding the pipes and the smoking mixture, were the essential accompaniments of the smoker (Vigié 1989: 66–8; Brongers 1964: 107–16). For the taking of snuff nothing more than the container, which could be miniaturized, was required.

Eighteenth-century European snuffing rituals and practices may seem a far cry from the Amerindian snuffers first witnessed by Europeans at the end of the fifteenth century. While European methods may seem peculiarly European, the ritual distance separating Europe from America was not as great as we imagine. Europeans and Amerindians snuffing tobacco were, as von Gernet has argued for pipe smoking, sharing a cultural language that was a composite of the individual experiences (von Gernet 1988). This language did not stop at snuff. It also involved the use of enemas. In so far as Amerindians employed enemas in their medical treatments before contact, the Europeans followed. And just as Europeans appropriated the consumption of tobacco by smoking pipes and cigars and by snuffing, so too they introduced the practice of using tobacco smoke in rectal devices known as clysters. To whom the credit for using tobacco smoke in this way is due is not certain but the practice was referred to in one of the first publications on the use of clysters written by Regnier de Graaf (de Graaf 1668; Brockbank and Corbett 1954). De Graaf was a physician who, among other distinctions, invented a syringe and its accessories which not only would be safe to administer but could be self-administered (Frieden-wald and Morrison 1940a: 83–8). The tobacco clyster, using either smoke or an infusion, depending on whose authority was being followed, was widely practised until well into the nineteenth century (Warren 1919; Friedenwald and Morrison 1940b). The possible erotic nature of the clyster injection did not escape some eighteenth-century artists; Fragonard, Lav-reince and Baudouin all represented the clyster in this way (Wagner 1990: 273; Friedenwald and Morrison 1940a: 110). Tobacco clysters were used in the hope of treating ailments of the colon and the bowel. They were also recommended in attempts to resuscitate drowned individuals, as well as those who had suffocated, had convulsions and fits, or were frozen (Society 1775: 91–3; Warren 1919: 14–17).

In the eighteenth century tobacco was used in several different forms. While the tobacco clyster may be said to be an example of a medical use, can one assume that for the most part, whether taken by smoke with a pipe or cigar, by chewing or by snuffing, tobacco had by the eighteenth century become a recreational drug? Many commentators on tobacco consumption in the seventeenth and eighteenth centuries were keenly aware that consumption was increasing and often warned about the excessive and non-medical use of the plant. But it is important to understand that their opprobrium was not strictly about the use of tobacco for recreation but, rather, they were concerned that this drug was being self-administered, rather than being dispensed by, or taken with the approval of, the medical fraternity. Certainly one can be forgiven for thinking that tobacco was a recreational drug given the vast amounts that were consumed in Europe, both absolutely and per head of the population. But this argument would be too hasty for the simple reason that in the eighteenth century the dividing line between consumption of commodities for leisure and for health was not as clearly drawn as it is now. To keep the body in humoral balance was the objective, and tobacco was clearly perceived as playing a key role in this. The recreational and medical use were reaffirming parts of a culturally specific view of the human body. It is impossible for us to say which was which. There is little doubt, however, that the taking of tobacco for medical purposes did decline in the nineteenth century. The transformation, however, was neither as complete nor as fast as normally understood; but that is the subject of the next chapter.

Europeans and Amerindians defined the culture of tobacco consumption. As Europeans were responsible for diffusing the tobacco habit around the globe, it is not surprising to find European methods of tobacco consumption wherever Europeans spread the news. Chewing, snuffing and smoking tobacco were all practised throughout Asia, the first embarkation point for tobacco as it circumnavigated the globe after its introduction into Europe. Smoking, however, appears to have been the primary form of tobacco consumption, as it was through smoke that Asians were first acquainted with the plant. Pipes and water pipes were the preferred devices throughout the region. As a previous chapter discussed, it was the Spanish, principally, who introduced Asians to tobacco through the Philippines, from as early as 1575 (Reid 1985: 535). The Japanese were apparently introduced to tobacco and pipe smoking by the Portuguese before the turn of the seventeenth century (Laufer 1924a: 2). After the expulsion of the Portuguese in 1638 it was left to the Dutch to keep the nicotian spirit alive and they possibly acquainted Japanese smokers with the latest European tobacco fashions in the seventeenth and eighteenth centuries. The Portuguese can also be credited with sowing the nicotian habit in India (Laufer 1924a: 11). Once Europeans had made the first strike for tobacco, the habit spread very quickly throughout Asia, though in this regard it was the Chinese, rather than Europeans, who were principally involved in diffusing tobacco consumption (Laufer 1924a: 15).

The Europeans introduced the pipe, presumably made of clay, as well as the cigar into Asia, but very soon after first contact and the exchange of the tobacco ritual Asian tobacco consumers were adapting the European methods of consumption to their own situation. Not only did they employ local materials – wood and metals, rather than clay, in the manufacture of the pipe – but the design of their pipes became quickly distinguished from the European variety. Indeed, if anything, many of the pipes extant of the period resemble more the Amerindian pipes of the period rather than the European ones (Samson 1960; Laufer 1924a). Moreover, it is clear that while European pipe smokers of the seventeenth century could do no better, generally, than a pipe made of white clay, Asian pipes were both simpler and then far more elaborate and imaginative in their choice of materials. The simplest pipe in East Asia was typically made of bamboo, either in one piece, that is bowl and stem, or two pieces, the stem of bamboo and the bowl of wood (Laufer 1924a: 21). More elaborate pipes had their stems made of bamboo but the mouthpiece of jade, ivory, metal or porcelain (Samson 1960: 22). When ivory was employed for the stem, the Chinese in particular frequently inserted a copper tube into the stem to protect the ivory from cracking (Laufer 1924a: 21). Though bamboo appears to have been the most common material for stems in China, specific woods, such as ebony and other black hardwoods, were also used (Laufer 1924a: 22).

Asian tobacco consumers also smoked their mixtures in pipes that were not of American or European origin. These were water pipes, known also as the hookah. One of the earliest, and possibly one of the best, descriptions of the hookah comes from Edward Terry in his report of 1616. Terry, chaplain to Sir Thomas Roe, the first English ambassador to the Mughal court, spent almost three years in western India, and witnessed the cultivation as well as the consumption of tobacco in the region. This is how he described the hookah:

They have little Earthen pots . . . having a narrow neck and an open round top, out of the belly of which comes a small spout, to the lower part of which spout they fill with water: then putting their tobacco loose in the top, and a burning coal upon it, they having first fastned a very small strait hollow Cane or Reed within the spout . . . the Pot standing on the ground, draw that smoke into their mouths, which first falls upon the Superficies of the water, and much discolours it. And in this way of taking their Tobacco, they believe makes it much more cool and wholsom.

(Samson 1960: 227)

The inventiveness of the hookah was extraordinary. John Fryer, who travelled throughout India and Persia in the last quarter of the seventeenth century, also described the hookah and confirmed very much what Terry described some sixty years earlier. Fryer, though, noted that the hookah was a crucial part of the coffeehouse and that it was the centrepiece of the ritual of the entire practice, including smoking, drinking and conversation, as we can read from his description:

They are modell'd after the Nature of our Theatres, that everyone may sit around, and suck choice Tobacco out of Cosy Malabar Canes, fastene'd to Chrystal bottles, like the Recipients or Bolt-Heads of the Chymists, with a narrow Neck . . . the Vessel being filled with Water: After this Sort they are mightily pleased; for putting fragrant and delightful Flowers into the Water, upon every attempt to draw Tobacco, the Water bubbles, and makes them dance in various Figures, which both qualifies the Heat of the Smoke, and creates together a pretty Sight.

(Fryer 1912: 34).

The hookah was used throughout the Asian world and even in Madagascar, where in 1638 Peter Mundy recorded its use (Mundy 1919: 384). Interestingly, Mundy's observations included the custom that the men he saw hung around their necks the mouthpiece for the pipe of the hookah, suggesting that the hookah was a shared smoking device.

Who invented the water pipe is not known but most authorities point to the Persians as the most likely candidates. And it was Muslims who were responsible for its spread, along the east African coast and throughout Asia, arriving in China by the turn of the eighteenth century (Laufer 1924a: 26–8). When the water pipe was invented is another mystery but it would appear to have succeeded the introduction of tobacco rather than the other way around. In other words, the water pipe appears to have been an invention of the Muslim world and was designed within the context of a shared social experience as it existed within the coffeehouse – unlike the Amerindian and European pipe the hookah is not easily portable (Birnbaum 1957: Hattox 1988). Some writers have persisted in the belief that the water pipe was used to smoke marijuana long before the introduction of tobacco, denying, as it were, the possibility of a spontaneous, and extremely imaginative, invention in Islam (Benet 1975: 48). Yet there is no evidence that marijuana was smoked before tobacco was introduced; eating or drinking an infused concoction were the most common methods of consuming marijuana, but the debate on the connection between marijuana and tobacco smoking continues (Abel 1980: 3–57; Philips 1983: 315–16).

Once it got into Chinese hands the hookah was refined to such an extent, that, according to Berthold Laufer, it 'was so convenient, simple, graceful, and artistic that it may be put down as an invention wholly their own' (Laufer 1924a: 28). In the same way, when at the end of the seventeenth and the beginning of the eighteenth century the Chinese became acquainted with snuff, they created a highly distinctive ritual involving not only exquisitely produced snuffs but also delicately designed bottles as containers. Whereas Europeans kept their snuff in boxes, the Chinese preferred bottles made from material ranging from gold, silver and brass to porcelain, jade, agate and coral (Laufer 1924a: 34–8). The reason why the Chinese were attracted to snuff was partly medicinal and partly, one may suppose, because of its theatricality. This is how a Chinese work of the early eighteenth century described the new use of the tobacco plant:

> Recently they make in Peking a kind of snuff which brightens the eyes and which has the merit of preventing infection. It is put up in glass bottles, and is sniffed into the nostrils with small ivory ladles. This brand is made exclusively for the Palace, not for sale among the populace. There is also a kind of snuff which has recently come from Canton and which surpasses that made for the Palace. It is manufactured in five different colours, that of apple colour taking the first rank.
>
> (Laufer 1924a: 33)

Snuff appears to have been confined mostly to China, with a small amount being consumed in Japan and India. By far the most popular form of tobacco consumption was smoking. The pipe, whether long-stemmed or in the form of the hookah, was the most widely adopted method of smoking but there were others. One of the most important was the *bunkus*, a cigar composed of shredded tobacco and wrapped in a dried leaf of banana or maize (Laufer 1924a: 20). These appeared in Java as early as 1658, having been introduced via the Philippines and the Moluccas earlier in the century. From Java the route of diffusion headed westwards towards Burma and India where they were being consumed by 1711, and called *cheroots* (Reid 1985: 536).

While smoking was universal throughout Asia and snuff confined primarily to China, tobacco chewing was confined mostly to those parts of Asia – India, the Malay archipelago and Indonesia – where betel chewing predominated. As far as we can tell, the tobacco was not chewed by itself, as was customary in Europe and America, but rather it was added to the betel leaves, much in the same way as other ingredients, principally areca nut, gambier and opium (Reid 1985: 536–7). The tobacco and betel chewing mixture appears to have been a relatively late practice, in Indonesia at least, possibly as late as the latter part of the eighteenth century or as much as a century and a half after the introduction of tobacco smoking: no one seems to know the reason for this (Reid 1985: 537).

Elsewhere, in Africa and colonial America smoking seems to have been the preferred method of consuming tobacco. In Africa the pipe predominated, as it did in North America, receiving impetus from Arab traders who diffused the water pipe along the east African coast and across the Sahara trade routes, and from Europeans who visited the western coast (Laufer 1930; Philips 1983; Samson 1960).

5

'THE LITTLE WHITE SLAVER'
Cigarettes, health and the hard sell

Before the nineteenth century snuff was the tobacco product that carried the characteristics of later forms of tobacco consumption. Both its manufacture and distribution can clearly be viewed as 'modern'. That is to say, snuff alone, of all tobacco products, can be considered in that class of goods that historians have identified as belonging to the first stirrings of modern consumerism in the eighteenth century (McKendrick *et al.* 1982). Even so, in general, tobacco remained a pre-modern consumer commodity, much more connected to the commercial, rather than the burgeoning industrial, system. As we shall see in Chapter 9, one of the ironies of tobacco was that, despite the fact that it was probably the first mass-consumed food-like substance, it was one of the last products to be mass-produced.

Usually snuff is viewed as an eighteenth-century phenomenon, an aberration in the history of tobacco consumption, with smoking considered as the norm. In fact snuff taking was the most popular form of tobacco, in Europe at least, well into the nineteenth century and, in a few cases, into the twentieth century. Far from being just the fashionable accessory of a few fops, it was the normal way of consuming tobacco in the early industrial age.

Despite the general lack of information on the type of tobacco consumed, that which is available presents a clear picture of the continuing importance of snuff consumption in the nineteenth century. Because of their centralized operations, the figures from the various European state monopolies furnish the clearest evidence. The strongest, and perhaps the most surprising, evidence comes from the sales of tobacco by the French state monopoly. On the eve of the Revolution (as pointed out in the previous chapter) the monopoly sold about 15 million pounds of tobacco of which more than 12 million pounds was either in the form of manufactured snuff or in *carottes*, from which individuals grated their own snuff (Rogoziński 1990: 96). Until the 1860s the sales of snuff, especially that already manufactured at source, rose considerably, reaching a maximum level of almost 28 million pounds in 1861 (Rogoziński 1990: 96). In per capita terms snuff consumption rose from around one-third of a pound per annum in 1789 to almost three-quarters of a pound per annum in 1861 (Rogoziński 1990: 96). It was not until after this date that snuff consumption began to fall for the first time since the early part of the eighteenth century. Even so, it was as late as 1925 before the total sales of snuff fell back to their level of 1789 (Rogoziński 1990: 96–7). In relative terms, however, the reign of snuff as France's most popular tobacco product ended in the 1830s when the sales of cut tobacco, for pipe smoking, overtook those of snuff (Rogoziński 1990: 94). What these figures suggest is that existing French consumers of tobacco continued to use snuff for most of the nineteenth century but that new consumers were increasingly introduced to tobacco through smoking.

Pipe smoking became much more popular in France during the first half of the nineteenth century than it had ever been before. Sales of cut tobacco, intended mostly for pipe smoking, though possibly used for rolling cigarettes, soared during the nineteenth century from around 6 million pounds in 1819 to 60 million by the turn of the twentieth century: this prodigious growth was, of course, accompanied by a marked increase in its relative popularity, rising from 28 per cent to 72 per cent of total consumption over the intervening period (Rogoziński 1990: 94–7). Cut tobacco accounted for over half of the monopoly's sales until after the Second World War: it was only at this point that the cigarette became, in terms of its relative position in the monopoly's sales, France's most popular form of tobacco consumption (Rogoziński 1990: 97). The image of the French cigarette smoker, so familiar in western culture, is, therefore, of only recent date.

The French attachment to snuff and its slow displacement in consumption patterns by smoking tobacco, whether in the form of the pipe, the cigar or the cigarette, was repeated elsewhere in Europe. In Italy, for example, sales of tobacco products by the monopolies in both Lombardy and Venetia show a rather similar pattern to that in France. In the former snuff accounted for at least 50 per cent of sales until the 1840s whereas in the latter the domination of snuff continued until the 1860s (Rogoziński 1990: 102). When the monopolies were reorganized in the aftermath of the unification of the peninsula the newly created Italian state monopoly was faced with a varied market, yet even so, across the country, snuff still commanded an important position. Even as late as 1880 over 20 per cent of the monopoly's sales was in the form of manufactured snuff, and this proportion was maintained until the end of the century (Rogoziński 1990: 103). During the second half of the nineteenth century cigars and cut tobacco gained prominence as the preferred form of tobacco consumption, a situation that was maintained until the eve of the First World War (Rogoziński 1990: 103). Cigars then began to lose favour while sales of cut tobacco remained fairly stable, but it was now the turn of the cigarette. Around 1930 the majority of Italians were cigarette smokers, and the figure continued to rise thereafter: by 1950, for example, almost 80 per cent of the state monopoly's sales was accounted for by cigarettes (Rogoziński 1990: 103). Sales by the Austrian state tobacco monopoly during the nineteenth and twentieth centuries confirm that the pattern of consumption in central eastern Europe was broadly similar to that of western Europe. Snuff consumption in absolute terms continued to maintain a significant position until the end of the nineteenth century, even though its relative position became eroded: cut tobacco and cigar consumption grew in the nineteenth century, as the most popular forms of tobacco consumption until the second and third decade of the twentieth century: then after the First World War, and then especially in the 1930s and 1940s, the cigarette eclipsed all other means of consuming tobacco (Rogoziński 1990: 90).

Of those European regions served by state tobacco monopolies, the exception to the general trend was the Scandinavian tobacco consumer, especially those in Sweden and to a lesser extent those in Iceland, Norway and Denmark. The case of the Swedes is the most exceptional and remarkable. In the first place snuff consumption grew steadily both absolutely and relatively from the late eighteenth century until the 1930s, while that of spun and cut tobacco fell, while cigars had a brief but unremarkable rise in consumption during the second half of the nineteenth century: snuff consumption finally gave way to the cigarette but not until after the Second World War (Rogoziński 1990: 105). The shift to smoking from snuffing, present in most of Europe in the nineteenth century, did not therefore occur in Sweden until well into the present century. The second difference between Sweden and other European countries is that the Swedes preferred moist oral snuff, placed behind the upper lip, to dry nasal snuff (Rogoziński 1990: 105–7). Smokeless tobacco, whether in the form of snuff or of chewing tobacco, still accounted for more than one-quarter of tobacco consumption in other Scandinavian countries, principally in Iceland and Denmark (Rogoziński 1990: 114).

In so far as Sweden presented an unusual example of a tobacco-consuming population committed to snuff rather than any form of smoking tobacco for far longer than elsewhere in Europe, the American tobacco consumer also displayed an unusual pattern of consumption. Rather than snuff Americans preferred chewing tobacco: even as late as 1900 chewing tobacco was the most popular form of tobacco consumption in the United States accounting for 44 per cent of total consumption (Rogoziński 1990: 111). The absolute growth of the consumption of chewing tobacco was reversed only just before the First World War, and though the fall in consumption was rapid the United States was consuming over 100 million pounds of chewing tobacco in 1940 (Gottsegen 1940: 34; Rogoziński 1990: 124). The consumption of snuff, taken orally rather than inhaled into the nostrils, actually increased from the late nineteenth century until after the Second World War and has recently been experiencing a revival, but, in relative terms, the consumption of snuff has been limited, representing at most 5 per cent of total consumption in the nineteenth and twentieth centuries (Rogoziński 1990: 124; Christen *et al.* 1982). Though the consumption of the cigarette increased sixtyfold in the space of thirty years between 1900 and 1930, it is often not appreciated that the cigarette continued to compete with all other forms of tobacco consumption, and that the rise of the cigarette to a dominant position in tobacco consumption was a long process (Gottsegen 1940: 28; Rogoziński 1990: 124): it was not until 1941 that 50 per cent of total American tobacco consumption took the form of the cigarette (Rogoziński 1990: 113).

Britain, too, entered the nineteenth century with more than half the tobacco market represented by snuff, but this declined very quickly through the nineteenth century to the point at which, by the turn of the twentieth century, snuff accounted for only 1 per cent of total British tobacco consumption: this proportion has been maintained for most of the century (Rogoziński 1990: 132). The pipe returned to the British tobacco scene with a vengeance in the nineteenth century. Already by the middle of the century the age of snuff had definitely passed. Some 60 per cent of British consumption near mid-century was accounted for by pipe tobacco, most, if not all of it, destined for smoking in the pipe. Moreover the pipe itself was changing during the nineteenth century, as it became manufactured of briar instead of clay (Alford 1973: 110–11). Interestingly, while most European and American tobacco consumers took some time to become cigarette smokers, the British consumer found this form of tobacco consumption highly appealing from a relatively early time. For example, in 1900, barely twenty years after the introduction of the machine-made cigarette, more than 10 per cent of all British tobacco sales were cigarettes (Alford 1973: 480). In most countries of the world the comparative figure was nearer 2 per cent (Rogoziński 1990: 111). By 1920, well in advance of most other countries in the world, more than 50 per cent of British tobacco sales was in the form of cigarettes (Alford 1973: 480).

While generalizing about the pattern of tobacco consumption in Europe and the United States during the nineteenth and twentieth centuries is difficult, and in some cases entirely misleading, certain features are clear enough. First, over the period there occurred not only an enormous increase in consumption in absolute terms, as one might expect, but also a quite astonishing rise in per capita use. During the nineteenth century per capita consumption was low by present standards, and within a range that was already established in the course of the eighteenth century. Levels of 1 or 2 pounds per capita were fairly common (Rogoziński 1990: 81–108). By the turn of the century only the United States and Denmark had per capita consumption exceeding 3 pounds: the former with a figure of 5.3 pounds per capita (Rogoziński 1990: 111). In 1950 consumption in most European countries exceeded 3 pounds per capita, many countries approaching 5 pounds per capita and a few above this figure: the American per capita consumption was still the highest at 7.5 pounds per capita (Rogoziński 1990: 113). Per capita tobacco consumption ceased to increase, in most western countries, in the 1960s and 1970s (Lee 1975: 4–5; Laugesen and Meads 1991: 1,346). Second, the increase in per capita consumption has occurred along with a changing pattern of consumption. Broadly, with the exceptions discussed above, the trend was towards smoking tobacco, and away from smokeless forms. And within the category of smoking tobacco there has occurred a vast transformation of the tobacco consumer from one with pipe in hand, or cigar or hand-rolled cigarette, to manufactured cigarette brands. Table 5.1 provides a picture of this phenomenon by comparing the relative position of cigarette consumption in 1950 with the year at which cigarette consumption accounted for 50 per cent of total tobacco consumption in a selection of countries.

Table 5.1 Cigarette consumption

	% of total tobacco consumption, 1950	Year of 50% of total consumption
Austria	76	1939
Belgium	44	1961
Denmark	44	1961
France	53	1943
Germany	37	1955
Netherlands	43	1972
Spain	31	1955
Sweden	49	1951
United Kingdom	84	1920
United States	72	1941

Sources: Rogoziński 1990: 113; Alford 1973: 480

The main point about Table 5.1 is that it shows a relatively late switch to the cigarette, in most countries after the Second World War. What the table also shows is that it would be inaccurate and misleading to relate the pattern of consumption, especially the switch to cigarettes, to any crude measure of economic development. Certainly any correlation between the level of industrialization or per capita income and cigarette consumption is not supported by the data.

In Asia and Africa the pattern of consumption in the nineteenth and twentieth centuries was as varied as in Europe. Japan and China, for example, were rapidly converted to cigarette consumption. In Japan this had certainly occurred by the 1920s and possibly earlier (Rogoziński 1990: 158). Japan was one of the first foreign markets to be targeted by W. Duke and Sons and the American Tobacco Company, the forerunner of British American Tobacco (Durden 1976). Japanese consumers were exposed to western commodities and culture in the aftermath of the Meiji restoration and this included the cigarette. British and American tobacco companies were quick to take a financial stake in Japanese cigarette manufacturing companies, but these were liquidated in 1904 when the state tobacco monopoly was formed. Japanese cigarette manufacturing technology was generally considered to be the best in the Far East (Durden 1976; Alford 1973: 218; Cochran 1980: 56–7). The cigarette also swept through China during the first half of the twentieth century. British American Tobacco played an important role in converting Chinese tobacco consumers to the cigarette and away from the traditional pipe and snuff. Before the arrival of BAT, tobacco was consumed on its own or in a mixture with opium (Spence 1975). Pressure was, however, mounting for the substitution of opium by tobacco. It is, therefore, difficult to determine whether Chinese cigarette consumption grew because of the decline of opium consumption or whether it was just a change in the pattern of tobacco consumption: evidence points in both directions (Cochran 1980: 27–8; Newman 1991). While the reasons for it remain unclear, the fact of the rapid growth of cigarette consumption in China is strikingly evident. In 1902 China is estimated to have consumed 1.25 billion cigarettes; in 1928 the figure stood at 87 billion and in 1988 China consumed nearly 1,500 billion cigarettes (Cochran 1980: 219, 234; Grise 1990: 22).

Japan and China stand out in Asia as the countries most thoroughly committed to cigarette consumption. In other parts of Asia the progress of the replacement of other forms of tobacco consumption by the cigarette has been far less complete. Traditional patterns have proved to be much more resistant to change than they were in Japan and China, though the resistance is showing distinct signs of crumbling.

In India, for example, manufactured cigarettes accounted for only 12 per cent of total consumption in 1950 despite the early intervention by British American Tobacco (Basu 1988). Chewing tobacco, smoking tobacco for hookahs, cheroots and *bidis* each accounted for about one-quarter of the market (Rogoziński 1990: 156). In 1979 India and Pakistan had amongst the lowest figures for per capita consumption of manufactured cigarettes, equivalent to 5 per cent of the annual per capita level in Japan (Tucker 1982: 186). Even so, the penetration of the Indian market by the manufactured western-style cigarette has progressed considerably. Between 1935 and 1965 the consumption of all tobacco products per adult fell marginally but cigarette consumption per adult more than doubled (Beese 1968: 4).

Indonesia, too, was targeted by British American Tobacco as a potential market. Its first factory manufacturing cigarettes was established in 1924, and output increased steadily until the 1970s (Reid 1985: 539–41). But the real success story in the Indonesian tobacco industry has been production of the *kretek*, a cigarette composed of locally grown, primarily dark, air-cured tobacco mixed with cloves, in a ratio of 3 to 2 (Akehurst 1981: 320; Reid 1985: 539–40). *Kretek* production has increased substantially since it was first introduced in the 1880s: in 1939 output was at a level of between

5 and 16 billion units – there are varying estimates of the output level – but by 1980 the figure had surpassed 50 billion units, almost twice the level of output of what are called 'white' cigarettes (Reid 1982: 539–41).

Other parts of Asia have experienced patterns of tobacco consumption somewhere between these extremes. The patterns are not static, however, and the tobacco marketplace continues to be transformed by the actions of tobacco multinationals and the responses to them (Chen and Winder 1990). Oceania, for example, has only recently become a battlefield between traditional and western forms of tobacco consumption. The Pacific Islands, at least those formed of coral, have never grown much of their own tobacco and have, therefore, been dependent on imported sources since contact with Europeans in the late eighteenth and early nineteenth century (Marshall 1981: 887). Most Micronesians, since the end of the Second World War, have been avid cigarette smokers, purchasing their supplies primarily from the United States and Japan (Marshall 1981: 889). The larger islands, such as Fiji and Papua New Guinea, now produce their own cigarettes from local tobacco as well as imported supplies: Fiji began its industry in the early 1960s, whereas in New Guinea, though plans were laid down for a cigarette manufacturing operation from as early as 1948, little happened before 1960 (Marshall 1987: 35). The impact of the establishment of a modern cigarette industry has been dramatic. Not only has tobacco consumption increased but, perhaps more importantly, the consumption of cigarettes has outstripped that of *brus*, a traditional tobacco product made from air-cured, as opposed to flue-cured, tobacco. Papuan New Guinea smokers first consumed more cigarettes than *brus* in 1969 but by 1979 sales of cigarettes accounted for 71 per cent of total tobacco consumption (Scrimgeour 1985).

A similar pattern of the increasing consumption of manufactured cigarettes, and the decreasing use of both local tobacco and traditional methods of consumption, is evident in Africa. The diffusion of tobacco growing and consumption in Africa was, in many ways, similar to its spread in Asia. In particular tobacco was incorporated into local specific cultures in such a way that there emerged great variation in how tobacco was used. Snuffing, chewing and smoking, primarily in pipes, was practised throughout the continent (Hambly 1930). Increasingly though, as the demand for manufactured cigarettes has risen, traditional methods of smoking as well as growing and curing methods have become less prevalent (Muller 1978). In Kenya, for example, cigarette consumption during the 1970s grew at an annual rate of 8 per cent – this compares with a global rate of only 1 per cent (Stebbins 1990: 229; Currie and Ray 1984: 1,132). Urbanization has been held partly responsible for shifting patterns of consumption, from traditional to western, in Africa, as it has in other parts of the world (Kaplan *et al*. 1990; Waldron *et al*. 1988; Finau *et al*. 1982).

The process by which the cigarette has come to dominate global tobacco consumption in this century has progressed at an extraordinary rate. If one compares the relative position of the cigarette in tobacco consumption in ten representative countries in 1900 and 1950, then the figures speak for themselves – in 1900, on average, cigarettes were consumed by 8 per cent of the population; fifty years later the figure was 57 per cent. Since the Second World War the figure has continued to climb. Towards the end of the 1980s, at least 80 per cent of the world's tobacco crop ended up in cigarettes (Grise 1990: 9–13). Despite the widely differing methods of consumption used throughout the world up to the end of the nineteenth century, the entire world has been converging in the twentieth towards one type of tobacco consumption, the cigarette. And, as the cigarette has come to dominate consumption, there has also been a convergence in its form, away from unfiltered to filtered varieties and from dark to light tobacco, from air- and fire-cured to flue-cured (Grise 1990: 10, 20–1; Tucker 1982: 38, 195).

How do we explain the emergence and dominance of the cigarette? The early history of the cigarette is obscure; there are several accounts, all at variance with each other. The story that the cigarette originated in the Near East, and was introduced into Europe by soldiers returning from the Crimean War where they were taught to smoke by Russians and Turks, is the least credible though most popular account. It may be that British soldiers were introduced to the cigarette by Russians and Turks, but it seems that the Russians and Turks learned of the cigarette from the French who in turn learned of it from the Spanish (Rogoziński 1990: 51).

This is not to suggest that the cigarette is a nineteenth-century European invention; like other forms of tobacco consumption, it originated in South and Central America. There the cigarette was commonly smoking tobacco crushed and wrapped in vegetable matter such as banana skin, bark, maize leaves, and reeds (Rogoziński 1990: 50; Wilbert 1987). Spanish cigarettes used maize wrapping at first, and in the seventeenth century fine paper was introduced (Perez Vidal 1959: 100–1). This latter method of enclosing tobacco for smoking the Spaniards called *papelate*: Goya portrayed in several paintings Spanish soldiers smoking *papelates* in the eighteenth century.

According to most accounts the *papelate* as well as the maize cigarette crossed into France around 1830, but it could easily have happened before that date. At any rate, what is important is that the French state tobacco monopoly began manufacturing cigarettes, the nomenclature that the French had given to the *papelate*, in 1845, during which year sales amounted to 7,000 pounds, or the equivalent of 3 million cigarettes (Rogoziński 1990: 96). And what is perhaps of even greater importance, especially for the subsequent history of the cigarette, is that at about the same time as the monopoly began manufacturing cigarettes French consumers were found to prefer American rather than French domestic tobacco in their cigarettes:

the verdict was that French domestic tobacco was too acrid (Tilley 1948: 505). Whether, as one authority suggests, the American tobacco was Bright tobacco is not clear, but certainly the early adoption of American leaf into the cigarette was of great significance (Tilley 1948: 505).

From France the cigarette, using American leaf, passed into Germany and Russia, where Turkish or Balkan tobacco was mixed with American leaf for the first time around 1850 (Tilley 1948: 505). The cigarette may have been introduced into England as early as the 1840s, therefore, antedating the supposed connection with the Crimean War, but there is little doubt that manufacturing did not begin in England until after that date (Alford 1973: 123). The problem with cigarettes at that time is best summed up in the words of the historian of W. D. and H. O. Wills and Company:

> These early cigarettes were wrapped in tissue paper with a cane mouthpiece attached. But they were very crudely made and it was necessary to pinch together each end to prevent the tobacco from falling out. Accordingly, at that time cigarettes had only novelty appeal. Moreover, dark air-cured and fire-cured tobaccos, most commonly in use, were generally too strong for use in cigarettes. During the 1860s cigarettes made from best quality Turkish tobacco and specially made fine texture paper came on the market, but even these, with their distinctive aromatic flavour, did not commend themselves widely to British tastes.
>
> (Alford 1973: 123–4)

Americans appear to have adopted the cigarette from England, it being relatively rare before the Civil War (Tilley 1948: 506–7). In 1869 production of cigarettes in the United States stood at the level of 2 million individual units, though it may have been much higher than that by the end of the Civil War (Tilley 1948: 507). It comes as no surprise to learn that the first manufactories of cigarettes were located in New York City and run and owned by Greek and Turkish immigrants (Heimann 1960: 206). One of the most important manufacturers of the time, the Bedrossian brothers, made an important contribution to the evolution of the cigarette for it was they who were the first to blend Turkish (or Balkan) tobacco with the recent innovation in American leaf production, Bright flue-cured tobacco (Tilley 1948: 507). A more detailed description of the evolution of Bright tobacco and flue-curing can be found in Chapter 8, but suffice it to say that Bright flue-cured tobacco produced much milder smoke than did the traditional dark air- and fire-cured varieties.

The adoption of flue-cured tobacco was of such great importance for the history of the cigarette and its cultural and pharmacological dynamics that it would be difficult to exaggerate its significance. Without getting into too much technical detail, flue-cured tobacco smoke is acidic while air- and fire-cured tobacco smoke is alkaline (Akehurst 1981: 647). Cigarettes are acidic and it is this chemical property which makes the smoke relatively easy to inhale. It is interesting to note that Turkish (Oriental) tobacco, which is sun-dried and highly valued for its aromatic properties, is more acidic than flue-cured tobacco; this might help explain why Turkish tobacco was adopted for use in cigarettes (Akehurst 1981: 579). Cigar and pipe tobacco smoke is more difficult to inhale. In addition nicotine is released gradually in acidic smoke whereas in alkaline smoke the initial release of nicotine is very fast but so is the decline (Akehurst 1981: 647). It seems that the relative ease of inhalation of flue-cured tobacco was critical in influencing new consumers who might have been put off by the adverse initial effects of consuming air- and fire-cured tobacco (Brecher *et al.* 1972: 229). This may be one reason why legislation against children purchasing and smoking tobacco was not required before the twentieth century. Recent studies of smokers who have access both to commercial flue-cured and traditional air-cured tobaccos certainly confirm that inhalation is much easier with the former (Vallance *et al.* 1987: 277; Mougne *et al.* 1982: 106).

The cigarette, even as it appeared in its relatively crude form at the end of the 1860s, nevertheless already had the hallmarks of a new form of tobacco consumption, one that could be exploited by tobacco manufacturers using new methods of production, employing technologies of manufacture from other industrial activities (see Chapter 9); and, more significantly, new methods of marketing, particularly packaging and advertising.

The origins of tobacco advertising are obscured but certainly the practice did not begin with cigarettes. One of the earliest forms of advertising was through branding. The idea behind branding was simple enough: it allowed consumers to identify with the name of the product rather than the name of the maker. Its historical trajectory is unclear but there is little doubt that branding evolved from trade cards used in tobacco distribution from as early as the first two decades of the seventeenth century (Brongers 1964: 125). Trade cards and wrappers from the late eighteenth century show clearly the images that manufacturers wished to convey. The ubiquitous American Indian together with hogsheads, tobacco leaves, snuff boxes and other accessories of tobacco culture were frequently portrayed (Scott 1966; Brongers 1964: 132–5). By the late eighteenth and early nineteenth century a brand name began to take prominence over the maker's name, as evidenced by extant trade cards and wrappings of the period. Not surprisingly, manufacturers of snuff were the first to exploit this early form of marketing because manufacturers prided themselves on providing snuff of different scents and colours (Alford 1973: 27–8).

Branding was important in presenting an image for the consumer as well as providing manufacturers with a movable asset. There is considerable evidence, for the United States at least, that both of these had already occurred by the middle of the nineteenth century. There were already brands of chewing and, to a much lesser extent, smoking tobacco that were famous in many parts of the county though they were manufactured predominantly in Virginia and North Carolina (Robert 1938: 218–19). Before the Civil War there seems to have been a preference for choosing the names of sweet fruits or famous people as the brand label. In the case of the former this was probably done in order to convey an image of the chewing tobacco as being delicious: in the latter case there was undoubtedly an effort to make the product seem prestigious. Some of the names used by Virginia and North Carolina manufacturers at the time included Cherry Ripe, Wedding Cake and Golden Pomegranate in addition to Lafayette and Pocahontas (Robert 1938: 219). The value of brand names as assets to the firm was shown in Richmond in 1852 after the death of Poitiaux Robinson, one of the city's most powerful tobacco manufacturers, when the issue of to whom the brand names should be ceded became the central concern of the administration of Robinson's estate (Robert 1938: 193).

In the United States branding tobacco was widespread since manufacturers were typically distant from their customers and distribution was handled not by the manufacturers but by agents and jobbers located for the most part in New York City (Robert 1938: 222–6). British manufacturers also turned to branding. Ricketts, Wills and Company began the practice in 1847 with brand names such as Best Bird's Eye, Bishop Blaze and Stansfields (Alford 1973: 97–8). The distribution of tobacco products by Wills in Britain, though operating on a different system from that in the United States, nevertheless allowed the firm to gain a national as opposed to a local reputation (Alford 1973: 100–7).

Branding preceded advertising and it involved the consumer and the manufacturer in an association through images (Mitchell 1992). Names were one kind of image but a more effective method was for the brand to be recognized by a symbol or sign. One of the first manufacturers to do this, and to do it very successfully too, was the smoking tobacco manufacturer W. T. Blackwell and Company of Durham, North Carolina who in 1866 launched the brand Bull Durham on a demanding public (Tilley 1948: 548). Not only did the firm place its product on the market with a trademark but, more importantly, they decided to adopt the bull as their motif on all wrappings and advertisements. One of the partners of the firm, Julian Carr, instigated a national advertising campaign in which the Durham Bull itself, rather than the tobacco, was portrayed in anthropomorphic situations, alternating between scenes in which the bull was jovial and boisterous and those where he was serious and determined. The campaign was in swing by 1877 but Carr, one of the first and principal exponents of advertising, had more tricks up his sleeve, two of which strike a particularly modern chord. One of these was to sponsor commencement exercises at the University of North Carolina by conveying guests from hotel to

100

campus with a livery of horses, each of which had attached to it a flag bearing the sign of the bull; the wagons dealt with the matter more directly – they had painted on them the sign 'Smoke Blackwell's Durham Smoking Mixture'. Each member of the band greeting the guests smoked a pipe containing Bull brand tobacco (Tilley 1948: 549). The other move was to sponsor the barbecue following commencement, to which all guests were invited: once having eaten, drunk and, one may safely suppose, smoked, the guests were returned to their hotel in Durham in the same fashion in which they arrived (Tilley 1948: 550).

That was in 1879, and just the beginning of W. T. Blackwell's experience of the power of advertising. The 1880s signalled Julian Carr's determination to announce the Bull to the whole country by contracting an advertising agency to take out space in both country newspapers and large dailies: at the same time he introduced the practice of offering premiums to customers: clocks, razors and soap were particularly favoured, as well as gifts to dealers handling the Bull brand (Tilley 1948: 550).

W. T. Blackwell and Company, and especially Julian Carr, set a new trend in the American tobacco industry not only in its use of advertising but also in understanding that this was a key to success in the marketplace. Other manufacturers of the time attempted to follow the Bull's lead but none seems to have had the imaginative approach of Carr, except that is for one manufacturer, James B. Duke. Duke not only was Carr's rightful heir in the field of advertising but also understood that advertising for mass consumption required techniques of mass production.

Like Blackwell's, Duke's business, inherited from his father, was founded on the manufacture of smoking tobacco, under the brand name Pro Bono Publico. Unlike the Bull, however, Pro Bono Publico had a limited market and, in frustration at the firm's inability to crack the market, James Duke made the fateful decision to withdraw from direct competition in the smoking tobacco business and launched into what was then a relatively little known field, the cigarette industry.

Virginia and North Carolina manufacturers concentrated for the most part on producing chewing and smoking tobacco: New York City and other large towns in the northern United States concentrated on cigar manufacturing and, later, on the cigarette (Cooper 1987). When Duke decided to produce cigarettes, therefore, he entered into a sector of the tobacco industry that not only was located in a distant part of the country but in which there were already some well established firms (Tilley 1948: 557). Duke launched his first brand of cigarettes, Duke of Durham, in 1881 and then went about creating a mass market, employing techniques reminiscent of those first used by Carr, as well as some of his own. He advertised nationally, offered premiums to consumers and special deals to jobbers, sponsored games at which free cigarettes were presented (to men only) and he introduced the cigarette card, using at first the picture of a popular singer of the period and in time extending to actresses, presidents of the United States, royalty and other important figures; at the same time he considerably cut the price of his brands of cigarettes (Tilley 1948: 558). The firm also had its own polo team, named after Cross Cut, one of the company's leading brands (Tilley 1948: 558).

It is clear from the extant information about the firm's operations that Duke's techniques of mass advertising were extremely successful in that until 1885 orders for cigarettes consistently exceeded the technical capacity of production (Tilley 1948: 558). Even his decision to open a factory in New York City in the previous year did not solve the supply problem. As the discussion in Chapter 9 will show, the solution came in the shape of the cigarette-making machine invented by James Bonsack but most fully exploited, in several meanings of the word, by Duke himself. Duke was not alone in using advertising to push his products: other cigarette manufacturers also pursued markets in this way and provided innovative ideas (Tilley 1948: 559–60).

However, Duke alone, in the first few years of using the Bonsack machine, suffered the reverse of the problem he had before 1885: the Bonsack machines raised the output of cigarettes from a level of 9 million in July 1885 to 60 million two years later (Tilley 1948: 559). Duke's solution was not only to advertise more aggressively, which he did, but to get closer to the consumer by circumventing the commission merchant and searching for global, as opposed to national markets. But before this could happen a technical hitch stood in the way and that was that cigarettes were poorly packaged, typically in flimsy wrappers, although by the 1880s there were some machines on the market that could package cigarettes in a sturdy container. Once more it was Duke who took the lead by presenting his version of the sliding box in the launch of a new brand, Cameo, in 1886 (Tilley 1948: 575). It was now possible for Duke to use the package itself as a form of advertising and to include in it a small memento, in the shape of a cigarette card. Cigarette cards, in particular, became a critical component of Duke's advertising as they quickly became desirable in themselves, the consumer being enticed to collect sets of cards which were issued in series. According to an authority on Duke's advertising techniques at this time, the most common theme was sex. In his words:

> The cigarette was used almost exclusively by a masculine clientele in the nineteenth century, and the cards . . . reflect the advertisers' keen awareness of the fact. Many sets of cards featured either photographs or lithographs of buxom young ladies in what must have seemed very daring, if not shocking, costumes. Usually these sets were labelled simply 'Actresses' or bore descriptive phrases such as 'Stars of the Stage', 'American Stars', or 'Gems of Beauty'. Since there was surely little personal identification by the purchaser with the stars,

who were usually unnamed, and since actresses were then accorded a low place in the social scale of polite America, it seems clear that such cards were designed for prurient attraction.

(Porter 1971: 35)

Advertising did not come cheap and Duke probably spent more on it, as a proportion of turnover, than did other manufacturers (Porter 1971: 43). According to Duke himself, the costs of advertising in 1889 accounted for 20 per cent of sales (Porter 1971: 41). In the same year it was estimated by a trade magazine that manufacturers could easily incur a cost of $250,000 by introducing a new brand on to the market (Porter 1971: 41).

Duke structured his business around the twin concerns of mass consumption and mass production, and in this lay his particular genius (Chandler 1977: 382). His advertising techniques targeted the market carefully, in the United States and abroad. In China, the most important foreign market for British American Tobacco, the successor to Duke's firm, advertisements followed a different pattern to those in use in the United States. In particular far greater use was made of outdoor advertising, handbills, wall hangings, posters and window displays, rather than newspaper advertising and cigarette cards (Cochran 1980: 35–8).

In Britain advertising also proved to be crucial to the performance of individual firms and to the industry at large. Firms like Cope's and Wills, while not spending as much on advertising as their American counterparts, nevertheless found themselves pursuing similar objectives and also innovating in their own right. Cope's, for example, not only advertised their products in familiar ways but also took the unprecedented step of publishing a magazine, *Cope's Tobacco Plant*, from 1870 to 1879, in which not only was the firm and its products advertised but also the delights of smoking (Seaton 1986: 16–22).

One of the major characteristics of the cigarette industry in both the United States and Britain before the First World War was the large number of brands available to the consumer. Shortly after the creation of the American Tobacco Company in 1890 James Duke stated that the company manufactured a hundred different brands of cigarettes (Tennant 1950: 42). Brands had particular followings, and advertising policy at the time reinforced this pattern, definitely in the United States and most likely in Britain too (Tennant 1950: 42; Alford 1973). But all this changed very dramatically in the immediate aftermath of the dissolution by legislation in 1910 of American Tobacco, over which Duke was the supreme head (for more detail see Chapter 9).

One of the effects of the dissolution was to divide up the trust's business and, particularly, its brands among the successor companies. The cigarette sector was split largely into three companies, the newly reorganized American Tobacco Company, Liggett and Myers, and P. Lorillard who between them accounted for just over 80 per cent of national output (Tennant 1950: 80). One manufacturer who did not produce cigarettes before the dissolution of the trust and consequently got none of the cigarette sector at the time was a chewing tobacco firm called R. J. Reynolds. In 1913 this firm, which had no experience of cigarette manufacture, launched on the market a brand called Camel, the output of which stood at just over 1 million cigarettes: the following year production surpassed 400 million cigarettes and in 1919 it surpassed the figure of 20 billion (Tilley 1985: 219). By that date, Camel accounted for just under 40 per cent of all the cigarettes sold in the United States.

The success of Camel cigarettes rested on three main factors, all of which revolutionized the cigarette industry and placed it on a trajectory which would continue until very recently. In the first place, Camel was a new product in that it was composed primarily of American leaf, a mixture of blended Bright tobacco with Burley tobacco (an air-cured mild tobacco grown mostly in Kentucky); Turkish tobacco was added for taste and aroma (Tilley 1985: 211). Reynolds redefined the cigarette in terms of its constituent composition. Before the introduction of the Camel brand American-made cigarettes were either pure flue-cured (termed Virginia) or pure Turkish, or blended flue-cured and Turkish (Tilley 1985: 210). Second, Camel cigarettes were sold only in packages of twenty, again breaking with a tradition of selling in units of ten or five. Finally, Reynolds decided to advertise their new product publicly, that is in newspapers and the like, and totally refrained from the use of premiums or prizes, so frequently used to promote the cigarette in Duke's day. Slogans proved to be especially important in this kind of advertising. In 1921, for example, the famous slogan 'I'd Walk a Mile for a Camel' was first used on billboards throughout the United States (Tilley 1985: 223). The cost of advertising soared in the first decade of Camel's existence, as did the expenses entailed in giving away samples for free, a widespread practice in the cigarette industry: in 1913, for example, the total expenditure on both these amounted to under $800,000; in 1924 the figure was over $8,000,000 (Tilley 1985: 224). But of equal importance was the advertising on the package itself. The refusal to offer premiums was turned to advantage by printing on the package the message: 'Don't look for Premiums or Coupons, as the cost of the Tobaccos blended in CAMEL cigarettes prohibits the use of them' (Tilley 1985: 214–15).

The other tobacco companies, naturally, responded directly and aggressively to Reynolds's challenge and, of course, its extraordinary success. Each of the major companies either pushed one brand already in existence or introduced an entirely new brand to compete directly with Camel. Liggett and Myers turned to the Chesterfield brand as their flagship, and, not surprisingly, modified the blending formula and packaged the cigarettes in units of twenty: the American Tobacco Company staked its future on a new brand, Lucky Strike, which it brought out in 1916: and Lorillard chose Tiger as its brand but, after failing to secure a reasonable market share, dropped it in favour of Old Gold, launched in 1926 (Tennant 1950: 78).

The impact of these responses to Camel's challenge was dramatic, as the cigarette market split into three main brands. Table 5.2 shows these market shares for Lucky Strike, Camel and Chesterfield from 1925 to 1949.

Table 5.2 Market share of leading cigarette brands, United States 1925–49

Year	Brand	Production (billions)	Market share (%)
1925	Lucky Strike	13.7	16.7
	Camel	34.2	41.6
	Chesterfield	19.7	24.0
1939	Lucky Strike	38.3	21.2
	Camel	42.8	23.7
	Chesterfield	33.0	18.3
1949	Lucky Strike	93.0	23.9
	Camel	97.0	25.0
	Chesterfield	68.5	17.6

Source: Tennant 1950: 88, 94

Needless to say, the amount of money expended by the companies to pursue what had come to be known as a cigarette war in advertising grew considerably. To hold on to its market share, for example, R. J. Reynolds continued to pour money into advertising. By 1925 advertising expenditure topped the $10 million level and until the outbreak of the Second World War the figure hardly slipped below $15 million: moreover, as a proportion of net earnings, the expenditure on advertising was enormous, averaging 70 per cent during the 1930s in the aftermath of the depression (Tilley 1985: 330). The American Tobacco Company with Lucky Strike also saw its advertising budget mushroom, tripling to just over $24 million in the period from 1925 to 1932, and similar changes were experienced by Liggett and Myers (Tilley 1985: 331; Tennant 1950: 165–6).

It was precisely in the period after the First World War that the consumption per capita of tobacco products other than cigarettes began to decline for the first time since the turn of the nineteenth century: by contrast, the consumption of cigarettes continued to increase, and by 1923, in these terms, cigarettes had become the single most popular form of tobacco consumption in the United States (Tennant 1950: 127). In the 1920s alone, cigarette consumption per adult more than doubled from 610 in 1920 to 1,370 in 1930; or, to put it another way, from less than two to just under four per person per day (Beese 1968: 62). The powerful attraction of the cigarette was a remarkable feature of this period considering that in per capita terms consumption of all tobacco products was virtually the same in 1940 as it had been in 1920 (Beese 1968: 62–3).

Men were abandoning the old tobacco comforts – pipe, cigar and chaw – and confirming themselves as cigarette consumers, and this certainly accounted for the greater part of the increase in per capita cigarette consumption. But a not insignificant component of this increase in the popularity of the cigarette was a change in the gender basis of tobacco consumption. As the previous chapter pointed out, the notion that women have started to consume tobacco only in the twentieth century is incorrect; on the contrary there is ample evidence that in the seventeenth and eighteenth centuries tobacco consumption was not gendered. The picture for the nineteenth century is blurred, but there is enough indirect evidence to suggest that some such process was under way: certainly in the nineteenth century the pipe, the cigar and chewing tobacco had an increasingly masculine image (Dunhill 1924; Seaton 1986: 22). Other than these general observations there is little hard information on the degree or extent of tobacco consumption by gender in the nineteenth century. However, there is scattered evidence in the early part of the twentieth century that might confirm the low incidence of smoking among women. In 1924 the editor of one of America's most important tobacco trade journals estimated that women accounted for only 5 per cent of national consumption (Tennant 1950: 136): in Britain there are no contemporary estimates for this period but a recent work estimates that women's consumption of tobacco for the same year was no more than 1.9 per cent of the national total (Alford 1973: 340).

During the 1920s, however, there is ample evidence that women became increasingly attracted to smoking and particularly to the cigarette. By 1929 women were estimated to consume 14 billion cigarettes in the United States, equivalent to about 12 per cent of total consumption; in 1931 the estimate had increased to 14 per cent and in 1935, in a survey carried out by *Fortune* magazine, the researchers were able to report that 26.2 per cent of women aged 40 years or more smoked cigarettes while for those under 40 the corresponding number was 9.3 per cent (Gottsegen 1940: 150–1). In urban areas the proportion regardless of age was much higher, between 31 per cent and 40 per cent, depending on the size of the city (Gottsegen 1940: 151). A similar pattern occurred in Britain where the proportion of total tobacco consumption accounted for by women doubled from 5 per cent to 10 per cent between 1930 and 1939 (Alford 1973: 362).

Tobacco companies were not slow to realize that changes were occurring, and through the 1920s their advertising became clearly targeted to this new and growing group of consumers. Whether the advertising was itself responsible, to some degree, for the increasing number of women smokers or whether it was capitalizing upon a discernible trend is a moot point. It is also unresolvable since the complex culture of tobacco consumption precludes any analysis in simple terms of cause and effect (Schudson 1985: 178–208; Waldron 1991). There is little doubt, however, that even if the rise in women's consumption of tobacco rested on other social and economic changes, the tobacco companies, in their cigarette advertising, provided new images to which women might aspire or be confirmed by smoking.

The targeting of women as cigarette smokers in the United States happened slowly and uncertainly. One of the first advertisements was for Helmar's cigarettes, a brand manufactured by Lorillard. The advertisement featured a woman with oriental features holding a cigarette between her lips (Tilley 1948: 614). This appeared in 1919 but it was not until 1926 that the first advertisement appeared in which women were portrayed in the role of accepting the challenge of smoking a cigarette. The advertisement for Chesterfield showed a couple in a romantic setting: the man is shown smoking while the woman, in a sensuous pose, pleads with the caption 'Blow some my way' (Marchand 1985: 97). In 1927 Philip Morris, one of the smaller cigarette manufacturers of the time, advertised one of its brands, Marlboro, showing a woman holding a cigarette with the caption 'Women, when they smoke at all, quickly develop discriminating taste' (Tennant 1950: 139). Lucky Strike entered the fray on two fronts: it solicited and printed testimonials from European artistes who informed the reader that they had discovered their favourite cigarette in Lucky Strike, a cigarette that was mild and mellow and because of a special process that treated the tobacco – 'It's Toasted' – Luckies protected your throat (Marchand 1985: 96). The makers of Lucky Strike, American Tobacco, pursued the advantages of smoking their brand with new hard-hitting messages in advertisements in 1928 and 1929 in which women were urged to smoke with the caption 'Reach for a Lucky Instead of a Sweet' (Tilley 1948: 614). This was backed up by testimonials from well-known personalities on the desirable effects on body weight and figure by substituting cigarettes for sweets (Marchand 1985: 99). And, if the point hadn't been driven home enough, in the next few years Lucky Strike adverts championed the svelte over the fat body with headlines such as 'Pretty Curves Win!' and captions such as: 'Be moderate – be moderate in all things, even in smoking. Avoid that future shadow by avoiding over-indulgence, if you would maintain that modern, ever youthful figure "Reach for a Lucky Instead".' (Marchand 1985: 101). Or the advertisement entitled 'The Grim Sceptre', in which a woman haunted by a double chin is urged, again, to reach for a Lucky instead (Tilley 1948: 614). Reynolds came back hitting hard in 1928 with their own version of the advertisement targeted directly at women, showing scenes of women alone as well as couples, the woman in each case getting closer and closer to smoking the cigarette (Tilley 1985: 340–1). In the following year Reynolds turned the first Chesterfield advertisement on its head when they showed a woman offering a Camel to a man who responds with a turn of phrase that must have warmed the hearts of the copy writers: 'I'd Walk a Mile for a Camel – but a "Miss" is as Good as a Mile' (Tilley 1985: 331). You've come a long way, baby, as the later advertisements put it.

What were the messages of cigarette advertising during the 1920s and 1930s and what was their connection with the culture of consumption? As Michael Schudson points out in his study of advertising and American society, the chief theme that advertising emphasized was mildness, a theme that integrated the ingredients of the cigarette (mild, mellow tobaccos wrapped in pure white paper) with the action of smoking the cigarette portrayed in refined terms and circumstances (Schudson 1985: 202). Presumably this message was designed as much to reinforce their confidence in the product as it was to wean pipe and cigar smokers and, presumably, tobacco chewers from their habit; after all, men were the smokers. The advertising of cigarettes to women, judging from the text of these advertisements, stayed on the edge of the social conflict which marked the rise of the woman cigarette smoker during the 1920s. It legitimized the results, and cultivated an image of the woman smoker that was complementary to the image of women over which the conflict arose in the first place (Schudson 1985: 187–91). The gender politics of the 1920s rested on the question of the access to power by women, in relation to men, as well as by women in the past. The cigarette was adopted as a symbol of the emergence of the new woman of the 1920s (Filene 1975: 148–9; Fass 1977: 9; Ernster 1985: 336; Waldron 1991: 994). The tobacco companies responded to this social change not by entering the debate but by adding their own fine tuning to the new woman image: slender, chic and mildly seductive (Ernster 1985: 338). As for men, as Schudson argues, the cigarette represented, both symbolically and actually, a convenience and refinement, in terms of pleasure, as opposed to the cigar and pipe, both of which came to be viewed as cumbersome and were therefore relegated to special occasions such as the after-dinner treat: cigarettes were more suitable to the work and leisure culture of the postwar era than were other forms of tobacco use (Schudson 1985: 198–202).

The cigarette was adopted by women as a badge of emancipation in the period following the post-First World War (Jacobson 1981). Since then the proportion of women smoking has increased continuously until just recently. In the United States there was a steady rise in the relative number of women smoking until the mid-1960s, after which it levelled off or decreased slightly (Waldron 1991: 989–90). A similar pattern can be observed in Britain (Jacobson 1988: 5). In Italy smoking among women was uncommon before the Second World War but increased very rapidly after, and continued to increase until the mid-1980s (La Vecchia 1986: 276). This is only part of the story, for while men have been decreasing their consumption of cigarettes, in the sense that the proportion of men smoking has been declining for a long time, women have only recently decreased theirs. If the cigarette is a symbol of equality, then men and women have only recently become equal, with some evidence that cigarettes have become appropriated by women more than men as a commodity in some parts of the world (Jacobson 1988: 5–12; USDHHS 1980).

Until the Second World War, the major tobacco companies in both the United States and Britain continued to be characterized by leading brands. Lucky Strike, Chesterfield and Camel accounted for the majority of the market share of the cigarette business, though in the 1940s the Philip Morris brand made significant inroads into the market share of the big three (Tennant 1950: 88). In Britain the picture was very similar: Woodbine, the flagship brand of Wills, accounted for as much as 70 per cent of all cigarettes sold by the company in the country and for as much as 30 per cent of all the cigarettes sold in total (Alford 1973: 478–9). Though advertising continued to target men and women in different ways, no specific brand was associated with any particular gender or class (Schudson 1985: 178–208).

Since the Second World War the cigarette market has undergone another major change involving three main developments: multibranding; filter-tipped and low nicotine brands; and gendered brands. These were not entirely separate developments though their chronology differed. Multibranding, for example, grew as a response to market conditions in the United States after the war. One of these conditions was the inability of the three leading brands to maintain, let alone increase, their share of the market: the sales of Camel, Chesterfield and Lucky Strike were virtually identical in 1949 to what they had been in 1946 despite an increase of just under 10 per cent in the size of the market (Tennant 1950: 88; Beese 1968: 63). Related to this was the other observation that unusual brands were finding a place in the market that was not only secure but actually growing. In particular there were mentholated brands, the most important being Kool, first marketed in 1933; there was a filter-tipped cigarette, under the brand name of Viceroy launched in 1936; and king-size brands such as Pall Mall (Axton 1975: 118). Viceroy and Kool were manufactured by Brown and Williamson, one of the smaller tobacco companies at the time, while the latter belonged to American. Between 1945 and 1952 Reynolds, Liggett and Myers, American and Lorillard brought out new brands to meet the competition (Tennant 1971: 228).

Multibranding was, therefore, well under way when cigarette smokers, in particular, were told by the prospective studies on cancer, carried out first in 1951, and published under the auspices of the American Cancer Society in 1954, that there was a causal relationship between cigarette smoking and lung cancer (Patterson 1987: 209). The consumption of cigarettes in the United States fell immediately by 24 billion units, or 6.4 per cent of the total manufactured, as the public reacted to the health scare

(Tennant 1971: 229). It was the first time in the twentieth century that cigarette consumption had fallen and shock waves rang through the tobacco industry.

Tobacco chiefs responded in a number of ways, including, not insignificantly, the creation of a powerful industry lobby and publicity organization called the Tobacco Industry Research Committee, and then in 1958 the more influential Tobacco Institute Inc., both of which, in their own way, sought to undermine the smoking–cancer equation (Patterson 1987: 211–12). Another much more important response was to manufacture a cigarette in which consumers would have more confidence and which, indirectly, would show that the industry was responsible and responsive to public opinion. The solution came in the form of the filter-tipped cigarette which had already been on the market but enjoyed only a small following. Viceroy sales doubled immediately; the sales of Kent, introduced only two years earlier, with an allegedly revolutionary filter system, accelerated and all the tobacco companies without a filter cigarette launched their own version (Tennant 1971: 229). Liggett and Myers's brand, L & M, came out in 1953 and in the following year R. J. Reynolds brought out its enormously successful Winston brand; in 1956 the company launched Salem, the first filter-tipped mentholated cigarette (Tennant 1971: 229; Tilley 1985: 496–503).

The speed with which the manufacturers and the public switched to filter cigarettes can be seen clearly from Table 5.3. While the data in Table 5.3 relate only to the United States, the trend towards the filter cigarette has been global, though the rates of transformation have varied considerably. Britain, for example, followed the American lead very closely though with a lag of some five years; in Japan, New Zealand and Venezuela over 90 per cent of cigarette sales were filter-tipped by 1970 while in India in the same year filter-tipped cigarettes accounted for less than 10 per cent of total sales (Alford 1973: 431; Tucker 1982: 195).

The filter-tipped cigarette was, in form, a refinement on the non-filtered type. It satisfied the consumer's need for a 'safer' cigarette, one which

Table 5.3 Filter-tipped share of the United States cigarette market (average %)

to 1950	0.5
1951–5	6.5
1956–60	42.1
1961–5	58.1
1966–70	74.6
1971–5	85.2
1976–80	90.6
1981–5	93.7

Source: Warner 1986: 23

filtered out 'irritating' elements in the smoke, and also drew the public's attention to the scientific research that had produced the filter: manufacturers were at pains to point out that their particular filter embodied the most advanced technology and materials. But there was also a large economic benefit in switching to filter-tipped cigarettes, as less tobacco was used in production.

This was not, however, the only cost-reducing manufacturing innovation of the postwar era. Since the early part of the century tobacco manufacturers had been attempting to reduce their costs of operations by reclaiming as much as possible of the waste products of manufacturing. Since tobacco leaf is the most expensive component of manufacturing costs, manufacturers looked for ways of using more of the tobacco plant other than the leaves. For technical reasons a successful method of reconstituting the waste products – the stems, scrap tobacco and tobacco dust – into a sheet, akin to paper, was not developed until after the Second World War (Tilley 1985: 488–94; Akehurst 1981: 656). But one cannot help feeling that there were other factors in the slow development of reconstituted sheet as a cigarette filling: the stem has a significantly lower nicotine content than the leaves and, while public tastes were formed around the high level of nicotine in cigarettes manufactured before 1945, any shift away from that position would undoubtedly hurt a manufacturer as consumers sought other brands for their taste (Davis 1987: 19). Once the cancer scare hit, manufacturers could confidently launch new tastes on the market largely because the filter-tipped cigarette was itself a novelty (Mann 1975: 100). The immediate impact of increasing reconstituted sheet, and the use of a filter, was to increase markedly the number of cigarettes manufactured from a given weight of tobacco. In the United States between 1939 and 1953 the number of cigarettes produced per pound of leaf was stable at a level of 324: in 1958 the corresponding level was 380 and in 1970 it was 467 – a 50 per cent increase over the figure in 1953, before the cancer scare (Johnson 1984: 65).

Besides the shift to filter cigarettes, another important change, especially since the publications of the Surgeon General in the United States and the Royal College of Physicians in Britain, has been the increasing importance of low-tar, low-nicotine brands. The United States was a big leader in the manufacture of these brands: between 1967 and 1970 cigarette brands containing 15 mg or less of tar captured hardly 1 per cent of the market but by 1981 the share had risen so quickly that six out of every ten cigarettes smoked were of this type (Warner 1986: 13). In 1980 the United States cigarette market was swamped with no fewer than one hundred new brands of the low-tar, low-nicotine cigarette (Taylor 1984: 185). Unlike filter cigarettes that have global markets, the low-tar, low-nicotine brands are mostly confined to the western industrialized countries, but even in these the market is not as large as in the United States: in 1982, for example, only 20 per cent of the British cigarette market was accounted for by low-tar cigarettes (Wilkinson 1986: 55; McMorrow and Foxx 1983: 302; Davis 1987: 17; Muller 1978: 53). The increasing consumption of low-tar, low-nicotine cigarettes has had an impact on manufacturing as well. Reconstituted sheet has been increasingly used as have other techniques designed to decrease the amount of tobacco leaf used per cigarette. One of the most important techniques of this kind is called puffing, in which dried tobacco leaf is processed in such a way that its size is increased to that of its green condition: by this method the filling capacity of the cigarette is increased by 40–50 per cent – the technique was first introduced in 1969 (Mann 1975: 100–1). Between 1970 and 1980 the number of cigarettes manufactured from one pound of tobacco leaf rose from 467 to 523 (Johnson 1984: 65). Both reconstituted sheet and puffing lend themselves to cigarettes of low-tar and low-nicotine delivery.

The filter cigarette and the newer low-tar, low-nicotine brands have pushed multibranding far beyond the levels envisaged when the practice first began after the Second World War. The process was further accelerated by the introduction and rapidly growing popularity of the female cigarette. The idea of a female cigarette was not new to the postwar period. The idea had been floated before and several attempts had been made by tobacco companies to get such a product established in the market: one such failure was a cigarette called Fems whose chief characteristic was its red mouthpiece designed, of course, not to show lipstick (Ernster 1985: 337); in Britain Wills had a proposal for launching a brand especially for women – its name was to be Rainbow – but this was declined in favour of targeted advertising (Alford 1973: 340). Marlboro, marketed by Philip Morris, was probably the only prewar cigarette that advertised itself as a luxury cigarette with a feminine touch (Ernster 1985: 336). Even in the 1960s the going was rough for the launch of a female cigarette as evidenced by the failure of Liggett and Myers's Eve brand (Jacobson 1988: 55). All the previous difficulties were swept away, however, in 1968 when Philip Morris launched Virginia Slims; in 1983 it was America's eleventh best selling cigarette in what was a very competitive market (Jacobson 1988: 56). The amount of advertising and promotional expenditures that went into sponsoring Virginia Slims and other female cigarettes is unknown but that this grew enormously during the 1970s is beyond doubt (Ernster 1985: 339; Jacobson 1988: 55–60). In 1984 cigarette advertising in the major women's magazines in the United States cost manufacturers many millions of dollars: *Family Circle*, to take one example, earned over $16 million from cigarette manufacturers, equivalent to 12.5 per cent of the magazine's total revenue for that year (Ernster 1985: 339).

Since 1950, therefore, cigarette brands have been placed on the market in bewildering numbers. These brands, at the time of their launch and subsequently, have been aimed at particular classes of consumers, each of which has an identity in the product, in terms of its packaging, design, colours and, of course, image. While the market is a mass one and the techniques of marketing reflect this, they also rest on the belief that while the market is segmented, within each segment the market is global.

One of the first brands to confirm this was Marlboro. From its launch in the 1920s until the 1950s, as stated above, Marlboro was first a luxury then a women's cigarette. In the wake of the cancer scare in the early 1950s, Marlboro underwent a total transformation: a filter was added, it was packaged in a flip-top, crush-proof box with distinct red and white colours, and it was unceremoniously taken away from women and given to men with the help of the masculine cowboy image (Lohof 1969: 443). The Marlboro advertising campaign has been one of the most successful of all time. Reflecting on the meaning inherent in the rugged images that the advertisements convey, one commentator has written as follows:

> The Marlboro image represents escape, not from the responsibilities of civilization, but from its frustrations. Modern man wallows through encumbrances so tangled and sinuous, so entwined in the machinery of bureaucracies and institutions, that his usual reward is impotent desperation. He is ultimately responsible for nothing, unfulfilled in everything. Meanwhile, he jealously watches the Marlboro Man facing down challenging but intelligible tasks ... Innocence and individual efficacy are the touchstones of the metaphor employed on behalf of Marlboro cigarettes.
>
> (Lohof 1969: 448)

'Power, status, success and confidence' as Bobbie Jacobson puts it; the appeal of Marlboro seems to cut across class and income (even gender) divisions by presenting an image which reinforces the self-awareness of the privileged, and satisfies the fantasy aspirations of the less privileged. In 1976 Marlboro became the best-selling brand in the United States (Tucker 1982: 80). Marlboro sales undoubtedly accounted in large measure for the rise of Philip Morris as a tobacco company: its share of the American cigarette market which from 1940 to 1960 hovered at around 10 per cent, reached 29 per cent in 1979, by which time it was firmly established as the second largest tobacco company in the United States. Globally, Marlboro is the world's most popular cigarette. In 1987 293 billion Marlboro cigarettes were sold worldwide (USDHHS 1992: 40). It is the market leader in many countries, especially, but by no means exclusively, in the Third World (Tucker 1982: 85–6; USDHHS 1992: 46).

Advertising has, as we have seen, always been of critical importance to the tobacco companies, even before James Duke raised the profile and use to which it was put. What the effects of advertising have been on consumption – especially on the extent of it – has been and still continues to be a hotly debated issue. Whatever its impact on total consumption, certain

elements of advertising are clear enough. One of these is that advertising
has probably been the most important expenditure by tobacco manufacturers. The twentieth century, especially the period since 1945, has witnessed an excessively large and increasing advertising budget despite the
bans on certain kinds of advertising taken by various different countries
throughout the world. Scattered data on advertising expenditure, in Table
5.4, reveal the extent of this trend. Though the data for 1970 to 1983 are
represented in current terms, even allowing for inflation advertising still
doubled in the period (Warner 1986: 45). Second, though there are
hundreds of brands available in many countries – in 1985, for example,
there were around 140 different brands of cigarette available in Britain –
there is a large degree of concentration in sales as there appears to have
been throughout the twentieth century (Jacobson 1988: 203–7). This confirms the remark about the segmented but mass aspects of the cigarette
market. In 1988–9 the top ten brands in Australia, Brazil, Canada, Italy,
France, Mexico, Germany, Britain and the United States accounted for at
least 70 per cent of total sales (USDHHS 1992: 40).

Table 5.4 Cigarette advertising expenditure,
United States, selected years 1939–83 ($
million)

1939	47
1959	148
1970	361
1975	491
1983	1,900

Sources: Tilley 1985: 331; Patterson 1987: 212;
Warner 1986: 45

The meaning of the cigarette has become intimately related to the image
and metaphor of its advertising. Smokers are reinforced in their choice by
the images presented and there is sufficient evidence available that brand
loyalty is extremely important – surveys in the United States in the 1970s
concluded that about half of cigarette smokers interviewed had never
changed their brand (USDHHS 1992: 41).

The culture of cigarette consumption, however, is not simply that of the
advertisement: rather it is a blend of the advertised images and metaphors
together with prevailing social customs. While there has been a powerful
transformation of tobacco habits globally towards the consumption of light
filter-tipped cigarettes manufactured by multinational tobacco companies,
the anthropology of tobacco use, as recent studies make patently clear,
remains complex (USDHHS 1992: 37–40, 46–7; Waldron *et al.* 1988;
Waldron 1991; Carucci 1987; Knauft 1987; Black 1984). In the western
industrialized world, however, there is a further element affecting the
culture of tobacco consumption, namely the increasing view of the cigarette

as a dangerous commodity, especially by the medical profession. In the words of the historian Allan Brandt:

> The new research agenda facilitated the ongoing process of delegit-
> imizing cigarette smoking in American culture. The cigarette – the
> icon of consumer culture, the symbol of pleasure and power, sexuality
> and individuality – had become suspect. The smoker would sub-
> sequently be redefined, in a process which we continue to see played
> out – from the independent Marlboro man or liberated Virginia Slim
> to a new vision of a weak, irrational, and now, addicted, individual.
> The stigmatization of the cigarette became a critical aspect of a revolu-
> tion in American values about personal health and behavior.
>
> (Brandt 1990: 168–9)

The medical significance of tobacco has been debated continuously since its first introduction into Europe in the sixteenth century. On the whole the judgements made about tobacco were favourable despite stern oppo-sition to consumption both by experts and by civil and ecclesiastical authorities. While the debates were most frequent and expressive in the seventeenth century, they did not disappear in the eighteenth century. The issues raised in the medical literature became more subtle, as exemplified in the debates concerning the comparative effects of tobacco in the form of snuff and smoke.

What is important to understand is that the discourse in the sixteenth, seventeenth and eighteenth centuries took place without anyone really understanding what constituted tobacco. Botanists were quick to establish the nature of the plant but other insights were lacking. There was some suggestion that tobacco was dangerous in that it contained dangerous substances. Francesco Redi, the Florentine scientist and physician, for example, published the results of his experiments in one of which he injected 'oil of tobacco' into a number of animals, all of whom died (Redi 1671). But the lack of knowledge about the constituents of tobacco ensured that many of the arguments pursued in the eighteenth century simply replicated those of the previous century without offering anything new. It is, of course, impossible to say precisely what impact the medical contro-versies had on the consumption of tobacco. However, with so much praise for the efficacy of tobacco from so many eminent physicians, it seems likely that favourable medical opinion did stimulate consumption. Certainly the survival of tobacco in folk remedies into the twentieth century suggests that conclusion (Kell 1965).

The tobacco discourse changed abruptly in the nineteenth century. The change came not from the medical profession but from the enormous growth in chemical investigations and insights. Of the many paths that chemists followed in the early nineteenth century to classify and understand the nature of matter, one with very special significance was to isolate the active, physiological principles of plant medicines. The interest in the pharmacology of plant medicines predated the nineteenth century but the techniques that would allow for an isolation of the active principle as well as the clinical trials necessary to confirm the connection were not generally developed until the end of the eighteenth century (Lesch 1981: 305–10; Smith 1979). In both Germany and France this problem seems to have captured the imagination of several eminent chemists and pharmacists. Initial interest focused on opium and cinchona but early work made little progress (Lesch 1981: 311). A real breakthrough came, however, in 1803 when Friedrich Wilhelm Sertürner, a pharmacist, discovered morphine, the active principle of opium (Schmitz 1985: 62; Lesch 1981: 312–13). Sertürner's work on opium was critical and it also raised hopes among chemists and pharmacists that there were other plant substances of an alkaline nature waiting to be isolated.

These alkali plant substances required classification as well as nomenclature. Joseph Louis Gay-Lussac proposed that substances in this class be nominated by attaching the suffix -ine to their name, and in 1818 the German pharmacist Karl Friedrich Wilhelm Meissner proposed the generic term 'alkaloids'. Both nomenclatures have remained ever since.

Whether codifying the discoveries in this way had anything to do with further breakthroughs is unknown, but in the aftermath of Sertürner's paper and Gay-Lussac's pronouncements, a host of new active plant principles were revealed. The French took the lead. Joseph Pelletier and Joseph-Bienaimé Caventou, both pharmacists working in Paris, declared in 1817 that they had discovered emetine, the active alkaloid of ipecacuanha root, a plant used in European and Amerindian medicine for its emetic properties; within the next few years they isolated strychnine, quinine, brucine and veratrine (Lesch 1981: 322–3).

Coffee was subjected to laboratory treatment in 1820, and caffeine was isolated; eight years later it was the turn of tobacco. Though the German physician Wilhelm Heinrich Posselt and his partner, the chemist Karl Ludwig Reimann, are credited with isolating nicotine in 1828, the attempt to find the active alkaloid of tobacco can be traced back to the experiments of Gaspare Cerioli in Italy and Louis-Nicolas Vauquelin, professor of chemistry in Paris, who produced an oil of tobacco in 1807 and 1809 respectively (Eiden 1976: 8). Experiments by Posselt and Reimann on dogs and rabbits confirmed what Francesco Redi had noticed more than 150 years earlier: that nicotine was extremely poisonous (Eiden 1976: 6).

There followed a spate of publications and experiments on nicotine from a chemical as well as a pharmacological viewpoint that filled the pages of scientific journals throughout Europe and the United States. Some of the work investigated the molecular structure of nicotine in the hope of synthesizing its constituent structure (Eiden 1976: 8–18). Other work investigated the physiological characteristics of nicotine from animal trials as well as from reported incidents in human consumption. A significant number of publications, however, fastened on the potential therapeutic uses for nicotine. A well publicized use for nicotine was as an antidote to strychnine, as well as other kinds of poisoning, including snake bites. The scientific journals reported on successful outcomes of nicotine for a wide variety of ailments including disorders of the nervous system, muscle contracture, genito-urinary diseases, haemorrhoidal bleeding (via a tobacco enema), strangulated hernia, infectious diseases (especially malaria) and tetanus (Silvette *et al.* 1958; Stewart 1967: 264–7).

The nineteenth century, like the seventeenth century, witnessed a controversy over whether tobacco was harmful or not but, for the time being at least, nicotine did not figure in it. As in the seventeenth century there was debate but no decisions, and, in fact, there had been little progress in the terms of the debate. Professor John Lizars of Edinburgh University in the sixth edition of his textbook on tobacco echoed the imprecise language of many critics of tobacco at the time. He not only listed an expansive catalogue of diseases and ailments caused by consuming tobacco – from vomiting and diarrhoea to ulceration, emasculation and congestion of the brain – that would have delighted any seventeenth-century nicotian critic, but spoke about tobacco consumption as a disease in itself. It was implicated, according to him, in the spread of syphilis, through the sharing of the tobacco pipe, and to national degradation, both physical, psychological and, of course, moral (Walker 1980: 393). Lizars was, perhaps, the most vehement critic of tobacco, but there were others in Britain, especially in the medical profession who made their views known in the *Lancet* and the *British Medical Journal* during the second half of the nineteenth century. In France, at the same time, a similar controversy raged (Perrot 1982; Nourrisson 1988).

While the terms of the debate remained unaltered from the seventeenth to the nineteenth century, there was an important change in how the debate was conducted. In the second half of the century anti-tobacco societies sprang up principally in Britain, France, the United States and, to a lesser extent, in other European countries. They all had similar objectives – to inform the public about the dangers of tobacco and to lobby for legislation against tobacco abuse – and similar memberships (Nourrisson 1988; Walker 1980; Troyer 1984). They also, at various times, associated themselves with the temperance movement, in a general crusade against intemperance, broadly defined. Furthermore, they were headed by charismatic individuals who imposed their own personalities on the societies. Despite having some outstandingly influential members – Pasteur, for example, joined the Société Française Contre l'Abus du Tabac in 1878 – most of the societies either fell into liquidation or ended their existence once the founders or the influential members lost interest or died. The two main French anti-tobacco societies were defunct by 1905, as were the chief British societies (Nourrisson 1988: 542; Walker 1980: 398–9). What they accomplished is hard to judge, but there is some evidence that the Children Bill of 1908, passed in Britain to prohibit the sale of tobacco to children and to ban juvenile smoking in public places, had enshrined at least some of the arguments of the anti-tobacco movement (Walker 1980: 401).

In the United States, however, the anti-tobacco movement had a much more direct and widespread impact, at least in terms of legislation. To speak of any early movement in the United States would be an exaggeration since the movement was perhaps no more than the publication, at irregular times, of the *Anti-Tobacco Journal* between 1857 and 1872. The main object of the attack was chewing tobacco and the main thrust was its uncleanliness but, in general, little seems to have come of the agitation which the editor of the journal, George Trask, hoped to inspire in his readers.

Once the cigarette became popular, however, an anti-tobacco movement coalesced around the problem of the cigarette. Publications of the period attacked the cigarette as a contaminating influence for both adolescents and women (Tennant 1950: 133). In the United States the argument connecting national decadence with cigarette consumption surfaced as it did in Europe. The favourite European bugbear was the ill-defined 'Turk', but in the United States, according to a writer in the *New York Times* in 1883, the finger was pointed directly at the Spaniard:

> The decadence of Spain began when the Spaniards adopted cigarettes and if this pernicious practice obtains among adult Americans the ruin of the Republic is close at hand.
>
> (Tennant 1950: 133)

The anti-cigarette campaign found supporters throughout the country and included some very influential people. Those in the front line, such as Lucy Page Gaston, who in 1920 declared she was running for President on a no-tobacco issue, were joined by such influential people as Henry Ford and Thomas Edison (Heimann 1960: 252). Ford, who carried out his own crusade against what he termed 'the little white slaver', solicited evidence from American notables about the inherent danger of the cigarette. One of them, Thomas Edison, in a letter to Ford, pronounced his verdict on the cigarette thus:

> The injurious agent in Cigarettes comes principally from the burning paper wrapper. The substance thereby formed is called 'acrolein'. It has a violent action in the nerve centres, producing degeneration of the cells of the brain, which is quite rapid among boys. Unlike most narcotics, this degeneration is permanent and uncontrollable. I employ no person who smokes cigarettes.
>
> (Tennant 1950: 135)

Edison's remarks are particularly interesting for two reasons. First, though he was wrong about what constituted the danger in cigarettes, he did attempt to provide a pharmacological and physiological basis for sustaining the argument that cigarette smoking in adolescence was particularly injurious. It should be pointed out that Edison was a committed cigar smoker. Second, his remark about narcotics should be seen in the context of what was perceived in the United States at the time as a virulent epidemic of drug taking, especially cocaine, opium and marijuana (Courtwright 1982; Courtwright 1991). (It is interesting that Edison should add cigarettes to the discourse on narcotics in light of our understanding of the powerfully addictive properties of nicotine.)

The tobacco reform movements culminated in a wave of legislation, beginning with the states of North Dakota, Iowa and Tennessee by 1897, and grew until 1921, when some form of legislation against smokers and the tobacco industry was on the statute books in twenty-eight different states (Gottsegen 1940: 155). The range of proscription was very wide but only two states – Idaho and Utah – passed specifically anti-cigarette legislation. In many cases, however, the repeal of the laws came as swiftly as their enactment: fiscal needs were often the reason. By 1930 little of the original legislation existed: the only prohibition that did remain in force was that on the sale of tobacco to minors (Gottsegen 1940: 154–5).

Most writers surveying tobacco consumption have treated those involved in the anti-tobacco crusades of the second half of the nineteenth and first two decades of the twentieth century in an anecdotal and largely unsympathetic manner. Certainly some of the pronouncements of those who carried the anti-tobacco banner border on the ludicrous. Dr H. A. Depierris, a guiding light of the *Association Française Contre l'Abus du Tabac*, claimed, for example, that the French suffered a humiliating defeat against German troops in the Franco-Prussian War 'because of the wreckage wrought by the narcotic plant . . . [they were] devoid of intellect, breathless, [with] emaciated legs and weak arms, they were incapable of taking up their rifles and marching towards the enemy on the day of the invasion' (Nourrisson 1988: 541). Dr Pidduck, in an article published in the *Lancet* at the height of the tobacco controversy in Britain, maintained that at an important London hospital the blood of smokers instantly killed leeches, and that fleas never attacked smokers (Walker 1980: 393). American commentators were equally colourful in their portrayal of the vile effects of tobacco in general, and the cigarette in particular. However, fanatics aside, there is a real sense of an underlying unease being expressed during this period. The difficulty, however, was to find a vocabulary to express it. Take for example the statement made in 1898 by the Supreme Court of Tennessee in support of upholding the state's anti-cigarette legislation:

119

Are cigarettes legitimate articles of commerce? We think they are not because they are wholly noxious and deleterious to health. Their use is always harmful; never beneficial. They possess no virtue, but are inherently bad, and bad only. They find no true commendation for merit or usefulness in any sphere. On the contrary, they are widely condemned as pernicious altogether. Beyond any question, their every tendency is toward the impairment of physical health and mental vigor.

(Tennant 1950: 134)

The quotation above, typical of many pronouncements against tobacco and cigarettes, demonstrates the problem. Moral condemnation alone is not enough; evidence is required and its source must be respected. An appeal to the medical profession was inevitable, given their remarkable rise as a professional group, but this was not without its difficulties. For one thing, most members of the profession did not consider tobacco as particularly dangerous. They were far more involved in other public health issues, especially the campaign against infectious diseases. The articulation and general acceptance of the germ theory of disease, helped by the enormous publicity given to scientists such as Robert Koch and Louis Pasteur in the 1880s, created an important shift in the public perception of health. In the United States, for example, but also in Europe and elsewhere, the medical profession focused on diseases such as tuberculosis which was, at the turn of the century, the greatest single killer in the West (Patterson 1987: 33). Though they were at the time unable to cure tuberculosis, the message from physicians (which was echoed in the popular press) was that the contagion could be arrested by attacking the germ through strict regimens of cleanliness (Burnham 1984; McClary 1980). It was difficult, if not impossible, to locate tobacco-induced disease within this paradigm. Interestingly, the practice of chewing was condemned by some doctors principally because it was believed that the spittle helped the spread of germs: this condemnation may also have helped the consumption of cigarettes.

Another problem was that many doctors themselves were smokers and thought positively about the substance. The Surgeon General of the United States Public Health Service, Hugh S. Cumming, while condemning cigarettes especially for women in a statement made in 1929, was accused of being half-hearted about it since he was a cigarette smoker himself (Burnham 1989: 3). Even as late as 1948 the *Journal of the American Medical Association* was arguing that the benefits of smoking – to reduce tension – outweighed any evidence of its harmfulness (Patterson 1987: 208). This attribute of tobacco was, of course, already part of the popular image of the cigarette, seized upon by tobacco companies and repeated in the popular press. Even as early as 1889 the *New York Times* carried the following articulation of the opposition of tobacco and tension:

> Whatever be its merits or demerits, one thing is certain – namely, that there is an ever-increasing subjection to the influence of this narcotic, whose soothing powers are requisitioned to counteract the evil effects of the worry, overpressure and exhaustion which characterize the age in which we live.
>
> (Tennant 1950: 141)

The third reason why the medical profession remained relatively silent on tobacco is that before the 1930s and 1940s there was no place for tobacco in the disease paradigm. Physicians who did speak out against tobacco did not adopt a scientific position and eventually resorted to rather shaky moral arguments. Ironically, in their attempt to distance themselves from the moralizers, many doctors found themselves tacitly arguing that tobacco was not harmful (Burnham 1989: 12–13).

Moreover, as the consumption of the cigarette increased in the United States and Europe, both relatively and absolutely, the cigarette became increasingly a cultural artefact that was resistant to carrying health associations. The parallel with the car is instructive. The rising mortality from traffic accidents was never attributed to the car itself, to its design, either internally or externally; this was above criticism. Traffic deaths were caused by reckless individuals not by cars. The car as a symbol of freedom from the tyranny of distance could therefore continue to develop without being confronted by traffic deaths (Flink 1988). Tobacco discourse had a similar structure. Victor Heiser, an influential American surgeon writing at the end of the 1930s, stated what many others thought: that 'tobacco has different effects on different people' (Burnham 1989: 12). The speed with which the cigarette was adopted as a means of personal expression (an expression of choice, loyalty and control) no doubt helped to prevent it being recognized as an agent of disease (Brandt 1990).

There is also little doubt that the powerful industrial and political forces that were building up around tobacco companies, the state and farmers were already having an impact on the medical discourse and deflecting interest. This is, of course, a very important factor and one which is dealt with in far greater detail in the next two parts of this book.

Finally, one should not overlook the fact that though we now associate smoking with cancer this link has been recognized only because of changes in the nature of the perception of cancer, and the understanding by medical researchers of what might possibly be dangerous in tobacco. Let us deal with the latter first: as we have already seen, the first real breakthrough in providing for a pharmacological definition of tobacco lay in the isolation and synthesis of nicotine by the end of the nineteenth century. The extreme toxicity of nicotine was also confirmed at the time. This led to two developments, one being the preparation of an effective insecticide – in general use until the production of DDT in the early 1940s – and the other a spate of work investigating the physiological action of nicotine in humans (Busbey 1936). While this research increased the understanding of nicotine, there was no evidence that in the doses taken by consuming tobacco nicotine caused any problem.

During the first two decades of the twentieth century medical researchers produced an enormous literature on the toxicology of tobacco. It is difficult to summarize this work because of its diverse character but, in examining the titles of these studies, certain patterns of enquiry emerge. First of all, there was a great deal of research on the effect of tobacco on intellect, efficiency and growth. This had more to do with the moral arguments about tobacco that were then current than any new departure in the scientific study of tobacco use. Second, there were many studies purporting to deal with tobacco consumption as a form of addiction, called variously tobaccomania and tobaccoism. Neither the physiological and psychological studies nor those on addiction called into question tobacco as a cause of disease. In fact the only area in which there was concern about tobacco in this respect was in relation to tobacco-specific disorders, in particular tobacco amblyopia and tobacco heart (Burnham 1989: 15–16, 21–3; Larson *et al.* 1961: 653–86; Dunphy 1969).

The rising consumption of tobacco, specifically cigarettes, and the public nature of the moral debate about tobacco, particularly in the United States, undoubtedly stimulated the increasing flow of research on tobacco until the 1920s. But in disease aetiology tobacco was generally not incriminated, until, that is, the western world began to understand that a new disease was in its midst – cancer.

Not just any cancer, but lung cancer in particular. While it is commonly thought that this is a twentieth-century disease, in fact not only was it a recognized disease of the nineteenth century but there was a considerable amount of research into the nature of the malignancy (Rosenblatt 1964). Already by the latter part of the century lung cancer was recognized as a significant type of respiratory disease. While studies describing the cancer abounded, there was little in the way of understanding what caused it. This was for several reasons: first, cancer was typically recognized only at the post-mortem, and so a search for causes was severely hampered; the diagnostic tools, especially the X-ray and the bronchoscope, did not exist until the twentieth century. Second, cancers, and particularly pulmonary malignancies, were understood to be occupational diseases (in those instances, at any rate, where there was an interest in such conditions). Actually, during the second half of the nineteenth century, interest in industrial or occupational diseases was increasing in Europe, especially in France, Germany and Britain. However, one of the effects of this interest was that pulmonary diseases became associated with specific occupations, rather than specific causes (Lecuyer 1983). In a study of mineworkers in Saxony published in 1879, the researchers, both of whom had considerable understanding of lung cancer, concluded that the reason why these workers suffered from fatal pulmonary malignancies was because of their long exposure to arsenic and other metals: ironically, in their search for the causative agent, both Hesse and Härting dismissed food and tobacco as possibilities (Rosenblatt 1964: 405). Finally, there was a significant body of opinion that held that the supposed rise in the incidence of lung cancer during the nineteenth century was apparent rather than real (Rosenblatt 1964: 413).

In spite of the diagnostic difficulties and sceptical voices, there was a growing body of evidence that the incidence of lung cancer, at least in relation to other cancers, was on the increase, especially during the second half of the nineteenth century; and, though aetiological factors were still kept in the background, references to possible respiratory irritants and previous infections such as influenza and tuberculosis were appearing in the literature (Rosenblatt 1964: 412). Yet, for all the medical work, the absolute level of deaths attributed to lung cancer was not only very small in absolute terms but paled into insignificance when compared to other known diseases. In the United States in 1900, for example, 48,000 people were reported to have died of cancer of which fewer than 400 cases were of lung cancer – while tuberculosis, the country's main killer, was responsible for 266,000 deaths (Patterson 1987: 32–3; Brandt 1990: 161).

While lung cancer contributed only marginally to overall mortality, cancer in general was rising in the United States during the second half of the nineteenth century from a mortality rate of less than 20 per 100,000 before 1850 to 64 per 100,000 in 1900 (Patterson 1987: 32). Some three-fifths of cancer deaths were of women (tumours of the breast, uterus and oral tissues were the most common form of malignancy) (Patterson 1987: 26–7). The increase in cancer deaths was attributed to many causes – stress and urban civilization, and greater longevity (since cancer was normally found in old, rather than young people) were the most popular explanations – but one proximate cause which found wide acceptance was that cancer could be caused by an irritant.

The idea of an irritant causing cancer logically followed from the observation that tumours were normal body cells that behaved in an anarchic and vividly fatal manner (Patterson 1987: 14–15). What made previously normal cells carcinogenic was unknown but the possibility that irritating a particular part of the body could cause tumours in that area was taken seriously. The definition of an irritant was, however, very broad and included, for example, bruises and cuts. In the case of tumours of oral tissue, mouth, lip and throat cancer, for example, doctors had little difficulty in ascribing blame to tobacco smoke because of its irritant qualities, normally perceived of as heat: at the same time the pipe was suspected of causing irritation by rubbing against the sides of the mouth and tongue, as were jagged teeth (Patterson 1987: 26–7). Snuff was also implicated in tumour development from at least the end of the eighteenth century when Antoine Fourcroy, the French chemist, linked nasal polyps to tobacco in powdered form (Körbler 1968: 1181).

Until the twentieth century lung cancer was not considered a serious problem, and the carcinogenic effects of tobacco were neither widely acknowledged nor understood. Over the next few decades several changes occurred which gradually led to the recognition that lung cancer was a twentieth-century disease of alarming proportions, and that tobacco smoke, delivered to the lungs directly by the cigarette, was increasingly suspected of contributing to the phenomenon. Yet in spite of all the growing evidence it was not until well after the Second World War, and in particular during the last two decades, that the connection between cigarettes and lung cancer, as well as other diseases, entered public discourse.

Of particular importance was the stark fact that deaths from cancer were rising while those from other diseases were falling. The figures speak for themselves. In the United States, for example, cancer deaths between 1900 and 1940 increased in the manner shown in Table 5.5. Within this pattern lung cancer began its inexorable rise: between 1930 and 1940, for example, the number of deaths from lung cancer increased much faster than deaths from other forms of cancer, rising from 2.3 per cent to 4.5 per cent of deaths from cancer (Patterson 1987: 88, 203). But of even greater importance was the growing suspicion that this phenomenon was being caused by tobacco smoke. In the United States and Britain a series of studies appeared in the 1920s, and 1930s, in which tobacco became implicated in the aetiology of lung cancer (Brandt 1990: 158; Patterson 1987: 205; Burnham 1989: 17; Cuthbertson 1968). Some of these studies, such as the ones conducted separately by Frederick Hoffman in 1931 and Dr Raymond Pearl in 1938 in the United States, were primarily statistical: that is, they demonstrated a link between smoking and mortality on the basis of a statistical analysis of the appropriate data. Pearl, for example, compared

Table 5.5 Cancer deaths, United States 1900–40

Year	Number (000)
1900	48
1915	85
1920	88
1926	95
1940	158
1950	170
1970	330
1985	462

Source: Patterson 1987: 32, 80, 88, 95, 159, 235, 301

mortality characteristics of smokers and non-smokers and came to the conclusion that those who smoked shortened their lives without, however, explaining the reason (Brandt 1990: 159). Hoffman was more precise in drawing a link between cancer and smoking, but the strongest advice he could give – as much a reflection of the problems of analysis as of the climate of the time – was to moderate the consumption of cigarettes (Brandt 1990: 159). Other studies reported observations by highly respected surgeons such as Alton Ochsner and Michael DeBakey which linked smoking and lung cancer even more closely (Burnham 1989: 18–19; Patterson 1987: 205). Then there were the experimental studies, conducted on laboratory animals, that attempted to induce cancers by painting likely carcinogens on skin – interestingly, but not surprisingly, nicotine, rather than other tobacco ingredients was suspected as being carcinogenic (Burnham 1989: 17).

Many, if not most, of the conclusions of these various studies were challenged on a variety of different grounds and, as far as it is possible to say, made little impact on public discourse and certainly none at all on consumption habits: in the United States annual consumption of cigarettes per adult rose from 1,485 in 1930 to 1,976 in 1940 and 3,552 in 1950 (Patterson 1987: 201, 207); in Britain over the same period consumption rose from 1,380 to 2,180 cigarettes per adult (Beese 1968: 60–1). Clarence Little, president of the American Society for the Control of Cancer, the forerunner of the American Cancer Society, reflected the lack of a clear commitment to the conclusions of the many studies on smoking and cancer when he stated that 'the more common use of tobacco is blamed by some for the frequency of lung cancer . . . It is impossible to say how accurate these opinions are' (Patterson 1987: 206). One of the first and most comprehensive studies of tobacco consumption in the United States, published in 1940, did not mention the word cancer even though the author covered the physiological effects of tobacco: again it is notable that the centrepiece of the discussion was nicotine (Gottsegen 1940: 81–105). While the work on cancer and smoking before the 1940s made little impact, it did, nevertheless, bring new researchers into the field using different techniques to prove or disprove a connection.

By 1950 lung cancer was accounting for 15 per cent of cancer deaths in the United States (Patterson 1987: 207). That year was the turning point for implicating tobacco smoke as a cause of cancer. The new studies, epidemiological in character, used new approaches: the one, termed retrospective, proceeded by interviewing patients positively diagnosed as having lung cancer about their lifestyle leading up to their diagnosis; the other, termed prospective, interviewed people about their lifestyles and then correlated this information later with the cause of their deaths. The major studies of this kind were those by Ernst Wynder and Evarts Graham, and E. Cuyler Hammond and Daniel Horn in the United States, and Richard

Doll and Austen Bradford Hill in Britain (Wynder and Graham 1950; Hammond and Horn 1954; 1958; Doll and Hill 1952; 1954; 1956).

Their conclusions were direct and clear. Prospective and retrospective studies were consistent with each other in the identification of tobacco smoke and the cigarette as a significant contributor to lung cancer (Steinfeld 1985). In the wake of these profound conclusions, medical researchers preoccupied with other big killers reviewed their understanding of the aetiology of these diseases. In particular, attention became focused on the role of tobacco in heart disease and emphysema. By 1960, in the United States, cigarette smoking had become officially implicated in coronary disease, in emphysema, in certain cardiovascular diseases and a host of other ailments (Burnham 1989: 20–3; Davis 1987: 19–22). But in both the United States and Britain there was a considerable lag between the publication of the major research findings and their general acceptance by the major institutions of the medical profession, who had considerable public power. It was not until 1962, in Britain, and 1964, in the United States, that the Royal College of Physicians and the Surgeon General, respectively, weighed up the evidence before them and declared that cigarette smokers exposed themselves to a very high risk of serious disease, and that this risk could be substantially reduced if they gave up smoking (RCP 1962; USDHEW 1964). But, as a later discussion will show, it took even longer for the public to react and to curtail consumption.

Seen over the long history of tobacco since the beginning of the sixteenth century, what happened in the 1960s was momentous. For the first time there was solid evidence that tobacco was a dangerous substance and that cigarette smoke caused fatal diseases. This had considerable impact in many different areas. First, tobacco was put on the political agenda as it became increasingly clear that there were powerful vested interests involved. Almost immediately an intense war broke out between the pro-tobacco forces, including tobacco companies, some government agencies, tobacco producers and some consumers, and the anti-tobacco forces, including consumer pressure groups and some other government agencies. The lines dividing the forces have never been entirely clear and shifted over time (Brandt 1990: 165–7; Patterson 1987: 211–30; Taylor 1984). The role of government came under close scrutiny especially since, on the one hand, it had a duty to protect consumers from potentially dangerous substances, while, on the other hand, it acted to protect its own interests, financial and electoral. In the United States, where tobacco growing is big business, the scale of the problem is enormous but even a few statistics from Britain, where there is no tobacco farming lobby, make the point clear enough. In 1980 the UK Treasury received around £3,000 million in taxes and duty from tobacco; tobacco sales from 350,000 retail outlets amounted to £4,300 million; cigarettes valued at £464 million were exported to 150 countries; overseas aid to Zambia, Malawi and Belize to develop and support tobacco growing amounted to £3,500,000 since 1974; and 36,000 people were employed in tobacco manufacture (Calnan 1984: 288).

Second, and as part of the tobacco war of the period, manufacturers responded to the medical threat by denying the conclusions and by launching a particularly aggressive counter-attack through advertising. To understand this phase of its history it is important to analyse tobacco in cultural terms. The medical history is only one aspect of the cultural significance of tobacco but one to which manufacturers did respond. When at the end of the nineteenth and early twentieth centuries the objection to tobacco was based on moral arguments, manufacturers and advertisers tried to project an image of respectability: when after the Second World War medical evidence condemned tobacco as dangerous and its consumption as risky, manufacturers and advertisers switched to projecting an image of individual choice and independence of authority. The attacks and counter-attacks amounted, in cultural terms, to a struggle over the image of tobacco's most refined form, the cigarette.

Part III

It is a culture of infinite wretchedness. Those employed in it are in a continual state of exertion beyond the power of nature to support. Little food of any kind is raised by them: so men and animals on these farms are ill fed, and the earth is readily impoverished.

Thomas Jefferson (1743–1826)

Marse ain' raise nothin but terbaccy, ceptin' a little wheat an' corn for eatin', an' us black people had to look after dat 'baccy lak it was gold. Us women had to pin our dresses up roundst our neck fo' we stepped in dat ole 'baccy field, else we'd git a lashin'. Git a lashin' effen you cut a leaf fo' its ripe. Marse ain' cared what we do in de wheat an' corn field cause dat warn't nothin' but food for us niggers, but you better not do nothin' to 'baccy leaves.

Quoted in Siegel (1987): 98

6

'WHOLLY BUILT UPON SMOKE'
The impact of colonialism before 1800

Despite the attention it received from European physicians, botanists and herbalists, tobacco retained the character of an expensive herbal medicine until the end of the sixteenth century. Exactly where it fitted into European perceptions of the New World economy is unclear. Despite the scattered references to tobacco cultivation in Brazil and in the Spanish-American possessions during the sixteenth century, there is very little hard evidence on quantities produced and amounts consumed. Historical sources pertaining to the imports of colonial commodities into Seville, the principal port for Spanish-American trade, are silent on tobacco before the first decade of the seventeenth century, despite a thriving commerce in other New World medicines and spices, such as sarsaparilla, canafistula and ginger (Lorenzo Sanz 1979: 605–13). It could of course be that tobacco entered Seville illicitly, that is without paying customs, but in the absence of any direct evidence its quantity remains wholly unknown. No other European country, with the possible exception of Portugal, had direct access to the New World's tobacco supply before the end of the century. Unless or until other evidence emerges we can conclude only that no regular commerce existed in tobacco before the last decade of the sixteenth century at the earliest. Tobacco was being cultivated in Europe before then but, as in the case of colonial production, nothing is known of its quantity (von Gernet 1988: 32–3, 61).

Searching through the documentary evidence one is struck above all by a general lack of interest in tobacco as a commercial commodity as opposed to its value as a miracle cure. Why this should be is not entirely clear. It may have to do with the fact that the sixteenth century resonated with the lure of gold and silver. Spanish conquistadors and colonists accumulated a vast treasure store of gold in the wake of their military conquests between 1520 and 1540 (Bakewell 1987: 203). The belief that more gold could be discovered, despite the fact that few considerable deposits were uncovered, continued to draw Europeans across the Atlantic. Following on from the discovery of silver deposits near Mexico City around 1530, and culminating in the most important strike of them all in Potosí in 1545, Spanish attention shifted to the problems of extracting, processing and transporting an increasing amount of silver across the Atlantic (Bakewell 1987: 206). Once other European powers, especially those hostile to Spain and Spanish influence, recognized that Spanish power rested on New World bullion, their interest became either to find their own deposits or to cut the Spanish supply lines (Elliott 1987: 98–9). During most of the second half of the sixteenth century European commercial interests in the New World revolved around silver and gold. Another reason for the lack of commercial interest in tobacco may have been that as long as the Amerindians controlled the supply there was little scope for its incorporation within European commercial capitalism. The problem here was not so much that Amerindians would not, and could not, respond to market forces – they managed it for other medicines such as sarsaparilla and guaiacum, and, of course, in North America, for beaver pelts – but that tobacco was sacred in the sense described in Chapter 2 (Amerindians, it turned out, were very happy to receive tobacco from Europeans, as they did increasingly in the seventeenth and eighteenth centuries; but giving it away on European terms was quite another matter) (Trigger 1991b). To transform tobacco from an Amerindian into a European commodity required Europeans to appropriate the means of production. This, in turn, required settlers who would be willing to migrate for a cash crop, a magnet with little appeal in an age of glitter (Boyd-Bowman 1976; Elliott 1985; MacLeod 1973).

Before the end of the sixteenth century Spanish colonists had managed to launch fledgling sugar plantations on Hispaniola, but these were in steep decline by the mid-1570s (Ratekin 1954: 12). By contrast the Portuguese were successful in their exploitation of the sugar cane on their Brazilian settlements; by the end of the sixteenth century Brazil was the world's largest producer of cane sugar (Schwartz 1987: 67–98). By 1600 at least 200,000 emigrants had left Spain for the New World (Slicher van Bath 1986: 25); in 1585 the European population in Brazil stood at around 30,000 (Johnson 1987: 31). The French, English and Dutch were hardly yet in evidence. The French had ventured into New World enterprises in several different ways, but before the end of the sixteenth century none was successful. Cartier's voyages in the St Lawrence River early in the 1530s did not lead to colonization partly because of the disappointment at not finding gold and partly because of the difficult climate and health hazards faced by the few colonists who attempted settlements in the St Lawrence Valley (Meinig 1986: 25–6). Huguenot initiatives in Brazil and Florida failed (von Gernet 1988: 26–9). Jean Ribault's 1562 settlement on Port Royal Island, South Carolina, also foundered, partly because of internal conflict and partly because it was successfully attacked by the Spanish (Meinig 1986: 27). Ribault's purpose was similar to that of Cartier – the search for gold, particularly the legendary Cibola, the Seven Cities of Gold – but none was to be found (Buker 1970). He also considered the settlement as a base for attacking Spanish treasure fleets on their return to Seville (Shammas 1978: 154). The account of Ribault's voyage circulated in the English court, and this clearly influenced English enterprises in the New World which, for most of the second half of the sixteenth century, were concerned to find gold and silver, to attack the Spanish treasure fleets and even to conquer all of Spanish America (Shammas 1978). Even the first but disastrous English settlement in North America, on Roanoke Island in 1585, was established with the express purpose of providing a convenient base for piracy and privateering, targeting the Spanish treasure fleets (Kupperman 1984). As the French were drawn by the legend of Cibola, so too were the English inspired by the legend of El Dorado, which was believed to be sited at the southern end of the Orinoco, and first investigated by the Spanish from New Granada (Hemming 1978; Lorimer 1989). As for the Dutch, their primary interests in the New World were very similar to those of the English, though settlements were not, at the time, on the agenda (Goslinga 1971; Israel 1989).

Through the next century the French, now under the inspiration of Samuel de Champlain, succeeded in establishing permanent settlements on the banks of the St Lawrence and Acadia, though both areas struggled to increase their numbers. By the end of the century the French population of North America was probably 15,000, almost 95 per cent of whom lived in New France (Davies 1974: 77). The Dutch, too, failed to make a significant impact on the settlement of the New World, though they managed to establish colonies in the Hudson River Valley, the Caribbean and on the north-eastern coast of South America. The largest Dutch colony in the New World, that of New Amsterdam, was lost to the English in 1664 after which 10,000 Europeans became the first residents of New York (Meinig 1986: 119).

While the French struggled, and the Dutch experienced varying levels of success, the seventeenth-century settlement of North America belonged almost entirely to England. The change in English attitudes towards the New World is one of the most important changes to occur in the history of colonialism and one of the most difficult to explain completely. Certainly there is clear evidence of a highly significant shift from privateering to legitimate trade. During the latter part of the sixteenth century, as the Spanish bullion fleets became larger and better defended, privateering became less remunerative, and the 'gentlemen adventurers' (usually funded by courtiers) became increasingly interested in other commodities, of which tobacco was perhaps the most important (Lorimer 1978; Shammas 1978). By the time the Anglo-Spanish truce of 1604 outlawed the preying on Spanish bullion ships, even the otherwise conservative London merchants were aware of the profits to be made from relatively risky, long-distance trading in 'exotic' commodities, an alteration from their preoccupation with Europe and the steady cloth trade (Shammas 1978: 164–6; Brenner

1972; Fisher 1976; Quinn 1974). The year 1600 saw the incorporation of the English East India Company, an event that reflected the change in mercantile ideology and outlook.

It was in this transformed political and economic climate that the settlement of Englishmen became a strategy in the competition for the New World. Yet, as the history of the first permanent colonies makes amply clear, settlement did not necessarily mean the construction of a society abroad. On 10 April 1606 a charter was granted in London to the newly-founded Virginia Company. In May 1607, just over a year later, 105 colonists from an expedition led by Captain Christopher Newport chose to establish themselves on a site on the James River that they named Jamestown, and the settlement of Virginia had begun (Quinn 1974; Morgan 1975). The primary aim of settling these colonists was to turn a profit for the shareholders. Only men were sent on the first voyage and many, if not most, of them were company employees. Women and children did not arrive in the fledgling colony until 1609 (Meinig 1986: 38–9). It is clear from the occupations of the first arrivals, as well as from the fact that women were not included, that the Virginia Company intended the colony to process the riches of Virginia, through extractive industries such as mining and glass making (Morgan 1975: 87).

Life in the new colony was very difficult, to say the least. Though the population of Virginia had grown to around 500 at the beginning of 1610, most of the colony's population starved in the winter, and in spring of the same year there were only 60 Virginians left (Morgan 1975: 63). One year later, in May, the new governor of the colony, Sir Thomas Dale, upon his arrival in Jamestown was appalled to find that the colonists had abrogated their responsibilities to subsistence and, instead of working in the fields, they were at 'their daily and usuall workes, bowling in the streetes' (Morgan 1975: 73). The population hardly recovered. In 1616 the colony was reported to have 351 European inhabitants (McCusker and Menard 1985: 118). Between 1607 and 1624 (when the Virginia Company was wound up) at least 6,000 people emigrated to Virginia: by 1625 only 1,200 Europeans were left (Kupperman 1979: 24).

Tragic as this was, there was a slight glimmer of hope for English society abroad in 1612. In that year John Rolfe, who is perhaps better remembered as the husband of Pocahontas, daughter of Powhatan, chief of the Pamankey Indians of Virginia, successfully grew the colony's first crop of tobacco. According to William Strachey in his account of Virginia in 1612, the type of tobacco grown there 'which the Saluages call Apooke . . . is not of the best kynd, yt is but poore and weake, and of a byting tast, yt growes not fully a yard aboue grownd . . . the leaves are short and thick . . .' (Strachey 1953: 122–3). Though he experimented with cultivating this variety of tobacco, known to have been *Nicotiana rustica*, Rolfe's success came with seed imported from Trinidad which, again according to

Strachey, was the best tobacco to have and was already growing to some extent in the colony (von Gernet 1988: 137; Strachey 1953: 38). Ralph Hamor, the secretary of the colony, was the first to credit Rolfe with establishing the cultivation of what we now know was the variety *Nicotiana tabacum* (Hamor 1957: 24). From whom Rolfe acquired his seeds is unknown though there was contact between Virginia and Trinidad through English merchants and seamen. One possible source was Sir Thomas Roe who was in Trinidad in February 1611 and had joined the Council of the Virginia Company in 1607, but there is no definite proof (Lorimer 1978: 141; Lorimer 1989: 37). Any of the considerable number of English and Dutch traders plying around Trinidad and the Orinoco delta and landing in Virginia could have conveyed the seeds to Rolfe, or someone else in the colony (Lorimer 1978; Kupp 1973). Rolfe's crop apparently arrived in England in July 1613 aboard the Elizabeth (Brown 1964: 639; Hillier 1971: 115, 410, 435). Rolfe was not alone in experimenting with tobacco. Others were clearly involved in the attempt to produce a marketable product by finding, in particular, new ways of curing and preparing it (Kingsbury 1933: 92–3).

Ralph Hamor saw the possibilities for the colony and in his account of Virginia in 1614 he emphasized to prospective emigrants the profitability of the tobacco plant. 'The valuable commoditie of Tobacco', he wrote, 'of such esteeme in England (if there were nothing else) which every man may plant, and with the least part of his labour, tend and care will returne him both cloathes and other necessaries. For the goodnesse whereof, answerable to west-Indie Trinidado . . . let no man doubt' (Hamor 1957: 24). In the same year the Elizabeth once again brought tobacco back from Virginia, perhaps as much as 170 pounds (Thornton 1921–2: 496). Two years later Virginia exported 1,250 pounds and in 1628, fifteen years after the Elizabeth's first deposit of Virginia tobacco in England, exports reached 370,000 pounds. As one historian has argued, Virginia's economy exploded into a boom, and wherever tobacco could be planted, it was (Morgan: 1971).

In more than one sense Rolfe's success with tobacco came at just the right time. Since 1592, in which year the island of Trinidad was settled by Spaniards, English ships had been calling either at the island or on the nearby mainland to trade for tobacco with the local Amerindians (Lorimer 1978: 125). Spanish settlers were cultivating tobacco on the Venezuela coast and New Andalucia. The high price of tobacco attracted the attention of many northern European ships, including English ones, who engaged in what must have been a lucrative smuggling trade. Even though tobacco cultivation on these Spanish settlements was increasing considerably, so much of the tobacco crop was traded illicitly that the King of Spain, in a royal *cédula* of 25 August 1606, forbade the planting of tobacco for a period of ten years (Arcila Farias 1946: 82). This effectively ended tobacco cultivation in Venezuela for more than a decade and it was not until 1620 that the trade in Venezuelan tobacco recovered to the level it had reached before 1606. Meanwhile, and in direct response to the destruction of the Venezuela plantations, the northern Europeans turned their attention to the small Spanish settlements in Trinidad and the Orinoco (Lorimer 1978: 129–31). The number of European ships calling at Port of Spain for contraband tobacco increased considerably over the following years: in the season 1608–9 there were twenty ships, thirty the following year, and in February 1611 alone fifteen ships were reported trading (Lorimer 1978: 132, 133–4). The trade, however, came to a climax in 1612. Spanish authorities sought to bring to an end the Trinidad operations, as they had previously ended those in Venezuela. In 1612 the resident investigator forbade cultivation and so there was no tobacco available for the next season (Lorimer 1978: 147). Though the illicit trade was not completely cut off, it appears that after 1612 only the Dutch continued to ply the waters for contraband tobacco (Lorimer 1978: 147).

There is no direct evidence connecting the events in Trinidad with John Rolfe. It may just be a coincidence that Rolfe harvested his first crop in the same year that the Spanish authorities brought tobacco cultivation in their possessions to a complete halt, but if so it is an amazing one. There was a fair degree of overlap between those merchants with an interest in the contraband tobacco trade and those with an interest in the Virginia Company (Lorimer 1978). The Virginia Company did not envisage tobacco as a possible crop for the colony; it was never mentioned in the list of potential commodities. The question as to why the Company did not seize upon tobacco as a crop in the early years, despite its proven profitability, may be answered partly by the existence of the contraband trade.

However, there is something even more significant for the history of tobacco in the experiences of Spanish-American tobacco cultivation. First, by the end of the sixteenth century, at the latest, Europeans had already learned from Amerindians how to cultivate tobacco. According to one estimate, the Trinidad and Orinoco plantations supplied as much as 200,000 pounds of tobacco annually in the early years of the seventeenth century (Lorimer 1978: 136). Bartering with Amerindians for the sacred herb was clearly a thing of the past. Second, Spanish cultivators were curing their tobacco in a more complex way than their Amerindian counterparts, a practice that resulted in a product with both a different appearance and a different taste. (Whether the Spanish developed this method on their own or, as is more likely, were imitating practices followed in Brazil, is unclear.) An English writer on tobacco, known to us only as C.T., contrasted the different curing practices of Spanish and Amerindian producers in a pamphlet published in 1615. From the description that follows it can be seen clearly that Spanish methods were not only more involved but were linked to other colonial enterprises in the New World:

The naturall colour of Tobacco is a deepe yellow or a light tawnie: and when the Indians themselves sold it us for Knives, Hatchets, Beads, Belles, and like merchandise, it had no other complexion, as all the Tobacco this day hath, which is brought from the coast of Guiana, from Saint Vincents, from Saint Lucia, from Dominica, and other places, where we buy it but from the naturall people; and all these sorts are cleane, and so is that of St. Domingo, where the Spaniards have not yet learned the Art of Sophistication . . . [their] Tobacco is noynted and slubbered over with a kind of iuyce, or syrope made of Saltwater, of the dregges or filth of Sugar, called Malasses, of blacke honey, Guiana pepper, and leeze of Wine . . . This they doe to giue it colour and glosse, to make it the more merchantable . . .

(C.T. 1615)

Both André Thevet's account of tobacco curing methods by Brazilian Indians and Girolamo Benzoni's account of the Taino of Hispaniola support what C.T. describes, namely that Amerindians simply dried their tobacco before consuming it (Dickson 1954: 119; Benzoni 1857: 80–1). Finally, the methods employed by the Spanish authorities to stamp out contraband and smuggling indicate that tobacco was a commodity to which governments had to give special attention.

The success of the Spanish plantations in Venezuela, together with the high prices that tobacco fetched in Europe, gave the plant commercial viability (Lorimer 1973: 270). The same factors were instrumental in linking tobacco with colonization in the minds of those Englishmen who established small settlements in South America in the years before and after the founding of Jamestown. Robert Harcourt, who maintained a small settlement on the Wiapoco River in Guiana from 1609, argued in his history of the colony that tobacco was a linchpin of successful colonization. The allusion to Spanish techniques of curing is important to note:

There is yet another profitable commoditie to bee reaped in Guiana, and that is by Tabacco, which albeit some dislike, yet the generalitie of men in this Kingdome doth with great affection entertaine it. It is not only in request in this our Countrey of England, but also in Ireland, the Neatherlands, in all the Easterly Countreyes and Germany; and most of all amongst the Turkes, and in Barbary. The price it holdeth is great, the benefit our Merchants gaine thereby is infinite, and the Kings rent for the Custome thereof is not a little. The Tabacco that was brought into this Kingdome in the yeare of our Lord 1610. was at least worth 60. thousand pounds: And since that time the store that has yeerly come in, was little lesse. It is planted, gathered, seasoned, and made up fit for the Merchant in short time, and with easie labour. But when we first arrived in those parts, wee altogether wanted the true skill and knowledge how to order it, which now of late we happily have learned of the Spaniards themselves, whereby I dare presume to say, and hope to prove, within few moneths, (as others also of sound judgement, and great experience doe hold opinion) that onely this commoditie Tabacco; (so much sought after, and desired) will bring as great a benefite and profit to the undertakers, as ever the Spaniards gained by the best and richest Silver Myne in all their Indies, considering the charge of both.

(Purchas 1906: 385–6)

In 1612 the settlers, together with others from Holland, were concentrating entirely on tobacco. Indeed tobacco was crucial to the survival of all the English, and Irish, settlements that appeared on the Wiapoco, as well as the Amazon. Philip Purcell's group of Irish settlers were sending tobacco to England and Holland by 1617 and, according to a Portuguese account, the English, Irish, and Dutch settlements in the Amazon were prospering with tobacco and, apparently, by 1623 shipping as much as 800,000 pounds to Europe (Lorimer 1989: 46, 57, 76). Thomas Roe inspired many of the settlements in the Amazon, together with merchants from Zeeland, in the Dutch Republic, as well as the Dutch West India Company after 1621 (Lorimer 1989: 75–6).

The speed with which European settlements were established and took to tobacco cultivation was remarkable. The choice of tobacco was deliberate, and the case of the colonization of Bermuda provides a clear case of this connection. Bermuda was discovered uninhabited in 1609 when a ship carrying Sir Thomas Gates, Virginia's new deputy governor, and Sir George Somers, together with a company of 150, ran aground just off the coast (Bernhard 1985: 57–8). Tobacco was listed as one of the possible crops to be grown in the islands, in accounts written at the time of the first venture. Experimental cultivation of tobacco was undertaken at about the same time as Rolfe was trying it in Virginia (Craven 1937: 353). It was already growing, to some extent, by 1613 (Ives 1984: 4). As the population of the islands increased, so too did the output of tobacco: exports to England totalled 30,000 pounds in 1617/18, 80,000 pounds in 1623 and 184,000 in 1628 (Craven 1937: 355–6; Williams 1957: 414). The case of Maryland, first settled in 1634, while different in important respects to Bermuda, nevertheless underlines the pivotal role of tobacco in settlement. George Calvert, the first Lord Baltimore, clearly perceived the importance of tobacco to a successful settlement in 1629, when he decided to abandon his settlement in Newfoundland and set sail for Virginia (Wroth 1954). For various reasons he was not allowed to settle his group of colonists in Virginia, and returned to England to press for the rights of settlement in the New World (Menard and Carr 1982: 175). He died before seeing his vision materialize, and his son, the second Lord Baltimore, inherited not only his father's estate but also his project. According to historians of early Maryland, the colonists were at first encouraged not to grow tobacco and, indeed, during the first year of settlement no crop was sown (Menard and Carr 1982: 187, 199). However, it was not too long before the tobacco infection hit the early Marylanders. By 1637 tobacco was already the colony's currency, and in 1639 the colony, with a population of less than 300, exported 100,000 pounds of tobacco to London (Menard and Carr 1982: 189, 198).

The English settlement of the Caribbean islands did not get under way until the 1620s, though one or two earlier but unsuccessful attempts had been made. Once again the crucial role of tobacco is clear, as is the connection with the earlier English settlements in the Amazon. The first of the Caribbean islands to be settled was St Kitts, founded by Sir Thomas Warner in 1624. In Warner's own words, the island offered great hope because it was 'a very convenient place for the planting of tobaccoes, which ever was a rich commoditie' (Harlow 1925: 18). Warner had first encountered tobacco cultivation on the English settlements in the Amazon, and was undoubtedly strongly influenced by this experience in assessing the potentialities of his newly founded island (Lorimer 1989: 70). Arriving back in London, he managed to persuade a group of London merchants to invest in the enterprise and in January 1624 about twenty men arrived in St Kitts and began cultivating tobacco (Andrews 1984: 301). The following year Warner returned to England with the colony's first crop of 9,500 pounds, the impressive sale of which confirmed that he had been right about tobacco. He returned to St Kitts in 1626 with a party of some 400 settlers (Batie 1976: 6–7). The island was divided between English and French settlers in 1627. Ten years later, in 1637, the English population of St Kitts stood at 12,000; in the following year nearly 500,000 pounds of St Kitts tobacco arrived in London (Gemery 1980: 223; Watts 1987: 158).

The settlement of Barbados was also planned with tobacco as the pivotal cash crop. The driving force behind this enterprise was Sir William Courteen, the leading figure of a wealthy Anglo-Dutch trading firm with interests in the New World. One of Courteen's associates, a Dutch Catholic named Aert Groenwegen, successfully founded a small Dutch settlement on the Essequibo River and began to grow tobacco (Goslinga 1971: 79, 81). It was from this Dutch settlement that the seeds and knowledge of tobacco culture were transplanted to Barbados (Innes 1970: 14). As in St Kitts, the population increased as tobacco production expanded. In 1638, with a population of some 10,000, Barbados exported 205,000 pounds of tobacco to London (Gemery 1980: 219; Watts 1987: 158).

In the founding and settlement of St Kitts and Barbados, as well as in the cultivation of tobacco in both Virginia and Bermuda, there is clear evidence of a link to Spanish experiences with tobacco in Venezuela. Once the English settlements were thriving on tobacco, further settlements appeared with tobacco as the cash crop. One such settlement was begun on Providence Island, off the coast of Nicaragua, first assessed by English vessels in 1628/9 (Appleby 1987: 233). The enterprise was headed by a consortium, formed in 1630 as the Providence Island Company, and included among its Puritan leaders those with a controlling interest in Bermuda (Kupperman 1988: 73). The governor of Bermuda, Philip Bell, was so convinced of tobacco's promise in a more tropical climate that, influenced by Sir Thomas Warner's success on St Kitts, he proposed to lead the first settlement on Providence Island and he became the colony's first governor (Batie 1976: 7). The settlements of Nevis in 1628, Antigua and Montserrat in 1632, all of which were initiated by Warner, were based initially on tobacco (Watts 1987: 169–70). Thus, as far as English colonialism was concerned, by 1640 the English population of the West Indies, Bermuda and Virginia, estimated at 40,000, was producing just over 1,250,000 pounds of tobacco (Gemery 1980: 212; Pagan 1979: 253).

English successes with colonization and tobacco cultivation did not go unnoticed in other parts of Europe, least of all by the French. Martinique was settled in 1635 by colonists from St Christophe, the French part of the island adjoining St Kitts, led by Pierre Belain d'Esnambuc, who had also founded the French settlement on St Christophe in 1627. Not surprisingly, the Martinique settlers, who were experienced tobacco growers, began tobacco cultivation on the island (May 1930: 87). On the other hand, Guadeloupe was settled directly from France in the same year, 1635, yet tobacco was again chosen as the cash crop. For both islands tobacco continued to underwrite their wealth until the 1660s (Schnakenbourg 1980: 54). Roughly one-third of the acreage of the two islands devoted to cash crops was accounted for by tobacco in 1671 (Schnakenbourg 1980: 49; Kimber 1988: 128). St Domingue, which became settled by the French during the second half of the seventeenth century, also established tobacco as the primary cash crop (Davies 1974: 147; Price 1973: 83).

The other major player in the tobacco enterprise in the seventeenth century was Portugal. Tobacco began to be grown commercially in the Bahian region in the north-east of Brazil by the end of the sixteenth century, though some historians give a later date (Nardi 1986: 15; Hanson 1982: 150–1; BM Add. Mss 20,846 fol. 167; von Gernet 1988: 149). Unlike the English and French for whom tobacco was the sine qua non of settlement, the Portuguese had already settled the region with sugar plantations (Schwartz 1985: 82–5). Even so, a considerable part of the increase in the population of the region can be attributed directly to tobacco cultivation (Flory 1978: 191).

Elsewhere in the Americas tobacco was also cultivated by the early colonists, though, compared with the Chesapeake colonies and Brazil, output was meagre. The Dutch were growing tobacco in New Netherland as early as 1629; Swedish colonists in New Sweden (in the present state of Delaware) were harvesting a crop early in the 1640s; and even the *habitants* in inhospitable New France were producing a small yield (von Gernet 1988: 149–86).

The seventeenth century was a formative but complicated period in the history of tobacco. Most historians have not fully appreciated the role of tobacco in the settlement of the New World, and the powerful attraction it held as a settler's crop. They have also not acknowledged the fact that the settlers were also heavy consumers of tobacco and treated it much in the same way as their counterparts in Europe. The addiction of the colonists to tobacco should not be dismissed as a possible factor in choosing tobacco as a staple crop (von Gernet 1988: 149–86, 188–9). Considering all the evidence, there is little doubt that initial settlement of the English, French, and Dutch in the Caribbean and Bermuda would not have been possible without tobacco, and without the knowledge that Spanish settlers had gained about harvesting and curing the product. There are several reasons why tobacco was the preferred crop on which to found colonies in the New World. The next chapter will explore the social and cultural factors but at this point the economic reasons for the choice of tobacco need some attention.

It should be stressed that both English and French seventeenth-century colonialism were primarily commercial in nature. The Chesapeake, Bermuda and Caribbean colonies were all funded by private investment. Tobacco had two major advantages over other crops: first, its growing cycle was short, on average nine months from planting to being ready for market; and second, it could grow in various soils and climates, with the result that no two consignments of tobacco were alike. Ships' captains, according to one account, on reaching the French Lesser Antilles were reported to have remained on the islands long enough to harvest a crop before returning to their home ports (Kimber 1988: 107). According to John Pory, writing from Jamestown in 1619, a man working by himself had made a clear profit of £200 while another, working with six servants, managed £1,000 (Kingsbury 1933: 221). Even by the 1640s, after a fall in prices, tobacco profits in Virginia could range from £225 to £300 per man (Morgan 1975: 110). In Barbados in 1628 even a planter producing a poor quality leaf could expect a profit on his annual output of between £35 and £56 (Batie 1976: 7–8).

While such windfall profits were sufficient inducement to plant tobacco in the New World, those who did not emigrate across the Atlantic were no less attracted by the economics of this cash crop. It is important to remember that Europeans first grew tobacco in Europe itself, though until the end of the sixteenth century it was, as previously argued, confined very much to physic gardens and not yet a commercial crop. The origins and early history of European tobacco cultivation are not very clear but some things are known. In France, for example, cultivation appears to have begun in the 1620s in the south-east of the country, in the Rhone and Garonne valleys, and in Alsace. By the 1640s the area of cultivation had extended throughout the Garonne valley and also westwards into the upper Dordogne (Price 1973: 4–5). Tobacco cultivation seems to have begun in Germany at about the same time as it did in France. The main area of tobacco culture was in Brandenburg and the Palatinate, but as the century advanced other areas were incorporated (Tiedemann 1854: 175). In Italy tobacco cultivation concentrated primarily in the north, in the Veneto and generally in the northern plain (Comes 1893: 93–105). During the eighteenth century cultivation spread to northern Europe: in 1724, for example, the Swedish government encouraged domestic cultivation (Roessingh 1978: 29).

In England tobacco cultivation began in 1619 when two London merchants entered into a partnership for growing tobacco around the town of Winchcombe in Gloucestershire (Thirsk 1974: 79). Precisely how one of the partners, Henry Somerscales, acquired the knowledge of how to cultivate tobacco is not known, but connections with both Virginia and Dutch merchants existed (Thirsk 1974: 78–80; Roessingh 1978: 26). Within ten years tobacco was growing more widely throughout Gloucestershire, Worcestershire and fitfully in Wiltshire. Until 1640 the Avon Valley and the Vale of Evesham were the principal centres of cultivation. Thereafter tobacco growing spread generally throughout England and Wales. In 1655 tobacco was growing in fourteen English and Welsh counties and in the next ten years a few more Welsh counties as well as Yorkshire and East Anglia were added to the list (Thirsk 1974: 94–5). Proof of tobacco's appeal to the small farmer is available in the shape of expected profits: in 1619 one acre of tobacco could have been expected to clear anywhere from £29 to £100, at a time when the average annual earnings of a farm labourer stood at £9 (Thirsk 1974: 86–7).

How much tobacco was produced in England is unknown, but whatever the level of output, it would have been no match for Dutch tobacco cultivation, probably the largest in early modern Europe. The beginnings of Dutch cultivation can be traced back to 1610 or 1615 in the provinces of Zeeland and Utrecht (Roessingh 1978: 23). Amersfoort, in Utrecht Province, became the principal centre of tobacco cultivation, retaining this position until the end of the eighteenth century. Even so, tobacco cultivation diffused to other parts of the Dutch Republic, into the provinces of Gelderland and Overijssel and the Duchy of Cleves (Roessingh 1978: 27–8). As cultivation spread and intensified, output rose substantially. Around 1675, the first year for which reliable figures exist, total Dutch tobacco output stood at between 5 and 6 million pounds. Peak production was reached in the first decade of the eighteenth century at a level varying from 17 million to 28 million pounds (Roessingh 1978: 42; Price 1961: 88).

Taking the New World settlements together with those parts of Europe where tobacco was cultivated, the seventeenth century was remarkable for its rapid incorporation of the plant into its agrarian structure. Yet as the century advanced the pattern of production changed considerably as many areas either stopped cultivation altogether or experienced a sharp decline in output. In the first category, the main casualties were the English Caribbean settlements, England, parts of France and Portugal. The timing and pattern of the abandonment of tobacco culture in the Caribbean colonies is unclear. In terms of tobacco shipments to London it is certainly the case that a maximum level was reached in 1638, when 675,000 pounds of Barbadian and St Kitts tobacco reached England; two years later the comparable figure was just over 200,000 pounds (Watts 1987: 158). By 1640 tobacco cultivation on Barbados was in severe depression, as economic resources were shifted into cotton and indigo production (Innes 1970: 16–19). But even as late as 1654, by which time we are told that the sugar revolution had swept all before it, there is some evidence that tobacco cultivation had not been abandoned, though compared to the acreage then under sugar the amount was small (Innes 1970: 16). According to an account of Barbadian exports between August 1664 and August 1665, 82 per cent of the islands' exports by value were accounted for by sugar and less than 1 per cent by tobacco (Puckrein 1984: 60). Whether and to what extent Barbadian tobacco entered intra-regional American, rather than transatlantic, trade is unknown, though for other commodities there was a brisk traffic (Morgan 1975: 139–40). On St Kitts tobacco cultivation certainly continued until the 1660s (Dunn 1973: 121–2). With the exception of Nevis, which had earlier converted to sugar, the other Leeward Islands continued to produce tobacco well into the 1660s, and probably later (Dunn 1973: 123). By the turn of the eighteenth century, England was still receiving almost 200,000 pounds of tobacco from the West Indies, but after 1706 the amounts were insignificant (Gray and Wyckoff 1940: 24–6). Why Barbados, in particular, shifted so wholeheartedly into sugar in such a short space of time is a question that is still strongly debated. It appears that there are several reasons, including the collapse of tobacco prices and the buoyancy of other, especially sugar, prices; the discouragement of tobacco growing on the island by order of the Privy Council to avoid over-production throughout British America, and the interest shown by the ever-present Dutch merchants in sugar, as well as in slaves (Batie 1976; Beckles 1985; Green 1988). Whatever the reasons, it is clear that the tobacco era in Barbados was crucial to the sugar revolution that so changed the island in the 1640s. Of the 175 largest sugar planters in Barbados in 1680 no fewer than 40 per cent had become established during the tobacco era (Puckrein 1984: 65). Meanwhile in England domestic tobacco cultivation became a problem for the Virginia Company who, in 1619, successfully persuaded the Privy Council to ban domestic cultivation (Thirsk 1974: 94). Enforcement over the next two decades was ineffective, and during the political turmoil of the 1640s tobacco growers were left alone. In 1652 the Council of State resumed the government's intent to ban domestic cultivation, but enforcement once again proved difficult. Finally, in 1688–9, the Privy Council succeeded where it had previously failed. With the help of the Royal Army, who had harassed tobacco growers over the previous two decades but now attacked them and burnt their fields, cultivation finally came to an end (Thirsk 1974: 95). The French government also sought to ban domestic cultivation, but this was not extended to the politically sensitive areas in the southwest, or to Alsace and Artois. The *arrêt* of 1676 and another in 1719 were successful in containing domestic cultivation (Price 1973: 143, 294). It was left to the Revolution to restore the rights of Frenchmen to cultivate tobacco wherever they wished. As for Portugal, it is clear that domestic cultivation was suppressed as early as 1647, though how successfully is not known (Lugar 1977: 35). There is little doubt, however, that once the state monopoly was reorganized in 1674 domestic production would have been extinguished (Lucio d'Azevedo 1947: 287; Nardi 1986).

In the French Caribbean recent research maintains that the growth of tobacco cultivation paralleled that of sugar until the 1670s (Schnakenbourg 1980: Petitjean Roget 1980). The pattern of output on the islands of Martinique and Guadeloupe is not known but evidence from Martinique certainly suggests that production grew until at least 1671 (Petitjean Roget 1980: 1144, 1396). On Guadeloupe the incursion by sugar seems to have begun earlier and to have been more revolutionary than in Martinique – that is, not very different from what happened on Barbados (Schnakenbourg 1968). In 1671, probably at the peak level of production, output on the two islands surpassed 1 million pounds, but as major producers of colonial tobacco neither island was of much significance after the 1680s. On St Domingue output probably peaked at about the same time as on Martinique and Guadeloupe but the decline was more protracted (Price 1973: 90–115). Still, by 1700, there were a few planters growing tobacco on St Domingue, as the island's economy was much more diversified than the other French possessions (Trouillot 1981; 1982).

Between 1600 and 1700 the tobacco market was severely shaken. Small, especially island, producers had abandoned the dream crop and almost all of them, with the notable exception of Bermuda, turned effectively to sugar. The eighteenth century was dominated mostly by large mainland producers: the Chesapeake colonies, and Brazil in the New World; Holland and Germany in Europe.

The graph below (see Figure 6.1) shows the eighteenth-century history of tobacco production according to the best figures available. The principal

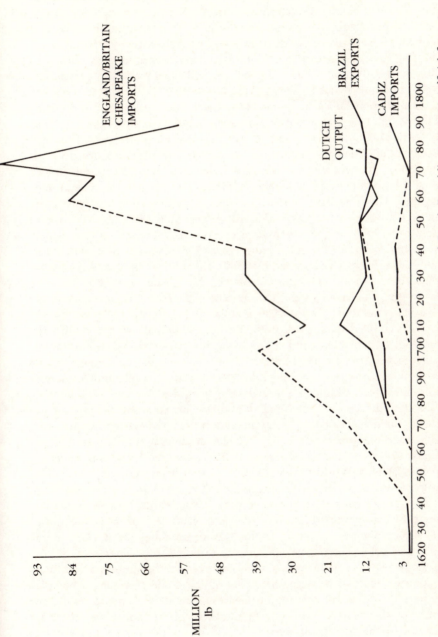

Figure 6.1 Tobacco quantities: Chesapeake, Brazil, Spanish America and Dutch Republic 1620–1800 (official figures). (Broken lines: incomplete data.)

Source: as in text

features of the period are clear enough. Dutch and Brazilian output remained fairly stable throughout the century, averaging around 8 million pounds of tobacco in each case. Spanish-American production, using imports into Cadiz as a proxy, was distinctly smaller, averaging around 2 million pounds annually. All of this was overwhelmed by production in the English colonies on Chesapeake Bay, from where tobacco exports to Britain soared from 37,166,000 pounds in 1700 to a maximum of over 100 million pounds in 1771 (USBC 1975: 1190).

At this point no further comment is needed on the Dutch and Brazilian figures, but the Spanish-American data require some discussion. It is important to note that by the end of the seventeenth century Cuba had become the principal source of tobacco imports into Spain. During the last decade of the century, for example, Cuba accounted for 83 per cent of total tobacco imports into Cadiz, in contrast to the situation around mid-century when the island's share of tobacco imports stood at only 2 per cent (Garcia Fuentes 1980: 369; Rivero Muñiz: 1964). Cuba's rise to dominance was at the expense, primarily, of Venezuela, whose export economy in the eighteenth century turned increasingly to cacao, indigo and hides (Ferry 1981; Fisher 1985: 60). In addition to Cuba's expansion, tobacco cultivation also grew in Mexico and Costa Rica, though in the latter case almost all of the tobacco entered into intra-American, rather than Atlantic, trade (Deans 1984; Acuña Ortega 1978).

Tobacco played a key role in the European colonial enterprise. Once the methods of cultivation, and curing, were both appropriated by Europeans, tobacco became rapidly transformed into an essential commodity of the transatlantic economy, and provided the economic foundations of successful settlement. Although tobacco was being widely cultivated in New World settlements, by the end of the seventeenth century most of the island planters ceased production, and turned their attention to other commodities, notably sugar. By 1700, and continuing for the next hundred years, Brazil and the Chesapeake colonies accounted for almost the entire output of New World tobacco, Cuba's output being relatively small. This pattern was, however, challenged in the following two centuries as tobacco cultivation spread rapidly throughout the world, partly because of its value for the new colonialism of the nineteenth century and partly because of its value to many countries in the world searching for a place in the international economic system, a theme explored in Chapter 8.

Settlements based on tobacco culture were the beginnings of an international circuit of tobacco that spanned the globe satisfying a complex pattern of demand. Europe controlled this circuit as it also accounted for most of the demand. The role of the merchant or trader was therefore critical to the success of tobacco in Europe at the same time as it was fundamental to the success of settlement and colonialism in the New World. The merchant and grower were natural partners in the exchange, but tobacco attracted other vested interests who at times were in conflict with one another. For reasons that will be explained later in this chapter the state quickly became involved in the exchange of tobacco. It also opened up a significant opportunity for smuggling, a trade that caused perennial concerns throughout the colonial period. Even without the interference of the state, the peculiarities of the colonial system as it operated in the seventeenth and eighteenth centuries brought merchants into conflict with growers. An uneasy, rather than a mutually beneficial, relationship characterized tobacco commerce.

Before the regular imports into England of Chesapeake tobacco in the latter years of the second decade of the seventeenth century, England received its consignments of tobacco from Spanish America either indirectly and officially, from Seville and Cadiz, or directly and unofficially, from the Venezuelan and Trinidadian plantations. Precisely how much tobacco reached the English consumer before the second decade of the seventeenth century is not known. Official figures refer only to the value, not to the volume, of imports, and those records that survive show that the imports of Spanish tobacco rose from around £8,000 in 1603 to over £44,000 in 1616 (Pagan 1979: 248). Other sources, however, give an indication of volume, showing a rise from an annual import of 25,000 pounds at the beginning of the century to between 50,000 and 75,000 at the start of the second decade (Gray and Wyckoff 1940: 18). Even though these figures cannot themselves be taken as indicative of the true level of tobacco imports, the trends they suggest are supported by the available figures on the export of tobacco from Spanish America. These show a considerable increase, from a level of 25,000 pounds at the end of the sixteenth century to a level approaching 60,000 pounds in 1610 (Arcila Farias 1946: 81; Chaunu 1956: 1032–3).

These are only the official figures, however. How much came through contraband is not known with any certainty, but there is little doubt that it was considerable, perhaps as much as 60 per cent of the English consumption in 1610 (Lorimer 1978: 136–7; Gray and Wyckoff 1940: 18). The temptation to contraband trade must have been considerable at the time. It has been estimated that English traders would have expected to pay about 3s. per pound at source directly from planters and, of course, clandestinely in Trinidad and the Orinoco plantations (Lorimer 1973: 271). This figure was not very different from the duty payable in England, which, itself, was not very different from the freight rate across the Atlantic (Lorimer 1973: 271–2). Once it hit the market, Spanish tobacco could fetch as much as 40s. per pound (Lorimer 1978: 137).

Those who chose the risky, but highly lucrative, path of contraband trade in the New World were not necessarily, as one might suspect, traders of dubious reputation. Though the background of most of them remains obscure, at least 10 per cent of them were merchants of considerable means, with commercial and financial interests, and a further 10 per cent were London grocers (Lorimer 1978: 137–8).

English contraband trade around the Orinoco virtually collapsed in 1612, as Spanish authorities clamped down on the illicit trade through various means, including the rather drastic action of forbidding the planting of tobacco, as happened in that same year (Lorimer 1978: 147). Once tobacco was cultivated for export in Virginia and Bermuda, however, important changes began to take place in tobacco's commercial system which, in turn, had important implications for all vested interests.

The commercial history of tobacco in the period until the 1630s is a complicated affair because it touches upon so many aspects of the English fiscal and colonial system that was in the process of being created (Beer 1959: 101–75). At the same time tobacco was also a focus for much of the political infighting that characterized this period of English history. To enter into this viper's nest would detract from the main point of this chapter and its details have been well documented elsewhere (Beer 1959; Craven 1932). For our purposes, however, some of the rough outlines of the controversies surrounding tobacco in these early years need to be covered, if only to provide a background to later developments. Possibly the best way to do this is to bring on the two principal actors. First of all there was the Crown, whose interests were fiscal and political. James I was granted the right to levy customs duties on both imports and exports by Parliament in 1604, and, as English trade grew in the early decades of the seventeenth century, the value of this revenue grew; between 1604 and 1625, for example, customs revenue increased by 50 per cent (Beer 1959: 103). At the same time, however, the prevailing ideology of trade considered imports as a drain on the country's wealth, in the sense of reducing its stock of precious metals. Tobacco was one of these imports since, before 1612, the bulk of the tobacco consumed in England was Spanish-American. Judging from his attack on tobacco in the famous *Counterblaste*, James I's solution to the problem was to prohibit the importation of Spanish tobacco by economic means through an excessively high duty – to have done it otherwise would have brought him into conflict with merchants, consumers and the Spanish Crown, with whom a peace treaty had been concluded in 1604. The total duty payable on tobacco shot up from 2*d*. to 82*d*. per pound though it fell back sharply in 1608 to steady at 24*d*. per pound by 1615 (Beer 1959: 108–9). Imports did not shrink: on the contrary, the total import of tobacco, primarily Spanish-American in origin, more than tripled. This was, of course, entirely unexpected, but so was the inflated size of the Exchequer's purse. Fortunately for the Crown, tobacco imports were relatively small when compared to the imports of manufactured goods, especially textiles, food and drink, particularly wine and brandy, and industrial raw materials, such as raw silk (Davis 1973: 55). These were the problem goods in the mercantilist's conception

of the commercial system. The economic potential of colonies lay principally in reducing the size of the import bill by producing the same commodities or substitutes for them (Shammas 1978). Both James I and
Charles I pressed the Virginia Company, and then the Governor and
Council of Virginia, to produce a commodity other than tobacco, and this
pressure continued until well into the century (Beer 1959: 90–1; Morgan
1975: 133–57, 180–95; Leonard 1967). Yet the Virginia (and Maryland)
colonists, as we have seen, could not be weaned from tobacco, and while
its production soared the dream of a diversified Chesapeake economy
faded, and finally disappeared. Precisely when the Crown began to accept
the fact that, as Charles I put it, the colony was 'wholly built upon smoke'
is unclear, but there was no avoiding the obvious: the revenue potential
of tobacco was overwhelming (Beer 1959: 149). One can easily recognize
the primacy of the Exchequer in the actions of the Crown, especially in
its relations with the Virginia interests in England whether in the first
instance in the shape of the Virginia Company or, after 1624, with individual merchants.

The second principal actor or actors were the Virginia interests. In the
guise of the Virginia Company, these interests were exempt from import
duties above the customary ad valorem tax that all goods had to bear (Beer
1959: 110). As Virginia and Bermuda tobacco imports soared after 1615,
the gap between the volume of imports and revenue began to widen and,
in response, the Crown entered into a long set of negotiations with the
Virginia Company, the results of which were extremely significant for the
subsequent history of tobacco. What emerged was an agreement that the
Virginia Company would pay twice as much in duties as they customarily
did: in return, the Crown would prohibit the cultivation of tobacco in
England (Beer 1959: 112). This happened in 1620, and signalled a victory
for the colonial interests in England over domestic interest groups. The
problem of what to do about Spanish-American tobacco was raised in the
following year when, once again, the Virginia interests put pressure on the
Crown to grant them a monopoly of the English market (Beer 1959: 114).
By 1623 the Virginia Company had convinced the Crown of its case, and
while the Crown did not wholly forbid the importation of Spanish tobacco
it did restrict it very severely (Beer 1959: 132). But the issue did not end
there, for the problem of Spanish tobacco continued for several more years
at the same time as the Crown accepted responsibility for governing Virginia in 1624 with the dissolution of the Virginia Company. In the end
the needs of the Exchequer for revenue clashed with the Virginia interests'
claim for a monopoly and the result was a compromise by which a certain,
but limited, amount of Spanish tobacco could be imported on which was
levied an import duty that was several times that on English colonial
tobacco. (It should be pointed out that Spanish-American tobacco enjoyed
a considerable following: to outlaw the importation altogether would have

antagonized these well-to-do consumers, at the same time as opening the trade to smugglers.) But the Virginia interests scored victories in other areas, particularly in getting the import duties on their tobacco reduced from a high level of 9*d*. in 1623 to 2*d*. for the period 1640 to 1660; in getting all tobacco shipped from the colonies to be landed first in England, regardless of its final destination; in establishing a drawback system whereby an importer would be reimbursed for a considerable proportion of the duties paid if the tobacco were re-exported; and finally in turning the attention of the Crown towards the retailers of tobacco as a source of revenue through a licensing system (Gray and Wyckoff 1940: 16; Beer 1959: 161–5; Menard 1980: 149).

The collusion between the Crown and the Virginia merchants resulted in a considerable flow of revenue into the Exchequer. In the 1660s, for example, tobacco duties from the Chesapeake colonies accounted for roughly one-quarter of total English customs revenues, and as much as 5 per cent of total government income (Morgan 1975: 193). In the last quarter of the seventeenth century, despite an enormous increase in re-exports, and a concomitant increase in drawbacks, tobacco income for the Crown is estimated to have quadrupled (Morgan 1975: 197). The Crown seems to have been content at the levy of around 2*d*. per pound of tobacco until 1685, when the duty was raised considerably to 5*d*. (Gray and Wyckoff 1940: 16). For the next seventy years the nominal duties on Chesapeake tobacco were raised at various times, reaching a level of just over 8*d*. per pound in 1758 (Menard 1980: 151). The legal imports of Chesapeake tobacco into England between 1685 and 1758 remained fairly stable, suggesting that Crown revenue grew in line with the rising level of duty. But the raising of duties also provided an incentive for smuggling, a problem of the tobacco trade that became of enormous concern to the state.

The first few decades of the commercial history of tobacco were dominated by London and by London merchants. There were many reasons why London should have taken this position but most important among them, as far as tobacco is concerned, is that London was the seat of the Virginia Company – as it was of all the chartered companies, such as the East India Company – and that the Crown, in 1624, had proclaimed that London should be the sole port for the importation of tobacco. This was done primarily in order to facilitate the agreements drawn up with the Virginia merchants over the import of Spanish tobacco and the collection of duties (Beer 1959: 197–9). The Crown's order was not entirely obeyed, however, and outports continued to trade in tobacco. Nevertheless, London did account for the greatest bulk of the import trade: available figures show that during the 1620s London handled, on average, 78 per cent of the tobacco import trade (Williams 1957: 418–20). Contributing to this concentration was the decline in the number of foreign merchants who imported Spanish tobacco and the relative fall in the volume of Spanish tobacco landed in England (Pagan 1979: 257–8).

Yet, while London monopolized the trade, the trade itself invited the participation of many merchants, far more than in any other branch of the colonial trade. In part this must have been because of the absence of a monopoly, following on from the dissolution of the Virginia Company and also because the trade itself did not require massive investment nor did it incur high transaction costs. The trade had a certain, predictable monotony about it in contrast to the other trades of the period to the Levant and the Orient where transaction costs were appreciably higher (Steensgaard 1974). A further incentive to wide participation came from the progressive decline in the costs of shipping; in the late 1620s and 1630s, transatlantic freight rates fell by more than half (Menard 1980: 147–8). This fall was primarily the result of improvements in packaging tobacco for export.

The evidence bears out the attraction of the tobacco trade to small operators, at least in the first few decades after the dissolution of the Virginia Company (Price and Clemens 1987: 4, 10). Soon enough, however, the trade became concentrated as a small group of large merchants increasingly dominated the commercial relations between the Chesapeake and the English and Scottish ports, notably London, Bristol, Liverpool and Glasgow (Pagan 1979: 259; Price and Clemens 1987). This 'revolution in scale', especially after 1680, was caused by a combination of a rise in duties, more burdensome administrative regulations and the easier access to credit and insurance by big firms (Price and Clemens 1987).

This increase in the scale of operations coincided with a profound shift in the spatial pattern of imports in the seventeenth and eighteenth centuries. There were two stages to this change. First, London began to lose its monopoly over importing tobacco especially after the 1660s (Gray and Wyckoff 1940: 18–20; Pagan 1979; Williams 1957: 419–20). Ports along the southern coast of England were the first to share in the tobacco import trade, but the most important newcomers to the trade were Bristol, Whitehaven, Liverpool, Hull and Newcastle (Gray and Wyckoff 1940: 21–2). By the mid-eighteenth century English ports other than London were handling half of the tobacco import trade (Nash 1982: 370; USBC 1975: 1190). The second change was even more striking. This was the rise of Glasgow as the second, and for the greater part of the second half of the eighteenth century the first, port of entry of Chesapeake tobacco into Britain (Price 1954a: 180–1). During the first half of the century Scotland imported no more than 13 million pounds of tobacco, but after mid-century the level reached as much as 47 million pounds in the few years before the outbreak of the American Revolution (USBC 1975: 1190). It was only after the American Revolution that London resumed its central role in tobacco imports, reflecting, in part, the decline of Glasgow (Price and Clemens 1987: 40–1).

Re-exports were crucial to the tobacco trade, and though reliable data do not exist before the mid-seventeenth century, figures for the second half of the century convey the scale of the operations: between 1677 and 1680 re-exports were, on average, 33 per cent of imports, and around 1695 53 per cent of legal imports were re-exported (Gray and Wyckoff 1940: 21–2; Nash 1982: 356). During the eighteenth century, however, the figure rose considerably, often exceeding 80 per cent of imports (USBC 1975: 1190). The main destination for re-exported Chesapeake tobacco was northern Europe, that is Holland, France and the Baltic countries; in the eighteenth century France, Flanders and Holland commonly consumed two-thirds of British tobacco exports (Gray and Wyckoff 1940; 21–2; Price 1973: 849). In terms of value the re-export trade in tobacco rose from £421,000 per annum at the beginning of the eighteenth century to £904,000 per annum at the outbreak of the American Revolution; at the same time tobacco re-exports accounted for as much as one-quarter of total re-export earnings (Davis 1969: 120).

The changes in the nature of the tobacco importing trade into Britain in the seventeenth and eighteenth centuries need to be understood in several important contexts. In the first place, there was a veritable explosion in the quantity of tobacco landed in Britain in this period. In 1620, for example, less than 60,000 pounds of tobacco was imported from the Chesapeake and, though the level of imports rose sharply over this decade and the following one, total imports around mid-century probably stood at around 1 million pounds (Menard 1980: 157–8). During the second half of the century the rate of expansion of tobacco imports accelerated very rapidly. By 1669 15 million pounds of Chesapeake tobacco were imported, reaching 30 million pounds by the turn of the eighteenth century (Menard 1980: 159–60). The trade continued to be characterized by enormous volumes after 1700 but the rate of growth in imports slowed considerably. During the first half of the eighteenth century an annual average of 45 million pounds of Chesapeake tobacco was imported into Britain, rising to an average of 76 million pounds during the second half of the century (USBC 1975: 1190).

Second, the huge increase in tobacco imports was part of a wider transformation in the nature of Britain's imports in which groceries, especially exotic commodities – tobacco, tea, coffee, chocolate and sugar – played a critical role. As a proportion of the total value of imports into England and Wales, groceries grew in importance from 8.9 per cent in 1559 to 34.9 per cent in 1800 (Shammas 1990: 77). Most of the growth occurred in the second half of the seventeenth century and the first half of the eighteenth century, reflecting, in good part, the growing importance of tobacco, as previously discussed, but also of tea and sugar. Unlike the other main exotic substances which were imported primarily for home consumption, tobacco was largely re-exported, especially in the eighteenth century, and entered into international circuits of trade (Shammas 1990: 81–6; Austen and Smith 1990: 99; Mui 1984: 12–14; Schumpeter 1960: 60–1).

Re-exported Chesapeake tobacco found its way principally to the Dutch Republic and France, though other markets were not trivial. The Dutch market was particularly important both because it took so much of the Chesapeake output and because Amsterdam was Europe's chief tobacco market. In the seventeenth century Holland accounted for about half of Chesapeake tobacco re-exports, and for most of the eighteenth century it remained the single most important re-export market, accounting for a proportion fluctuating between one-third and one-half (Price 1964: 500–1; Price 1973: 845–8). Holland had been the most important foreign market for Chesapeake tobacco from the beginning of its successful exploitation in the colonies. Dutch merchants effectively competed against English merchants in Virginia by offering relatively higher prices for the leaf than their English counterparts, but for various reasons, most notably that the main London merchants importing tobacco had extensive land holdings in Virginia, the bulk of the Chesapeake output flowed not to Amsterdam but to London (Pagan 1982: 485–6; Pagan 1979: 260–1; Kupp 1973). Their control of the tobacco trade was not guaranteed and had to be fought for. The Dutch were unable to make any headway in the early 1630s, but the tide began to turn towards them in the latter years of the decade. Two big changes occurred. First, in the matter of just a few years, Chesapeake output at least doubled: English imports of Chesapeake tobacco rose from 500,000 to just over 1 million pounds between 1634 and 1640, and prices collapsed by the same proportion; second, the outbreak of the Civil War in 1642 disrupted regular maritime communications between England and the colonies (Menard 1980: 157; Pagan 1982: 486). London was swamped with colonial tobacco, and everything was done in the city to shift the excess supply. Re-exports, which stood at just under 50,000 pounds – or 12 per cent of imports – in 1634, rose to just over 400,000 pounds – 42 per cent of imports – in 1640 (Pagan 1979: 255). Though we have no information on the size of the annual tobacco harvest in the Chesapeake from which to calculate the amount which was bought by English merchants, it is nevertheless clear that a proportion was left behind, and that it was the Dutch merchants who moved in to dispose of it. One Amsterdam merchant family put down roots in Virginia, and in 1640 they exported at least 60,000 pounds to Amsterdam, more than any London merchant handled (Pagan 1982: 487–8). Dutch merchants continued to profit from direct contact with Virginia planters until the 1650s, when a combination of parliamentary legislation (which made it statutory to send colonial products on to England, on English or colonial ships) and the First Anglo-Dutch War made it extremely difficult for the Dutch to trade directly with the Virginians (Pagan 1982: 489–97). For the next few years some Dutch traders undoubtedly circumvented English maritime law, but as the price of forfeiture increased, especially after the passage of the Navigation Act of 1660 and the Staple Act of 1663, the Dutch (and other nations, for that matter) were excluded from direct commerce with the colonies. After this date Virginia tobacco was, to all intents and purposes, English tobacco.

Even though Dutch merchants relinquished their place in Virginia's commerce, there were several compensations. Not least of these was the fact that Amsterdam was, and had been for a while, Europe's chief (possibly only) tobacco mart. There is nothing surprising in this, since Amsterdam was Europe's premier mart for an enormous range of goods, as well as the focal point of European trade and financial information (Smith 1984). Into Amsterdam flowed tobacco from the Chesapeake, Spanish America, Brazil, and from other parts of Europe, from the Dutch countryside, as well as from Germany (Roessingh 1978: 42–3; Price 1961: 6–7). How much tobacco flowed in and out of Amsterdam is unknown with any certainty but a reasonable estimate would be in the order of 25 million pounds at the turn of the eighteenth century (Roessingh 1978: 42). The amount of commercial income that this flow generated must have been substantial. The compensation for the loss of Virginia tobacco did not end there. Besides growing in importance in importing and disposing of tobacco, Amsterdam also became important in the processing side of the tobacco trade. Processing consisted of two stages, blending and spinning, both of which were established in Amsterdam at an early date, possibly 1631 (Barbour 1963: 63). Helped by the import of Chesapeake tobacco, either directly or indirectly through London, by the import of Brazilian and Spanish-American tobacco and by the prodigious expansion of Dutch cultivation, the Amsterdam tobacco industry increased enormously; in 1700 the city could boast more than twenty tobacco workshops where the blending of Dutch tobacco with colonial imports was carried out with typical Dutch ingenuity: at one point in 1670 they were accused by the English of falsifying tobacco products by, for example, selling their own tobacco blended with Virginia as Spanish (Price 1961: 7). Processing doubled the value of tobacco (Barbour 1963: 63).

London and Amsterdam were the twin pillars of the international circulation of tobacco, both colonial and European. Merchants in both cities were locked into a permanent search for markets, spreading tobacco consumption to the periphery of the European heartlands. The heartland was mostly represented by France, to which we will return later in the chapter. For the moment, however, it is revealing to examine how vital it was for both English and Dutch merchants to find outlets for their ever-increasing supplies, and no part of Europe provides a better example of a potential market than the north, both Scandinavia and Russia.

The northern European market was important for both Holland and

England. Both countries bought considerable quantities of raw materials in the region, including timber and iron and naval stores, but the problem was that there was little that either Holland or England could sell in the northern markets. So trade balances were adverse for Holland and England and could be balanced only by the export of bullion (Johansen 1986). Russia, in particular, was seen as an extremely promising market by London mercantile circles and the Crown itself, even though the consumption of tobacco was forbidden by law until 1697 (Price 1961: 17–20). For various reasons the dream of a vast market was never realized, for the English at least. Official figures for the export of tobacco leaf from England to Russia show a peak export of just under 1.5 million pounds in 1700, but for the rest of the century the figure barely reached 2,000 pounds annually (Price 1961: 101–2). By contrast the Dutch were extremely successful in the northern market, particularly in Sweden. One estimate places Dutch exports of tobacco, both leaf and processed, to the north at 15 million pounds (Price 1961: 89). Even though the Dutch had managed to gain more than a foothold, they were not able to retain it for the whole of the eighteenth century. The Russian market, for example, became increasingly difficult to furnish because, in true mercantilist fashion, tobacco cultivation was itself increasing in the Ukraine; in the late 1760s, for example, Russia was exporting Ukrainian leaf at a level of around 1.7 million pounds per annum (Price 1961: 95). It was only in its manufactured state, increasingly in the form of snuff, that Russia continued to import tobacco from the West (Price 1961: 95). The Swedish market also collapsed, and, though in absolute terms the import of tobacco from Britain remained stable over several decades, it never amounted to more than a minor element in British tobacco commerce (Price 1961: 101–2). The Dutch were the ones who suffered most from the collapse of the Swedish market because not only did Sweden begin to cultivate its own tobacco, once again under mercantilist pressure, but it also began to manufacture its own products (Price 1961: 89, 103; Roessingh 1978: 46; Israel 1989: 385, 388).

While both the Dutch, and particularly the English, struggled in the northern market for a foothold, much closer to home there was a market that not only was considerable – compared to both England and Holland – but that required far less expenditure in energy. The French market, from 1674 to 1791, had one buyer only, the state monopoly, and it preferred to purchase its supplies from Britain. This was, of course, a most fortunate state of affairs since the balance, if not the conduct, of trade with France was a source of constant worry to the English Crown, especially since French exports to England consisted of luxuries rather than essentials.

While the course of the commercial history of tobacco in England may correctly be interpreted as a victory by the Crown and the Virginia interests over the consumer, France, by contrast, presents a very different picture. The French tobacco market was already from its first beginnings much more varied than the English market. Before cultivation began in the French Caribbean colonies France was not only importing considerable quantities from Brazil and Venezuela but also producing substantial quantities in the south-west and the east of the country (Price 1973: 4–5). Once cultivation began in St Christophe, Guadeloupe and Martinique, and then later in St Domingue, French consumers were possibly the best supplied in Europe with a range of tobacco types and tastes – only the Dutch could match the range of the French tobacco market. The tobacco trade did not come directly under the French Crown's jurisdiction until 1621, when an import and export duty was first placed on the movement of tobacco (Price 1973: 11). Over the succeeding half-century the rate as well as the administration of these duties changed as new interests groups, such as the French West India Company, attempted to get special treatment for their participation in French tobacco commerce. Notwithstanding changes in the details of French duties on tobacco, until 1674 these were, by comparison with those of England, small. The French Crown did not earn as much from tobacco as it might have done had the duties been higher, or had there been a different method of collecting them (Price 1973: 11–16).

In 1674 the entire method of collecting duties, and the entire structure of French tobacco commerce, changed beyond recognition when the state established a tobacco monopoly that was to last, with a few interruptions, until its downfall in 1791, along with other *ancien régime* institutions. The monopoly was responsible not only for the commerce in tobacco but also for its cultivation, manufacture and sale. Monopolies over the control of tobacco were not uncommon in Europe, and by the time the French went in this direction there were already in existence in Spain, Portugal and the Italian peninsula institutions of this sort, though no two monopolies were identical (Price 1973: 17; Comes 1900; Gray and Wyckoff 1940: 4–13). The French monopoly operated under a system which was common in many parts of continental Europe, namely that the state did not directly administer the monopoly but rather farmed it or leased it out to whatever private company offered the state the highest price for the farm. In the event the French tobacco farm had few lessees over its history, allowing for a remarkable continuity in the monopoly's policies regarding purchasing and manufacture (Price 1964: 503).

The tobacco monopoly was a business that had to cover the price of the lease granted to it by the Crown. At the same time it sought to be the sole provider of tobacco to the French consumer. One of its first decisions was to control the sources of supply to prevent leakage through smuggling. Tobacco cultivation in France itself was severely restricted, though not abolished altogether as was the case in England; supplies overland from Holland and Germany were discouraged, and, as cultivation in the French colonies began to decrease, the monopoly turned increasingly towards England, then Britain, for its supplies. Even so, while the monopoly would have been happy to buy all of its supplies from across the Channel, it could not forget the French consumer who still hankered after the particular tastes of both Brazilian and Spanish-American leaf.

We can follow the purchases of the French monopoly with some degree of precision. During the first few years of the monopoly's existence French domestic and French colonial tobacco predominated, though just under 20 per cent of the tobacco supply was Brazilian; Chesapeake tobacco accounted for less than 5 per cent of the total, and purchases from both Holland and Germany were very small, in line with the monopoly's overall policy (Price 1973: 174–5). The main change in the structure of purchasing occurred around the turn of the eighteenth century when, because of the decline of tobacco cultivation in St Domingue, difficulties in getting increased supplies from Brazil and a growing demand among French consumers for Virginia leaf, English tobacco imports into France started to move ahead rapidly (Price 1973: 177–81). By 1708, for example, French domestic tobacco purchases by the monopoly had fallen, in relative terms, to around a third of total purchases, compared to a figure of 80 per cent at the time of the monopoly's founding; Virginia tobacco represented the single largest purchase, accounting for nearly one-quarter of the entire year's supply (Price 1973: 174, 187). Seven years later the proportion of Virginia tobacco in the purchases by the monopoly rose to 60 per cent of all imports, and by the same time France had become England's second market, surpassed only by Holland (Price 1973: 189). Once established to this extent, the place of Britain in the tobacco purchases of the monopoly remained paramount; at times in the eighteenth century, until its dissolution, the monopoly received virtually all its tobacco from Britain (Price 1973: 390–1). This trend towards British tobacco supplies was also reflected in British re-exports, where France, for most of the eighteenth century, vied with Holland as the major market (Price 1973: 849).

The relationship between the rising purchases of Chesapeake tobacco by the French monopoly, and the changing French preference for tobacco as snuff rather than in smoking or chewing form, has been discussed in Chapter 4. While French consumers and the purchasing strategy of the French monopoly clearly benefited the British tobacco trade, the market power of the monopoly directly affected the nature of the tobacco mercantile system in Britain as well as the Virginia planter.

The French monopoly bought in bulk through agents they placed in the main tobacco markets; in the seventeenth century these agents concentrated in Amsterdam and Lisbon, and in the eighteenth century they were in London, Bristol and especially Glasgow. By the time that the French monopoly was turning its attention across the Channel for tobacco supplies, the tobacco import trade in England had been transformed from a trade characterized by easy entry and numerous participants to one with high entry costs and concentrated business power. This situation was, of course, perfectly suited to the French monopoly whose aim was to buy as cheaply as possible in the British market, and without any appreciable difference in price across the market the monopoly could settle on doing business where transaction costs were at a minimum. The bigger the importing merchant in London, for example, the more likely he was to make a deal with the monopoly. That this was more than possible is borne out clearly by comparing the size structure of the London tobacco importing business and the purchases of Chesapeake tobacco made by the French monopoly in one year, 1719. In that year France imported just over 6 million pounds of tobacco from England. This quantity could easily have been supplied by only five of the leading London tobacco importers (Price 1973: 380; Price and Clemens 1987: 11).

For the French buyer it was not, however, simply a matter of the size of the business offered by the English merchant that was the decisive factor. The kind of tobacco business also mattered. In the tobacco trade at the turn of the eighteenth century there were two main kinds of importing merchants, those who purchased Chesapeake tobacco on consignment from the planter, and those who purchased Chesapeake tobacco on their own account. The French monopoly preferred to do business with the latter. This preference had a considerable effect on the entire British tobacco trade, as well as on life in the Chesapeake. To understand why this was so we have to explore the nature of both consignment and direct purchasing systems in greater detail.

It is not clear when the consignment system first took root in the Chesapeake. It was established in the trade to the Caribbean in the early seventeenth century, and there is some evidence to suggest that in the Chesapeake it began as early as the 1630s (Davies 1952; Price and Clemens 1987: 4). The foundations of the consignment system in the Chesapeake trade lay in the practice of having factors resident in Virginia, who consigned tobacco to a number of merchants in London, superseding a system in which tobacco was entrusted to ships' captains on their way to England (Price and Clemens 1987: 4–6). Most of the growth in the consignment trade seems to have occurred after the 1690s, and lasted right through the eighteenth century, though its significance, in terms of its relative share of the import trade, declined. At the height of the consignment system perhaps as much as 50 per cent of the tobacco trade was handled in this manner, but on the eve of the American Revolution the figure had slumped to 25 per cent (Price and Clemens 1987: 6; Breen 1985: 84; Bergstrom 1985: 198–9). The essential features of the consignment system were that the Virginia planter entrusted his tobacco to a London merchant who arranged to have it sold in the British and European market and returned to Virginia with British and European goods purchased with the revenue from tobacco sales. The system also involved the London merchant providing credit for the Virginia planters (Rosenblatt 1962; Price 1980; Breen

1985: 84–175). In its operation the consignment system integrated the Virginia planters and London commission merchants with the tobacco trade and the consumer goods market (Breen 1985: 118–22; Bergstrom 1985: 163–79).

Virginia planters, as the following chapter will explain, prided themselves on the individual quality of their tobacco, and no two hogsheads were considered to be the same, even if the tobacco they carried came from the same planter. This aspect of tobacco culture was carried forth into the consignment system, and manifested itself in minutely differential pricing, by type of tobacco, to reflect supposed or actual quality differences. To make the system work, therefore, consignment merchants were under an obligation to treat the sale of each hogshead as a separate event (Price 1964: 507). The Virginia planter and the consignment merchant were required to be loyal to each other, and, as a result, the relationships between the two ran far deeper than business alone (Rosenblatt 1968: xvi; Breen 1985).

When the French monopoly started purchasing Chesapeake tobacco in increasing amounts, it had no choice but to trade with the consignment merchants, principally in London, because it was only they who handled sufficiently large quantities of tobacco, even though the problem of individuality remained (Price 1973: 592). Dealing with the French monopoly was considered as a crucial adjunct to the British–Chesapeake trade because of the cash facilities which the French buyers extended. As the authoritative historian of the trade has aptly described it, 'the French were thus the liquidity grease which kept in motion the entire sluggish mechanism of the British–Chesapeake trade' (Price 1973: 660). Anyone who entered the tobacco trade after 1700 must have been aware of this role of the French monopoly and none capitalized on it more than the Scottish merchants.

Scottish merchants had been trading in Virginia from as early as the mid-seventeenth century but it was not until after the Act of Union, in 1707, that Scottish merchants had free access to Virginia, the British domestic and the re-export market (Price 1954a: 182–3). Records of the import of Chesapeake tobacco into Glasgow, the Scottish port of entry for tobacco, show an increasing share of the British tobacco import trade from 1707 and, though growth was slow in the first few decades after Union, by the 1740s Scottish imports were already 20 per cent of the total and rising; by the 1750s, the proportion had risen to 30 per cent and by the 1760s to 40 per cent (USBC 1975: 1,190). Even more striking, by the late 1750s Glasgow had become the chief tobacco port in Britain and for most of the years leading up to the American Revolution tobacco imports into Glasgow exceeded those into London (USBC 1975: 1,190; Price 1954a: 180). These figures refer only to legal imports, and, though there is no certainty about the degree of smuggling, recent work suggests that, once illegal imports are taken into account, the import of Chesapeake tobacco into Scotland was much higher in the years immediately following Union than originally thought (Nash 1982: 364). Yet, even when the revised figures are taken into account, the chronology of the rise of Glasgow as described above remains in force.

The rise of Glasgow in the tobacco trade obviously opened another sector in this branch of commerce and offered the potential of an alternative first to the outports Whitehaven, Liverpool and Bristol, and eventually to London itself. What made Glasgow a real alternative were two distinguishing but interrelated features. First, Glasgow possessed advantages in terms of the costs of transport from the Chesapeake, since the route to Glasgow was considerably shorter than to London; second, it had a well-developed financial and commercial system in which the 'tobacco lords' held special prominence; and finally, Glasgow merchants could offer exceptionally low freight rates (Price 1954a: 187–90; Devine 1975). It was the latter feature of Glasgow's tobacco trade that most influenced the price of tobacco at the port. The price of tobacco in Glasgow was generally below the London price (Price 1964: 508).

The French monopoly became increasingly attracted to buying tobacco from Glasgow. The Scottish share of the French market increased substantially after 1740; on the eve of the Revolution the share of Scottish tobacco in the monopoly's purchase of British tobacco stood at 72 per cent (Price 1973: 592). Price differentials were obviously one of the reasons why the French agents found themselves increasingly in Glasgow rather than London, but they were not the only reason. Mention has already been made of the needs of the French monopoly for bulk purchases and of the constraints imposed upon this by the special relationship that existed between the London commission houses and the large Virginia planters. Scottish merchants totally bypassed this because they operated on a completely different system in purchasing Chesapeake tobacco.

The consignment system placed the Virginia planter in a web of financial relations across the Atlantic that were continuous and based on close, familiar, interpersonal commitments. Consignment was as much a social as a commercial system and typically it was the large Virginia planters that constituted the colonial side of it. Small planters, on the other hand, did not have the social stature to place them in personal contact with the impressive London commission houses and, therefore, it was fairly useful for these planters to sell their tobacco to the large planters before the lot was sold on commission (Breen 1985: 36). Into this system came Scottish traders who offered an alternative to the consignment system that was particularly attractive to the small planter.

There is little doubt that Scottish traders were operating in Virginia in the latter part of the seventeenth century but it was not until the opening-up of the Virginia piedmont to settlement and tobacco culture that these traders came into their own. The following chapter will outline the dynamic aspects of tobacco culture in the Chesapeake as the frontier of cultivation moved westward, pushing further from the inlets where the large planters had their holdings. This migration, initially of small planters, was of little interest to the London commission houses partly because they produced small crops and their credit-worthiness was generally unknown and partly because of the difficulty in transporting tobacco from the interior (Devine 1975: 56–7; Kulikoff 1986: 92–9).

In contrast to the consignment system, Scottish merchants bought tobacco directly from the planters, and sold them European and West Indian goods from country stores that they established in the Chesapeake hinterland. Many of the country stores were actually part of a chain of stores owned by the three main Glasgow tobacco firms, the Glassford, the Cunninghame and the Speirs groups (Devine 1975; Devine 1976; Devine 1984). These stores were very sophisticated businesses that, in addition to their buying and selling activities, offered a substantial provision of credit to medium and small planters (Soltow 1959). The social and economic development of the Virginia piedmont, and especially southside Virginia, was intimately linked to the presence of these stores, and to the credit that they disbursed (Kulikoff 1986: 122–31; Farmer 1988; Price 1973: 666–71). The Scottish merchants offered a service that was highly attractive to small planters, but the cutting edge of the Scottish presence was that they offered relatively high prices (Devine 1975: 58). They could do this because direct marketing, especially purchasing tobacco in advance, substantially reduced operating costs; and, because the Scottish merchants could bring tobacco into Britain in hogsheads that were undifferentiated and could be sold in bulk, they distinguished themselves to the French buyers. As the tobacco was landed in Glasgow, the title was passed immediately to the French monopoly and money changed hands (Price 1964: 508). And so it went on to the next planting season in the Chesapeake.

Alongside the London consignment and the Glasgow direct purchase system operating roughly from London and Glasgow respectively, there were other means of purchasing tobacco, as well as other players in the field, including merchants from Liverpool, Bristol and Whitehaven (Price and Clemens 1987: 24–31; Tyler 1978: Price 1986). Until the American Revolution, however, London and Glasgow and their purchasing methods dominated the British tobacco trade. London reclaimed the ground it had lost to Glasgow in the years after the Revolution (Price and Clemens 1987: 39–40).

Though the British colonial system and the place of tobacco in it evolved over the seventeenth and eighteenth centuries, certain features stand out as relatively permanent. For one thing, the tobacco trade seems to have been sustained and promoted by a combination of state and mercantile interests; when, for instance, planters complained of extremely low prices because of overproduction and suggested a moratorium on planting, it was the Virginia merchants in London who successfully rejected the idea

(Hemphill 1985: 93–7; Olson 1983). Second, trade was severely constrained by the legislation of the period excluding non-British merchants from purchasing Chesapeake tobacco directly, forcing all Chesapeake tobacco, regardless of its final destination, to be landed first in Britain and to be shipped in British (or colonial) vessels. This may seem to have been drastic, but it made perfect sense given the contemporary economic discourse. Forcing colonial exports to Europe, on their first leg of a possible international circuit, was the policy of most European states, but in the history of tobacco there was one very important exception, that of Portugal.

The amount of tobacco that was exported from the New World was far in excess of the demands of the home country. For example, British consumers would have had to consume more than 10 pounds per head of the population in order to clear the Chesapeake output – at no time in British history has per capita consumption exceeded 5 pounds. It is not surprising, therefore, that the vast majority of New World tobacco was re-exported from the home country.

By comparison with the Chesapeake tobacco trade, commerce in Brazilian tobacco was complex. Part of this complexity was established in the seventeenth century, precisely in 1644, when King João IV authorized direct trade between Bahia and the Mina Coast, on the Bight of Benin to the east of the River Volta in West Africa (Hanson 1982: 153; Van Dantzig 1980: 118). This was not only a bold but a highly controversial step since it acted entirely in opposition to the prevailing political economic ideology of the time: namely that colonies were there to serve the home country, and all trade between colonies was to be conducted through the home country, and not directly. What inspired the royal decree was the capture of Angola by the Dutch in 1642, thereby forcing the Portuguese into other parts of Africa for their source of slaves, and the unease that Brazilian tobacco planters felt at the re-establishment of the tobacco monopoly only several weeks earlier (Hanson 1982: 153).

The direct trade between Bahia and the Mina Coast was essentially the Brazilian slave trade. Though in existence in the sixteenth century, it was not until the seventeenth century, and especially during the second half, that the slave trade to Brazil began to grow; according to one estimate almost four times as many slaves were imported into Brazil during the second half than during the first half of the seventeenth century (Curtin 1969: 119). During the eighteenth century the growth in the slave trade accelerated dramatically; nearly two million Africans were imported into Brazil between 1700 and 1800 as opposed to 560,000 between 1600 and 1700 (Curtin 1969: 216). Brazil was participating in a trade which was generally on the increase during the eighteenth century, but in the case of Brazil the main reason was the discovery and opening of gold mines in the interior (Lovejoy 1982: 497; Russell-Wood 1987). In this considerable trade Bahian tobacco played a critical role.

Thanks to its rich soil and climate the Bahian Recôncavo could produce three successive tobacco harvests, the first in September of each year (Antonil 1965: 313–15). The first growth was destined for the European market; it was considered strong and suitable for smoking, chewing and snuffing (Antonil 1965: 315). The second and third grades, however, did not find a market in Europe and their import into Lisbon was illegal (Verger 1964: 354). Their main market was the Mina Coast. Bahian planters prepared tobacco for Africa in a distinct manner, especially drenching it in molasses to give it a particular taste and aroma (Verger 1964: 354). Why West Africans were so dedicated to Brazilian inferior-grade tobacco is unclear, but their passion for it was remarked by many writers. It was this passion for it that created a barter economy of large proportions.

No one knows when tobacco ships began their regular voyages from Bahia to the Mina Coast. In 1678, the first year for which the record of such voyages exists, only one ship made the journey (Verger 1976: 578). Twenty years later twenty-one ships made the same trip. Over the follow-ing century the number of tobacco ships to the Mina Coast fluctuated around an average of maybe ten per year and by the end of the century over 1,300 ships are known to have been involved in this trade (Verger 1976: 578–9).

As the number of ships plying the South Atlantic increased so too did the export of tobacco. At the turn of the eighteenth century tobacco exports to the Mina Coast stood at no more than 160,000 pounds (Flory 1978: 162). By 1704 the amount exported had risen to 425,000 pounds; by the 1720s it had reached 2 million pounds and by mid-century perhaps as much as 4 million pounds (Flory 1978: 162; Hanson 1982: 160; AHU, Baia, doc. 3305). In the ten years from 1743 to 1753 just under 330,000 rolls of tobacco, yielding a total weight of 29 million pounds, were shipped (AHU, Baia, doc. 748). Tobacco shipments to the Mina Coast throughout the eighteenth century far exceeded those to any other parts of Atlantic Africa and, in many years, far exceeded shipments to Lisbon (AHU, Baia, doc. 748, 3305, 4667, 6589).

For the first half of the eighteenth century Lisbon dominated the export tobacco trade from Bahia. In 1698, for example, a year for which there are comparative figures, just over 5.5 million pounds of Brazilian tobacco were received in Lisbon (Hanson 1982: 157). During the eighteenth century Lisbon had to compete with the Mina Coast, or, to put it another way, with Bahian merchants. Even so, Lisbon's import trade grew substantially during the period. In 1756 the figure reached 6.1 million pounds and in 1784, a very good year for Bahian tobacco planters and merchants, Lisbon accepted almost 12 million pounds of first growth tobacco (AHU, Baia, doc. 3305; Lugar 1977: 49).

No exports from Bahia ever compared to those destined for Lisbon and the Mina Coast. However, it should be pointed out that there were other primary destinations for Bahian tobacco even though their quantity was never very large. On the African coast itself Bahian tobacco was landed in Angola, São Tomé, Principe and Benguela. In 1753, for example, exports to the last three destinations amounted to 3,500 rolls representing 1.2 million pounds' weight of tobacco – though large in absolute terms, these values were only about a quarter of the exports to the Mina Coast (AHU, Baia, doc. 748). The available evidence suggests that shipments to parts of Africa other than the Mina Coast fluctuated substantially from year to year, undoubtedly in response to the slave trade in these areas. Other than the African coast, Bahian tobacco also found its way into the Asian market through Goa, Macao and Timor. Little is known of this trade and even less about its size. Here and there in the documentation one finds references to the Asian trade. In 1778, for example, two ships were bound for Goa laden with almost 130,000 pounds of tobacco; a similar quantity was exported four years later (AHU, Baia, doc. 9733, 10944). The tobacco entering the Goan trade was different from that used in the African trade, the difference being in curing methods. The curing method for the Goan market was a Cuban innovation first introduced into Bahia in 1757. The first shipments of tobacco using the new process, weighing a total of 32,000 pounds, were sent at the end of the same year (Lugar 1977: 43; AHU, Baia, doc. 3275). As for the trade to China, it is not clear whether the tobacco was shipped direct from Bahia, and therefore in an unprepared form, or whether it came from Lisbon already prepared, typically as snuff. The Bishop of Macao writing to Martinho de Melo e Castro, the Secretary of State, in 1775 left little doubt that tobacco was the principal commodity of the colony – this was probably tobacco in powdered form manufactured in the tobacco monopoly's factories in Lisbon and of the most expensive variety (AHU, Macau, caixa 10, doc. 19; Nardi 1986: 19–20; Dermigny 1964: 1254). Chinese preference for Brazilian tobacco was widely acknowledged (AHU, Macau, caixa 10, doc. 11; Blanchard 1806: 473).

Bahian exports to Lisbon did not remain long in the capital. During the second half of the eighteenth century at least half of the imports were re-exported; in the 1780s the figure reached almost 70 per cent (Pinto 1979: 202; Lugar 1977: 46–9). This re-export trade consisted of unprocessed Bahian tobacco. The main markets were Italy, Spain, France and Hamburg (Lugar 1977: 46). The tobacco that remained in Lisbon, and which was destined for the Portuguese market, was processed into smoking and, increasingly in the eighteenth century, into powdered tobacco (Nardi 1986).

Brazilian tobacco was, as we have seen, in demand in Europe, Africa and Asia, supplied either directly from Bahia, or from Lisbon. There was also a strong, but quantitatively unknown, market for Brazilian tobacco in Amerindian communities in North America (von Gernet 1988: 204–16, 221–39). Like their West African counterparts, eastern woodland Amerindians had a strong preference for Bahian leaf and firmly rejected other types, including Virginia (von Gernet 1988: 204). The Hudson's Bay Company was alarmed in the early 1680s to find many native traders dealing with the French because they could supply them with Brazilian tobacco. One official of the Company wrote the following in 1685, reflecting a distinct policy change, a reaction to the leverage of the consumer:

> We are sorry the Tobacco, we last sent you, proves so bad, we have made many yeares tryall of Engelish Tobacco be severall persons, and whiles we have traded, we have yearly complaints thereof. We have made search, what Tobacco the French vends to the Indians, which you doe so much extoll, and have this yeare bought the like (vizt.) Brazeele Tobacco, of which we have sent for each Factorey, a good Quantety, that if approved of we are resolved in the future to supley you with the like, as you have occasion . . .
>
> (von Gernet 1988: 207)

The Company was thereby forced to purchase its supplies from Portugal with whom Britain had particularly good relations during the eighteenth century. Even the London pipe makers got in on the act as the Amerindian consumer preferred Bahian leaf in a clay pipe (Oswald 1978: 346; von Gernet 1988: 284–8).

It hardly needs repeating that the pattern of tobacco exports from Bahia was unusual in the context of early modern European political economy. The ability of Bahian merchants to break down mercantilist ideology was impressive; petitions from other Portuguese colonies for direct trade in tobacco to other colonies were not so successful (AHU, Timor, caixa 3, doc. 20). The British commercial and colonial system, by contrast, did not allow for any exceptions.

7

'TOBACCY'S KING DOWN HERE...'
Planter culture to 1800

The best smoker looks for the best cigar, the best cigar for the best wrapper, the best wrapper for the best leaf, the best leaf for the best cultivation, the best cultivation for the best seed, the best seed for the best field. This is why tobacco-raising is such a meticulous affair... The tobacco grower has to tend his tobacco not by fields, not even by plants, but leaf by leaf. The good cultivation of good tobacco does not consist in having the plant give more leaves, but the best possible... the cultivation, processing, and manufacture of tobacco is all care, selection, attention to detail... Everything having to do with tobacco is hand work – its cultivation, harvesting, manufacture, sale, even its consumption.

So wrote Fernando Ortiz, in his justly celebrated work *Cuban Counterpoint* (Ortiz 1947: 24, 26, 39). Ortiz was appealing to his readers to understand that the history of Cuba is deeply inseparable from the history of the island's two main crops, sugar and tobacco. Out of an analysis of these two crops, especially of the contrast, born of distinct botanical features, Ortiz was attempting to lay bare Cuba's complex web of social structures and cultural differentiation.

Ortiz has had his critics; the most serious charge against him points out that the contrast between tobacco and sugar (so central to his analysis) is not only exaggerated but presented as a moralistic argument; tobacco is good and sugar is bad (Hirschman 1977: 94–5). Yet, despite reservations, Ortiz does provide an insightful point of departure for this chapter simply because, as he and other historians have put it, there is a culture to agriculture (Breen 1984: 250).

This chapter has two main purposes. The first is to understand the transformation of tobacco from the grower's point of view: how it has been affected partly by changes in the labour organization of cultivation, broadly from plantation to smallholder; partly by changes in the nature of the tobacco crop; and partly by changes in marketing and manufacturing. Second, the chapter considers the symbolic value of tobacco to grow-

166

ers, and how this has altered historically. While historians and anthropologists agree that forms of cultivation, and plants themselves, influence culture, it is important to appreciate that such concepts as 'tobacco mentality' are specific in both time and place.

Chapter 2 argued that tobacco held special significance for Amerindian societies. This significance pertained to both consumption and production. The former has already been discussed in detail and there is no need to repeat what was said there. What does need reiterating is that the symbolic value of tobacco in consumption was mirrored in production. Proscriptions in consumption were echoed in cultivation. This took the form of laying down who, specifically, could cultivate tobacco, whether it should be men or women of a particular status, or those specifically chosen for the task. The Crow Tobacco Society is simply the most stark example of this; but throughout North and South America, wherever tobacco was cultivated, special rules existed for cultivation. Moreover, within agricultural societies, tobacco was usually grown separately from other crops and, as we have seen, tobacco was often the only crop grown by fishing and hunting societies.

Europeans, not surprisingly, eschewed Amerindian proscriptions, though they learned a great deal about the plant from them. Yet the European involvement with the plant was far from neutral and, though the cultural impact of tobacco was not constructed within a specific cosmology as was the Amerindian, it had deep significance nevertheless. One key to this lies in what I will call the labour history of tobacco.

Little is known about the early years of the transition in the cultivation of tobacco from an Amerindian to a European crop. Certainly it was rapid and there is little doubt that in these years, and in places such as Trinidad and Venezuela Amerindians and Europeans worked side by side (Lorimer 1978). Not only was the transition period rapid, it was extremely short as the previous chapter showed. By the time tobacco began its rise in the Chesapeake the Amerindian connection with tobacco was both severed and forgotten, and its association with Europeans firmly established. The rapid transformation of tobacco from an Amerindian to a European commodity was reflected in the rapidity with which Europeans reversed the original direction of the tobacco exchange and began, increasingly, to dispense 'European' tobacco and 'European' smoking instruments to Amerindians (von Gernet 1988: 187–318).

There was, however, nothing predetermined about tobacco's early connection with Europeans. That is to say, there was no particular characteristic of the plant that made it European, in contrast to sugar which, from its early beginnings in the New World, was inexorably linked to African slave labour. The contrast between tobacco and sugar in ethnic or cultural terms is one of the great and enduring themes in the history of the plant, and it needs explaining.

167

Two main factors can account for tobacco's Europeanness. The first is economic. There were no economies of scale in tobacco cultivation: that is to say, any increase in the area of land under tobacco demanded a proportional increase in labour and capital. The economic size of the tobacco holding could therefore vary quite widely. Smallholders were not at an economic disadvantage as they were, for example, in sugar cultivation. Tobacco cultivation could thus be embedded within a European mode of agricultural production, typically the peasant or independent farmer. It is not surprising that when tobacco was grown in Europe in the seventeenth and eighteenth centuries it was grown by the peasantry and the independent yeomanry (Roessingh 1978; Thirsk 1974; Price 1973). There is no reason to doubt, therefore, that in principle the same kind of labour system would have prevailed in New World tobacco cultivation. Indeed Dutch tobacco growers were invited to migrate to New Netherland in the seventeenth century for this very reason, and all the available evidence confirms that tobacco cultivation in the colony was similar to that in Holland (von Gernet 1988: 171–6). The problem for the Chesapeake, however, is that the colony, especially in its formative years, did not attract these kinds of people, and labour shortage undermined the colony's future prospects. Not only was the flow of people to the Chesapeake slow – 1,700 between 1607 and 1616 – but mortality was so high as to make the settlement precarious: death rates in Jamestown varied from 46 per cent to 60 per cent per annum between 1607 and 1610 (McCusker and Menard 1985: 118). The combination of open land and short free-labour supply provided fertile ground for solving the colony's problems by coercing labour through some sort of bound contract. It is at this point that the Chesapeake faced conditions that prevailed throughout the colonies further to the south and were solved there by resorting to the importation of African slaves. Here, then, is the second factor. Rather than turning to Africa, England turned to its own people. In England a system of servitude existed typically involving men and women aged between 13 and 25 (Galenson 1984: 3). The servant lived in the master's household under a contract normally lasting one year. The Virginia Company looked to this institution to solve its problems of labour recruitment (Diamond 1957–8). The indentured system in the Chesapeake was transformed by stages between 1609 and 1620 by which time it had elements specific to the conditions in the colony as well as the changes taking place in the relationship between the immigrants, the planters and the Virginia Company (Galenson 1984: 3–6). Indentures lasted anywhere from four to seven years and, after the servant had repaid the cost of passage, he was, in principle, free to establish himself as an independent planter, for example.

Whether in the Chesapeake or later in Bermuda and the West Indies, indentured servitude, settlement and tobacco cultivation were inextricably linked. The flow of indentured servants to Chesapeake increased rapidly as the tobacco economy began to boom. Between 1617 and 1623, for example, at least five thousand English people emigrated to the Chesapeake (McCusker and Menard 1985: 118–19). In the 1630s over ten thousand emigrated and the upward trend reached its high point in the 1650s, when an estimated 23,100 immigrated, at least two-thirds of whom were bound in servitude (Gemery 1980: 215; McCusker and Menard 1985: 242). After 1660 the migration of indentured servants fell back to a level 20 per cent below the peak of the 1650s, but thereafter the pool of English people willing to migrate in indenture began to shrink considerably despite efforts to attract these people to the colonies (McCusker and Menard 1985: 135; Gemery 1980: 215). Nevertheless, this flow, together with an appreciable decline in mortality, was responsible for a surge in the colony's population. In Virginia population grew from a level of 1,300 inhabitants in 1625 to almost 63,000 at the turn of the eighteenth century (Morgan 1975: 404). Maryland's population also grew enormously from 600 inhabitants in 1640 to 34,100 in 1700 (McCusker and Menard 1985: 136).

In the seventeenth century the cultural composition of the Chesapeake colonies was essentially white European; in 1670, for example, only 2,500 Africans lived in the Chesapeake, a mere 6 per cent of the total population (McCusker and Menard 1985: 136). And since the colonies' main, and it could be argued only, staple was tobacco, as the population increased, the association between Europe and tobacco deepened. What kind of society was being constructed around this crop?

The first thing to remember is that the kind of society that both the Virginia Company in Virginia and the Lords Calvert in Maryland hoped to establish and see prosper never materialized. Their vision of a hierarchical society modelled on the English pattern remained a paper vision and much of the reason for this failure lay in the labour problems encountered in both colonies (Kulikoff 1986: 26–7). Rather than hierarchy, what materialized in the Chesapeake was a society based on the dynamics of the indenture system located within the tobacco economy. High tobacco prices induced planters to recruit servants and allowed ex-servants to accumulate enough capital to become independent planters themselves and procure their own servants (Horn 1979: 92). Depressions had the opposite effect. Immigration fell, profits were diminished and recently freed servants, finding their entry into tobacco cultivation blocked by the lack of resources, either tried their hand at some other activity or departed for other shores. The booms and slumps of the tobacco economy, therefore, provided a powerful context for the evolution of Chesapeake society (Menard 1980; Wetherell 1984).

The Chesapeake colonies were transformed considerably during the seventeenth and eighteenth centuries by economic, social and political forces but, in general, what continuity there was derived from tobacco cultivation. Despite changes in labour organizations and significant moves towards economic diversification, the tobacco crop continued to stamp the region with a distinct culture. That culture, and the society in which it was located, derived first and foremost from the pattern of work that tobacco cultivation demanded.

Tobacco cultivation consisted of a series of distinct stages that took the crop from seed to market. Unlike other staples of the period, tobacco was unique in that a sustained labour input was required constantly, different kinds of skills were called for at different times of the year, and each stage had to be accomplished with the utmost care. Tobacco cultivation did not lend itself naturally to gang labour, though as the Chesapeake became a slave society that aspect did change (Galenson 1981b: 151). Nevertheless, the general features of tobacco cultivation remained unaltered.

The first stage in tobacco cultivation was to plant seed in a seedbed and not in the open field. The seedbed method of cultivating tobacco is nearly universal, and is still the common practice today, suggesting several possible advantages over planting in the open field. The first is that in the seedbed not only does the plant has a better chance of survival than in the open but in a fairly concentrated space there is less chance of weeds and better opportunities for detecting diseases and pests. These are all important factors in determining the intensity of labour use. The seedbed could also be more intensively fertilized – with an appreciable cost benefit – than the open field. Second, where the growing season is short it would be impossible to grow a successful crop if the seedling growth was added to the time in the field. Finally, there is evidence to indicate that the act of transplanting the young plants from seedbed to open field produces certain physiological effects that actually benefit growth (Akehurst 1981: 98). Though the last factor was probably not realized at the time, there is little doubt that Chesapeake planters would have been aware of the first two.

December and early January were typical times for planting seeds on an area normally not exceeding one-quarter of an acre. In common with agricultural practices followed elsewhere to minimize the overall risk of loss, plant beds were frequently kept separate over some distance. The beds were protected from the frost and winds and, with luck, the young plants made an appearance in March and were generally ready for transplanting to the open in April (Breen 1985: 46–7).

The seedbeds required constant surveillance to give the plants the optimum chances of survival. Even so, because of an inevitable rate of loss, more plants were sown than required. The seedbed stage was not particularly onerous on labour, but the transplanting to the open was. Before the actual transplanting began, it was necessary to choose the date. This was crucial for various reasons but most importantly because the lifting had to be done when ground conditions were most favourable, that is, just after a heavy rain when the root system was fairly free of the surrounding soil. The entire operation of transplanting normally began in late April and the work spread over many weeks, frequently spilling over into June. This stage required all hands and they needed to be available at just the right moment. The fields needed to be prepared and the actual process of transplanting had to be done fairly quickly as the plants could not be allowed to dry out. It is important to realize that transplanting was done plant by plant. Plant populations, though not known precisely, could easily have run into several thousand per acre (Akehurst 1981: 201). Landon Carter, one of the biggest Virginia planters, cultivated more than one hundred thousand plants on his estate (Breen 1984: 257). In the seventeenth century land was not ploughed and, therefore, the tobacco was transplanted into hillocks of prepared soil set in partially cleared land – the hillocks were set at a predetermined distance apart, usually 3 feet, and in parallel, a method of cultivation clearly inherited from Amerindian practices (Middleton 1984: 111; Carr and Menard 1989: 415–16; Herndon 1967).

Once in the field, the plants began to grow leaves at the same time as weeds appeared around them. This caused problems, and the tobacco fields had to be visited and worked upon constantly. Weeding absorbed an enormous amount of time and energy but the growing plant needed a lot of attention too. The tobacco plant needed to be topped, that is the top of the plant had to be removed. Topping stopped the plant from producing flowers, so that all the growing energy was put into the leaves. But as soon as the plant was topped it produced secondary shoots which had to be removed to allow the plant to process nutrients in the leaves. Suckering, as it is termed, and topping were done plant by plant (Breen 1985: 48–9). Once again, the precise stage at which topping was done was crucial to the final product. Late topping would have had particularly serious results on both leaf quality and yield. Modern research shows that, on average, each day's delay in topping can result in a yield loss of about 15 pounds per acre (Akehurst 1981: 217).

To say that the preliminary stages of seed planting, transplanting, topping and suckering were all critical to the viability of the final product is no exaggeration, but in terms of the culture of the field the success at these stages depended on an interaction between human and natural factors. Obviously, despite the most careful attention to detail, natural conditions could easily overwhelm all the human input. But, come September, when the tobacco plant reached maturity, the planter had to make the most difficult choice of all: when to cut. After this point nature took a back seat, and all the pressure fell on the human agent.

Cutting, naturally, ended the biological life of the plant at the same time as it began its transformation into a transactable commodity. Early cutting was avoided as the plant would not cure properly and would be worthless in the marketplace. Waiting for the right moment also had its problems because climatic conditions in September were typically uncertain. Frost, for example, could easily wipe out a crop (Breen 1984: 257). When to cut, therefore, was a decision fraught with anxiety and, according to an authority on the subject, there was little precise guidance available (Breen 1984: 258). William Tatham's treatise on tobacco culture gave his readers the following rule-of-thumb: 'The tobacco, when ripe, changes its colour, and looks greyish: the leaf feels thick, and if pressed between the finger and thumb will easily crack' (Breen 1984: 258). This may sound rather vague to modern readers, and there is little doubt that it had the same ring to both the experienced and inexperienced seventeenth-century planter. Experience alone, even today, is the key to success.

After cutting, the autumn months were spent in producing the final product, beginning with the curing stage. This was accomplished in barns where the tobacco stalks with their leaves were allowed to dry out naturally. Again it was an anxious period. Since air-curing depended upon environmental conditions, themselves subject to daily change, the point at which the tobacco reached a satisfactory point of dryness was highly uncertain. Too much moisture and the tobacco would rot; too little and it would turn to dust. Once the decision had been taken that the tobacco was cured, the leaves needed to be stripped from the stalk and the main stem of each leaf had to be removed. Stalking and stemming could easily be a twenty-four-hour operation given the size of the plant population. Once this was completed, the leaves were packed into hogsheads in a process called prizing. This was usually not accomplished until well into the next calendar year, and it would be a further two months before the prized tobacco was ready for export.

Moreover, if that was not enough, by the time prizing was completed the seedbeds needed to be well under way and the next production cycle begun. Not surprisingly, the calendar of the whole year moved to the rhythm of the tobacco plant, as did the labour of those on the land.

Recent research has clarified many aspects of the labour input required to bring tobacco from seed to commodity. The tobacco plant has a voracious appetite for soil nutrients, particularly potash, calcium and nitrogen (Akehurst 1981: 138). In modern production methods these minerals consumed by the growing plant are replenished by the application of large amounts of chemical fertilizer. In the Chesapeake, however, the only form of fertilizer available was farm manure but, because animals were not as a rule penned in, the supply of this was very limited (Carr and Menard 1989: 409, 417–18; Carr, et al. 1991: 55–75). The fertility of the soil was, therefore, restored in a system of long fallow. The fallow period in the Chesapeake region was twenty years. Each hand could tend no more than 3 acres successively for three years: consequently, a tobacco planter needed 20 acres of land per worker to turn a tobacco crop annually (Kulikoff 1986: 47). It was the amount of labour, paradoxically, that determined the size of the holding (Carr and Walsh 1988: 150). Twenty acres was, however, a minimum figure and did not include any provision for land to grow food or to grow timber for heating, constructing fences, barns and making tobacco hogsheads. When these requirements were added into the tobacco land, it was generally felt that planters needed 50 acres per worker (Kulikoff 1986: 48).

Tobacco cultivation in the seventeenth century absorbed about half of the year's working time. The slackest time of the year was between January and early April during which time the seedbeds were being made and tended. In these months, typically the only time in the year when workers could afford some leisure time, only about 10 per cent of available working time was used on the fledgling plants. In April, however, labour demands rose enormously. Transplanting absorbed virtually the entire working schedule in both April and May. Some respite came in June when weeds began to appear in the fields, but in July and August topping, suckering and weeding left workers with no time for leisure. September, October and November were somewhat less demanding, but even then cutting, stripping, stemming and prizing absorbed, on average, 50 per cent of the entire working schedule (Carr and Menard 1989: 414).

Tobacco cultivation was often merciless with labour. The planter, as we have seen, did not escape lightly. Even if he did not work in the fields – a rare event in the Chesapeake with the possible exception of the very big landowners – his consciousness, if not his hands, was always involved in critical decisions (Breen 1985: 69). Although the discussion so far has tended to treat labour as homogeneous, in reality this was not so. Different kinds of skills were displayed throughout the year within an unchanging regime of extreme careful handling. The most skilled work occurred in cutting and in prizing. The cutter did not just perform an intensive task – each plant had to be cut separately and handed carefully to someone who would gather several plants at a time – but he had to make decisions in the field as to which plants to cut and precisely where to make the incision (Tilley 1948: 58). Prizing was an activity that required considerable judgement and acquaintance with the materials. The object of prizing was literally to stuff a wooden barrel with as many layers of tobacco leaf as possible without rupturing the container. These hogsheads of tobacco, as they were called, were exceedingly heavy. In the seventeenth century the weight of a hogshead varied between 400 and 800 pounds, the latter figure commonly for sweet-scented tobacco; in the eighteenth century hogsheads containing sweet-scented tobacco weighed from 950 to 1,400 pounds, while those with oronoco tobacco, though typically smaller, nevertheless could weigh as much as 1,150 pounds (Middleton 1984: 113). What came out of the hogsheads on the other side of the Atlantic was the result not only of the curing stage but more importantly of prizing. Planters were under a strong incentive to pack the hogsheads to breaking point since freight rates were reckoned on the number not the weight of hogsheads (Breen 1985: 51).

Tobacco cultivation gave the Chesapeake region a particular work rhythm that bound the colonies in a singular pursuit. Small variations in this rhythm could be observed as slightly different soil and climatic conditions meant adjusting the work schedule. Moreover, by the end of the seventeenth century, tobacco planters had more or less settled on growing two varieties named respectively sweet-scented and oronoco, the former in the York basin and parts of the Rappahannock and the latter in most other parts of Virginia and Maryland (Walsh 1989: 396). These varieties had somewhat different work schedules as oronoco was never stemmed whereas sweet-scented normally was (Walsh 1989: 397).

Time was, however, not the only dimension of Chesapeake culture that was profoundly affected by the tobacco plant. The entire human and material geography of the region was shaped by the demands, or perhaps lack of demand, of tobacco. In the first instance, the large size of holdings relative to labour meant that settlement was dispersed. In Arundel County, Maryland, during the second half of the seventeenth century, 78 per cent of the farms had no more than one bound worker (Carr and Menard 1989: 410). Robert Cole, a 'small' Maryland tobacco planter of the mid-1650s, had an estate of 300 acres, of which 90 to 100 acres was planted in tobacco; yet he had no more than five bound workers (Carr and Menard 1989: 411; Menard et al. 1983: 186). Second, tobacco cultivation required little in the way of auxiliary activities. Fixed capital took the form of a barn and a ramshackle house built of wood and a few farm implements, mostly hoes and axes (Morgan 1975: 185; Carr and Menard 1989: 409). Beyond the farm tobacco needed little save hogsheads and these were often made on the farm. Since the settlements were on a river frontage; and the marketing, as shown in the previous chapter, was a metropolitan monopoly; and processing, to the extent that it was necessary, was done in England; and Europe was the eventual consumer, all the planter had to do was roll his hogsheads to his own wharf, or lay them into small craft. If neither of these methods of transport was available, then the hogsheads could be rolled down specially designed roads straight into the hold of ships waiting to return to England (Middleton 1984: 113). To put it succinctly, what need was there of any services, other than those provided by the farm? There was not enough time in the working schedule on the farm for manufacturing and, at any rate, better-quality goods could be imported directly from Europe. For all these reasons, Virginians and Marylanders did not live in close communities and hardly any towns sprang up in the seventeenth century (Rainbolt 1969). Even as late as 1770, with a total population accounting for as much as 30 per cent of British North America, Virginia and Maryland could count only two principal towns, Norfolk and Baltimore, each of which had only 6,000 inhabitants, compared to 30,000 in Philadelphia; but both had, in fact, relatively little to do with the tobacco trade (McCusker and Menard 1985: 131). It is perhaps without parallel elsewhere that both the Virginia and Maryland Assemblies, dismayed by the lack of urban growth, legislated towns into existence. Between 1655 and 1705 the former passed legislation six times, while the Maryland Assembly, probably in even greater desperation, passed ten town acts between 1668 and 1708 (Earle and Hoffman 1976: 14). All to no avail. By 1725 only seven small villages had materialized, containing between fifty and a hundred inhabitants, some merchants, more innkeepers and a doctor (Kulikoff 1986: 105).

Until the 1680s Chesapeake society was typified by what some have called the 'age of the small planter'. The foundations and the reasons for this are not hard to find, and they were the result of partly the economic structure of tobacco cultivation and partly the demographic regime and immigration pattern of the region. The Chesapeake tobacco economy began its life in the second decade of the seventeenth century when tobacco prices were extremely high. Between 1618 and 1624 tobacco prices were in the range of 30*d*. to 36*d*. per pound with prices as high as 60*d*. and 90*d*. being recorded (Menard 1976: 404). Profits from such high prices were substantial and attracted both investment and immigrants to the colony (Menard 1980: 114–42). After 1625, however, prices fell sharply as output grew rapidly, hitting the low level of 1*d*. per pound in 1630 and barely scraping 3*d*. or 4*d*. until the 1660s (Menard 1976: 405–8). Though the tobacco economy swung from boom to slump over this period, in the long run tobacco continued to be a profitable crop. Most of the explanation for this rests in the ability of tobacco planters to increase labour productivity. Although the data are scattered and not by any means conclusive, what is available shows a distinctly clear rise in productivity in all the Chesapeake tobacco regions from as early as the 1630s; more than a doubling of output per worker was recorded on Maryland's lower western shore, perhaps a tripling in the lower James basin – productivity gains on Maryland's upper eastern shore may have been even larger, though the information available is too sparse for confident conclusions (Walsh 1989: 395). Rising labour productivity increased the demand for servants, especially male workers who, with little previous experience or training, could be put to work in the tobacco fields. Recent estimates suggest that only 25 per cent of the servants who emigrated to the Chesapeake in this period were of yeoman stock (Kulikoff 1986: 32).

Those servants who survived the 'seasoning' and paid off their passage could look forward to becoming planters and, in time, buying their own servants. That this degree of economic and social mobility was a feature of early colonial life in the Chesapeake is beyond doubt. Ex-servants accumulated as much as £2 sterling each year they were free (Kulikoff 1986: 36). Ex-servants held political office and enjoyed other advantages, and though poverty and poor households existed there was far less distance in economic fortunes between the lowest and highest landowners than there was in other places and would be in the future (Kulikoff 1986: 36–7; Bernhard 1977).

The 'age of the small planter' is an accurate label for Chesapeake society in the first eight decades of its existence. It was short-lived, however. Beginning in the 1680s and then accelerating at a fantastic rate, Chesapeake society was transformed from a society based on servitude to one based on slavery; from relatively egalitarian to rigidly hierarchical; and from ethnically European to ethnically African. This change has, naturally, attracted a great deal of scholarship and, though the debate is by no means settled, it is certain that the market for coerced labour was undergoing a considerable transformation at the time (Eltis 1992). In particular it seems that the supply of English indentured servants was shrinking and their price rising relative to African slaves (McCusker and Menard 1985: 133–8, 238–45; Kulikoff 1986: 37–44; Menard 1977). Though tidewater planters seem to have preferred a European labour force, the economics of the marketplace did not accommodate their needs. It was not simply a change of labour supply, for what happened had a profound impact on the course of Chesapeake, and American history, and also on the relationship between planter and plant.

The total population of the Chesapeake colonies increased substantially during the eighteenth century from a level of 98,000 in 1700 to 786,000 in 1780 (McCusker and Menard 1985: 136). The black population, which at the end of the seventeenth century was small in comparison to the white population, rapidly overshadowed the growth of the latter. Whereas blacks accounted for only 13 per cent of the population in 1700, the figure reached nearly 40 per cent in 1780 (McCusker and Menard 1985: 136). Both a high rate of slave immigration and a high level of reproduction led to this diverging experience between whites and blacks (Kulikoff 1977).

Tobacco farms were transformed into tobacco plantations. Instead of servant labour, planters turned to slave labour, and wealth became vested in the number of slaves on the plantation. There was a steady change in the organization of tobacco cultivation in several ways, but in the end the main result was a generally increasing size of unit, on the one hand, and a growing complexity of management on the plantations, on the other. Plantation records are sparse but their message is quite clear. In the first place, over the eighteenth century it was increasingly unusual to come across tobacco farms where there was no slave labour: in Anne Arundel County, Maryland, for example, the proportion of farms without slave labour decreased from 62 per cent to 32 per cent between 1658 and 1777; at the same time those with more than three bound workers increased their share from 13 per cent to 40 per cent over the same period (Carr and Menard 1989: 410). In the middle of the eighteenth century several parts of the Chesapeake had a preponderance of large plantations. Around half of the plantations in Anne Arundel County, in Lancaster County and in the tidewater had more than twenty-one slaves, but on the big plantations the number of slaves could be huge; Robert Carter, one of the biggest plantation owners, owned 734 slaves in 1733 (Kulikoff 1986: 331–2). A second feature of the changing human geography of the Chesapeake in the eighteenth century was the expansion of 'quarters', plantations separate from the main or home plantation. These, even more than the home plantations, showed an increasing concentration of slave labour.

With the emergence of slave labour, tobacco cultivation itself became socially complex. Unlike the seventeenth century, when planters cultivated tobacco under fairly similar labour regimes, and with roughly similar acreage, the eighteenth century witnessed cultivation under distinct social relations.

At the one extreme were the great planters, men such as Robert and Landon Carter, William Byrd, even Thomas Jefferson, whose stately mansions began to appear along the banks of the Chesapeake tidewater (Breen 1985: 35). These families, and others like them, made up the Chesapeake gentry, a social class which was in embryonic form in the previous century (Kulikoff 1986: 261–313). Slavery cemented social relations while tobacco provided the gentry with material and symbolic existence. These men, as a recent study has brilliantly portrayed, were obsessed by tobacco (Breen 1985). The reasons are not difficult to find. First, tobacco continued to be their primary staple, though on their large plantations some amount of land was devoted to food crops, a proportion of which was marketed (McCusker and Menard 1985: 128–31; Clemens 1975; Walsh 1989). Second, their personal material world, their homes, clothes and food, all of which were culturally European, were accumulated through the consignment system operated by English merchants (Breen 1986; Breen 1988; Shammas 1990). In rough outline, the planter sold his output to an English merchant who would sell the tobacco back in England and return to the Chesapeake with manufactured goods and European foodstuffs. As Thomas Breen puts it: 'this marketing device became a badge of class, a means of distinguishing great planters from those of lesser status' (Breen 1985: 36). Finally, tobacco was pivotal to a culture of debt that was endemic among the great planters and which, during the second half of the eighteenth century, would begin to undermine the structure of the Chesapeake economy (Breen 1985: 160–203). On a symbolic plane, as plantations grew in size, as the bound labour force expanded and as the demand for supervision increased, planters became the repository of the almost mystical understanding of the ways of tobacco. In contrast to the way in which tobacco lore was passed to servants in the seventeenth century, this lore passed from planter father to planter son in the eighteenth century (Breen 1984). If slave-owning was a gauge of wealth, then tobacco was a measure of esteem. Individual planters prided themselves on the quality of their output, which typically carried their owner's name to market (Breen 1985: 64–7). Failure to harvest a fine crop in adverse climatic conditions was viewed not as bad luck but as bad management. A planter who achieved the finest results from his fields was given the highest praise as a crop master (Breen 1985: 61–3). In the same way as tobacco was the medium of exchange in seventeenth-century Chesapeake economy, so the tobacco leaf was representative of Chesapeake planter culture (Morgan 1975: 177). Of course, though they were present on their main plantations, the Chesapeake gentry were gentle-men and not labourers (Kulikoff 1986: 280–300). Field work was the responsibility of the overseer and the slave labourers (Carr and Walsh 1988; Morgan 1988).

All of this was in stark contrast to tobacco cultivation in areas distant from the tidewater. Up the many rivers flowing into Chesapeake Bay lay the Virginia frontier. White farmers, many of them ex-servants finding their access to tobacco cultivation in the tidewater blocked by rising labour costs, began to seek opportunities in this area. Thousands emigrated to the piedmont between 1740 and the eve of the Revolution. At first, the region was populated by poor families, cultivating tobacco on smallholdings with-out slave labour (Kulikoff 1986: 141–53). In contrast to their counterparts in the tidewater region, small planters in the piedmont did not sell their tobacco on consignment to English merchants but rather directly to Scot-tish merchants who, as detailed in the previous chapter, established country stores in the tobacco region (Kulikoff 1986: 122–4, 226–7; Price 1954; Soltow 1959; Farmer 1988). By involving themselves with Scottish and not English merchants the small planters of the piedmont retained an independence from the gentry class to the east and north while at the same time being firmly locked into the transatlantic economy (Price 1964). But the Virginia frontier was not static and, as the century progressed, many large planters from the tidewater bought land, and slave-holding on a large scale became just as common there as in the tidewater. In Amelia County in Virginia's southside the percentage of households owning slaves increased from 23 per cent to 76 per cent while the median number of slaves rose from two to six between 1736 and 1782 (Kulikoff 1986: 154). This drift to tobacco slave societies was typical of other counties in the southside and in central parts of the colony. But the frontier continued to move westward and the small tobacco planter after the 1780s was exploring opportunities in Georgia and Kentucky (Kulikoff 1986: 161).

Chesapeake society was the first European society to develop on the basis of tobacco. In many ways its history exemplifies the diverse cultural arrangements that accompany tobacco cultivation. Though plantation society swept through the region from the late seventeenth century, the ideal of a yeoman tobacco planter continued to be pursued, albeit on the fringes. It was these planters, together with the freed slaves after the Civil War, who carried the cultural history of tobacco forward into the nineteenth century; but before looking at that we need to retrace our steps to other tobacco-growing regions of the New World and Europe.

As Chapter 3 showed, tobacco cultivation was attempted by Europeans in many parts of the New World. Many of these areas did not pursue tobacco cultivation for very long, yet their short-lived history is nevertheless significant in terms of the theme of planters and tobacco societies.

With the exception of Jamaica, the Caribbean islands colonized by France and England were, as we have seen, based on tobacco cultivation (Dunn 1973: 149–87; Zahedieh 1986). The transatlantic servant trade that featured so prominently in the early history of the Chesapeake was just as evident in the Caribbean and, indeed, for several decades the islands rather than the mainland were the primary destination for the indentured (Gemery 1980). Over the seventeenth century about 60 per cent of an estimated British emigration level of 378,000 was bound for the Caribbean, compared with around 30 per cent for the Chesapeake; a little under half of the Caribbean immigrants arrived there in the two decades after 1630 (Gemery 1980: 215). Death rates were so appalling that, despite heavy immigration, population growth was miserably low: between 1630 and 1660, for instance, an immigration of some 144,000 English people managed to increase the total population by no more than 43,000 over the same period (Gemery 1980: 197).

All of the British-controlled islands in the Caribbean followed a roughly similar pattern in white population changes. Growth was particularly rapid until the 1640s, followed by a much slower rate of change reaching a maximum level around 1660, and then declining well into the eighteenth century (Gemery 1980: 212). Within the region there were some marked differences in demographic experience. Nevis, Montserrat and especially St Kitts, which experienced a drop in its white population from a maximum of 12,000 to 1,000 in 35 years, sustained a considerable decrease in their white population; Barbados, with the largest white population among the islands, experienced a smaller and slower decline, though by 1712, there were half as many whites on the island as was the case seventy years before; Bermuda and Antigua (also Jamaica) were the only islands to keep their white population stable over the seventeenth century (Gemery 1980: 219–25).

What is significant about the white population pattern is that, in general terms, the initial and often quite dramatic rise in population, and the subsequent equally dramatic decline, were accompanied by, or even caused by, the growth of and subsequent collapse of tobacco cultivation. Barbados is a case in point and one for which there is relatively good information. Settlement and tobacco cultivation went hand in hand, a situation which was quite different from what happened in the Chesapeake colonies. Grants of land to settlers were fairly large, on average 100 acres between 1628 and 1629 (Innes 1970: 9). Land usage was organized along plantation lines and in the very first years of settlement five such plantations came into being supported by a labour force of 150 people: within a few years, these five were reconstructed as thirteen plantations while the number of settlers had risen to 1,850 (Innes 1970: 4). The point about these figures is that under Barbadian political and economic conditions tobacco cultivation followed a path wholly distinct from that in the Chesapeake. Unlike the planter with a servant or two, typical of Virginia at the time, the Barbadian tobacco planter had a substantial army of labour under his control, almost all of it bound. Unfortunately there is no precise measure of the size of the labour input on a Barbadian tobacco plantation. One indirect piece of evidence, namely the labour structure on 15 plantations – probably cotton, indigo and some tobacco – between 1639 and 1643, i.e. before the sugar period, shows that all had at least 2 servants and several had more than 20 (Beckles 1985: 34). That was in general terms, but the actual method of cultivating tobacco was for smallholders to lease from 5 to 30 acres from the estate (Innes 1970: 16). As in the case of the Chesapeake, the motive for accepting an indenture was surely a hope of becoming a tobacco planter. On Barbados it was clear from the way the land was parcelled out that the chances of a freed servant becoming an independent small-holder were small, and became increasingly unlikely within a decade or two of initial settlement (Innes 1970: 10–11). Rather than hope to join the planter class, freed servants were faced with a less satisfactory set of alternatives: 'Their choice was to try one of the less congested Leeward Islands, or return home, or stay as wage laborers in Barbados' (Dunn 1973: 53).

Whatever they decided to do, one thing is clear: until the 1640s, tobacco was the most profitable staple in the English Caribbean. Output increased substantially and population levels soared. Between 1628 and 1638 the total amount of tobacco cultivated on Barbados and St Kitts rose from 100,000 pounds to 675,000 pounds: total population on the two islands increased from 2,400 to 25,000 (Beckles 1985: 25; Gemery 1980: 219, 223). Even though the price of tobacco fell by 50 per cent between these dates, rising output offset the effects of this price fall, in income if not in rate of return (Menard 1980: 157). One historian has estimated that yearly incomes for tobacco planters on Barbados, given the prevailing levels of productivity, averaged from £37 to £56 from the sale of leaf that each planter raised (Batie 1976: 8). Income, not rate of return, was the important goal since investment in tobacco cultivation was insignificant. At such high levels of income expectation it is small wonder that thousands flocked to the islands and, just as in the Chesapeake, were captivated by the tobacco plant. Descriptions of the material culture of Barbados in the early years of settlement bear a striking resemblance to those of Virginia. Here is how Sir Henry Colt summed up Barbadian agriculture in 1631: 'your grownd & plantations . . . lye like ye ruines of some village lately burned, . . . all ye earth couered black wth cenders nothinge is cleer. What digged or weeded for beautye?' (Harlow 1925: 65–6). What was an affront to Colt's eyes was nothing more than the attempt by the early planters to adapt Indian agricultural practices to the particular characteristics of the tobacco plant (Dunn 1973: 6).

Barbados, St Kitts and many of the lesser islands were planted in tobacco. Despite the growing populations and output there is a sense of desperation on the one hand, and addictiveness on the other hand, about tobacco cultivation that percolates through the surviving literary evidence. It is hard to make complete sense of it, but there is little doubt that tobacco growing on the English islands was becoming increasingly difficult; but giving it up was not easy. The English attempt to colonize the island of Providence off the coast of Nicaragua provides some evidence of this double-edged experience. The island was settled in 1630 by a select group of Puritan grandees headed by the Earl of Warwick and Sir Nathaniel Rich, both of whom were responsible for the settlement of Bermuda (Kupperman 1988). Philip Bell, then Governor of Bermuda, was attracted to the opportunities of planting tobacco in tropical climes, reacting partly at least to the success of tobacco plantations on both Barbados and St Kitts (Batie 1976). Writing to his financial backers, Bell was clear on the wealth that would accrue: 'in short time [it would] be made more rich and bountiful either by tobacco or any other commodities than double or treble any man's estate in all England' (Newton 1914: 33). Tobacco is what they settled on but the going was never good. Partly this was because the enterprise itself ran into trouble and partly because in the 1630s the glut on the market was eroding the commercial viability of the crop, especially in the Caribbean (Kupperman 1988). William Jessop, the Secretary to the Providence Island Company, put it succinctly in a letter written in 1635 to a settler: 'Tobacco', he wrote, 'sells now in London for 15d. and the market beyond sea is so little above it that men have no great stomach to ship it out' (BM Add. Mss 63,854B, f. 127). In 1634, to put the matter into perspective, the Chesapeake colonies shipped around 400,000 pounds to London, the Barbados and St Kitts colonies perhaps another 200,000, and some more (exact amount unknown) was shipped from Bermuda and Spain (Pagan 1979: 254; Beckles 1985: 25; Williams 1957). The little that the Providence settlers could export would have paled into insignificance when placed alongside the main producers. The Providence Island directors were fully aware of the market conditions for tobacco and tried to convince settlers to cultivate other commodities. The alternatives ranged from silk grass, cotton and sugar cane to pomegranates and figs; juniper berries were the flavour of 1634 (Batie 1976: 10; BM Add. Mss 63,854B, f. 10). But tobacco had a hold on the settlement, and though prices collapsed after 1634 tobacco was still being planted as late as 1637 (Kupperman 1988: 81;

BM Add. Mss 63,854B, f. 236). Whether this was for export, for their own consumption or both is unknown.

It was not just falling prices, and increasing competition from the Chesapeake, that produced problems for Caribbean producers. The quality of their product, to judge from contemporary accounts, was inferior. Condemnation of Barbadian tobacco in terms of quality was voiced as early as 1628 when John Winthrop, upon receiving his son's consignment of tobacco in London, wrote back to him that it was 'very ill conditioned, foul, full of stalks and evil coloured, and your uncle Fones taking the judgement of divers grocers, none of them would give five shillings for it' (Innes 1970: 15). It is just possible that the quality problem reflected nothing more than the fact that this tobacco was among the first that Barbados exported, but later comments suggest otherwise. Archibald Hay, the principal proprietary trustee for Barbados, was quite certain that the difficulty in disposing of Barbadian tobacco lay in its poor quality. In a letter of 1638 to Peter Hay, the Receiver General for the proprietary estate in Barbados, Archibald Hay complained: 'Your Barbados tubaco cannot expect to come to a good Marcat any where it hath the reputation to be so bad w[hi]ch the planters must help by making lesse and better' (Bennett 1965: 15). Captain Daniel Fletcher, a local planter, concurred. Tobacco, he argued, was 'A Commodity of Noe Better Estimation, not worth Any thinge, for it is the worst of all tobaccoes and I am p[er]swaded Never will be worth one farthinge token' (Bennett 1965: 15). Archibald Hay entreated his agent to encourage planters to try cotton, and though they would do so in the next years to come, it is instructive to hear Peter Hay on the reluctance of local planters to abandon tobacco. In his reply to Archibald Hay he wrote as follows: 'You desire us all to plant cotton, w[hi]ch is a thing the planters can hardly doe, because they are indebted, that if they leave planting of tobacco they shall never be able to pay thare debts . . .' (Bennett 1965: 16). But the poor quality continued to plague Barbados, and probably other islands. Richard Ligon, an especially eloquent observer of life on Barbados, dismissed its tobacco in summary terms by calling it 'the worst I think that grows in the world' (Batie 1976: 12).

Whatever the truth about the inferior quality of Barbadian tobacco, the planters were handicapped by their lack of resources. Specifically, they could not compete with the Chesapeake, for the simple reason that they had limited land and given that there was no alternative to a fallow rotation system, there was a physical limit to how much could be produced. Also there is a strong possibility that Barbadian planters were not achieving increases in productivity as evidenced in the Chesapeake. All of these factors undoubtedly contributed to the demise of tobacco cultivation on Barbados, which in the 1640s shifted to other staples and especially sugar cane production (Green 1988). And with that shift there was a corresponding shift away from the use of indentured English servants to African slave labour (Beckles 1981: 237, 243; Beckles and Downes 1987: 228–9; Dunn 1973: 68). The size of the slave population soared, and that of the white population, both free and unfree, began to shrink: in 1660 half the population of the island was white, the other half black; by 1712 African slaves outnumbered whites by more than three to one (Dunn 1973: 87). Tobacco planters on the Leeward Islands, especially on St Kitts, Antigua and Montserrat, continued to grow tobacco, though by the 1670s the pattern established earlier on Barbados and on Nevis began to be followed. Planters consolidated their holdings, shifted to sugar and to a black labour force: white servants and smallholders started looking elsewhere for a livelihood (Dunn 1973: 117, 122, 131, 141).

Bermuda did not share the experience of the other English islands. In the first place, tobacco cultivation began on the islands as early as 1613, according to one piece of evidence, but certainly by 1615 (Ives 1984: 4–5). Being thus contemporaneous with Virginia, there was less of a feeling on Bermuda that planters there were in direct competition with an established producer, as was clearly the case on Barbados and St Kitts. In the early years, judging from the correspondence surviving in the papers of the Rich family, there was considerable confidence among the islands' settlers about the ability of tobacco to underpin the local economy. There were no problems, as far as one can tell, about the quality of the output such as faced by planters in Barbados, but other problems associated with tobacco cultivation did emerge in Bermuda as elsewhere. There was, for example, the perennial problem that planters did not put their efforts into building settlements in the material sense of the word. Governor Tucker pointed this out to Nathaniel Rich in a letter of 1617 where he advised settlers on 'not spendinge most of their tyme in Making Tobacco but to manuer, plant and sett the Islands, for thereby the Plantacion would be made to flourishe and yield more pleasure to the Inhabbitants' (Ives 1984: 97). There was also that other perennial problem, namely that, as a petition to the Commissioners of Bermuda so succinctly put it, 'the planting of Tobacco doth suck out the hart of the ground' (Ives 1984: 267). Nevertheless, neither of these problems seems to have deterred the planting of tobacco in Bermuda. For one thing the islands were flourishing, as is apparent from population figures. In 1628, after 16 years of settlement, the island supported a white population of 2,000, a level slightly larger than that of Virginia at the same date (Gemery 1980: 225; Menard 1980: 157). By 1656 it had reached 3,000 inhabitants and by 1679 its maximum for the century at 4,000 (Gemery 1980: 225). During all this time the main staple was tobacco. Output grew slowly, reaching perhaps 500,000 pounds in the 1680s, a far cry admittedly from the 20 million pounds from the Chesapeake, but respectable nevertheless (Gray and Wyckoff 1940: 22). What all this means is that Bermudans, though living in an economy where both land and labour were scarce, were remarkably successful in dealing with problems of soil fertility (Bernhard 1988). How they coped is not certain, but there is some evidence that Bermudan planters fertilized their ground with marl, and used crop rotation; they also used their lands for growing food and were, therefore, not dependent on external supplies (Ives 1984: 267). Though other cash crops were frequently put forth as alternatives to tobacco, none of these, in the seventeenth century at least, made any real headway. Sugar cane was one such crop that was heavily promoted by the absentee landowners but found little favour with the tenant farmers. As was common in the Chesapeake and in the Caribbean in the early days of settlement, tobacco was also used in Bermuda as the medium of exchange (Ives 1984: 49, 229, 382).

In many respects what happened on the English Caribbean islands also happened on the French Antilles. These French possessions, specifically Martinique, Guadeloupe and St Christophe, were settled with tobacco as the staple crop, and though the date of the change from tobacco to sugar varied, it produced changes in the organization of labour, land structure and population profile very similar to those the English experienced. Guadeloupe was the first of the French islands to abandon tobacco cultivation; in 1671, according to the official land use survey, tobacco, in terms of the area under cultivation, was the least important cash crop, behind sugar and ginger – indeed, the area under tobacco was no more than 3 per cent of the area under sugar (Schnakenbourg 1980: 49). By that date the white population of the island stood at 3,112 while the black population stood at 4,627. The critical years for the change in the nature of the economy and the social structure of the island were from 1656 to 1671 during which time the white population shrank from a level of 12,000, and the black population increased by more than 50 per cent from a figure of 3,000; the number of sugar mills rose from 10 to 107, and tobacco cultivation fell precipitously (Schnakenbourg 1968: 300, 302). The white population began to increase again towards the end of the seventeenth century but its growth rate was substantially inferior to that of the black population (Davies 1974: 82). Martinique had a similar experience except that the economic and social transitions happened later, and at a slower pace, than in Guadeloupe. In 1671, for example, even though sugar was clearly the most important cash crop in the settlement, tobacco cultivation still occupied 15 per cent of the arable land and as many as 65 per cent of the *habitations* grew tobacco (Kimber 1988: 128; Petitjean Roget 1980: 1396). The white population of the island fell substantially between 1652 and 1685, though the extent of the fall is disputed because of uncertainty surrounding the accuracy of the 1652 figure (Kimber 1988: 115). Nevertheless, by 1660, blacks outnumbered whites (Kimber 1988: 115). As in Guadeloupe, the white population did grow again in the eighteenth century but not at the rate of the black population (Davies 1974: 83).

It is clearly the case that the abandonment of tobacco on the French islands was accompanied by the kinds of profound changes that had already occurred on the English islands. The exception was that the initial collapse in white population in the French islands was not followed, as it was on the English islands, by a long period in which the white population remained more or less stable (Gemery 1984: 322; Wells 1975: 260–8). The collapse in the white population was, in both the French and English islands, caused primarily by the abandonment of tobacco, by the shift towards cheaper African slave labour, and by consolidating landholdings. The pattern of indentured servitude was also similar on both sets of islands, though the timing of the decline of the servant trade came naturally at a later date on the French islands than it did on the English islands (Mauro 1986: 99–101). If the experiences of tobacco planters on Martinique were in any way general then it seems that the market for indentured servants dried up primarily because of a lack of demand on the part of the planter, but also because of the decreasing ability of the indentured servant to pay back the cost of the indenture in a reasonable time. On Martinique tobacco prices plummeted from 150 livres tournois per pound in 1636 to between 10 and 15 livres tournois in the early 1660s. The effect of such a depression in tobacco prices was, on the one hand, to double the cost of living calculated in tobacco and, on the other hand, to make it increasingly difficult for an indentured servant to pay back the cost of his passage within the period of his indenture (Petitjean Roget 1980: 1149–51). The diverging experiences after 1700 were primarily the result of the introduction of other, non-plantation cash crops, such as indigo, ginger, coffee and cocoa into the French Caribbean economies, as well as the continuing emphasis on food crops (Kimber 1988: 128; Schnakenbourg 1980: 49; Davies 1974: 190–2; May 1930). Tobacco cultivation on St Domingue also suffered the fate of other producers in the Caribbean, though the change happened much later. In 1674 St Domingue was the largest producer of tobacco in the Caribbean area with an output of between 2.5 and 3 million pounds. But this level was not sustained, and output began to decline, reaching 50 per cent of its 1674 level in 1714 and more or less disappearing thereafter (Price 1973: 83–115). St Domingue was becoming a sugar island, with the typical pattern of the concentration of land into the hands of a few large planters, and an enormous build-up of slave labour. On the eve of the Revolution St Domingue had a white population of 40,000 set against a slave population exceeding 450,000 (Stein 1988: 42–3). Free whites who were not absorbed within the sugar economy withdrew to the hinterland where, as small planters, they pursued coffee culture (Dupuy 1985: 92; Trouillot 1981; Trouillot 1982).

Planters who abandoned tobacco on the French islands must have felt that an important stage of the settlement of these islands was over. There is little doubt that on both Martinique and Guadeloupe, as on Barbados, tobacco cultivation provided the economic foundations upon which the vast sugar empires were built. But the passing of the tobacco culture did not go unnoticed. Father Jean-Baptiste Labat, the French missionary who travelled extensively throughout the Caribbean region towards the end of the seventeenth century, was clearly dismayed by what had happened to the society of Martinique and Guadeloupe. In his account of his voyages he reminded his readers in forceful language that French island society was made possible by tobacco cultivation. He maintained that the considerable progress of sugar cane came with a high price. In the first instance, it led directly to the depopulation of French people and the consequent erosion of a French cultural presence in the region. A typical sugar plantation supported no more than four or five Frenchmen whereas the same land area could support fifty or sixty tobacco planters. African slaves, he admitted, were fine for work, but they would not defend the rights of France against its enemies, nor would they necessarily go about their activities in a peaceful manner as French *habitants* did. The presence of so many slaves would, he predicted, lead inexorably to protest and revolts, and the fact that by the time he arrived in the area there had been no such unrest was nothing short of miraculous. Second, tobacco cultivation on the islands had repercussions on the metropolitan economy that sugar did not. Specifically, Labat was thinking here that French islanders would demand all of the fine manufactured goods and provisions that France could offer whereas African slaves, as he put it, 'need no more than 4 or 5 yards of linen and a bit of salted beef' (Labat 1742: 334). The white depopulation was deplorable, Labat argued. The only way to get French people to emigrate was to renew tobacco culture. Fresh lands were available on some of the least populated islands such as Marie-Galante, St Martin, St Barthélemy and Grenada, but also in still uncultivated parts of Martinique and Guadeloupe. As a pioneer crop tobacco had no competition. Sugar was capital-intensive and, according to Labat, indigo and cocoa were not much better since returns on the latter, for example, would not materialize for five or six years (Labat 1742: 328–37).

As we have seen, after Labat wrote, the white population in the islands did increase without a return to tobacco culture. Despite his incorrect predictions, Labat's plea for tobacco cultivation must be understood in its proper context as a powerful reminder of the tobacco mentality. It is interesting to note that most of Labat's section on tobacco is dedicated not to a polemic about tobacco culture but rather to a description of how tobacco is cultivated. Judging from the close similarity between Labat's description and that of later writers, both in the United States and in Brazil, there is little doubt that tobacco culture for French island planters was as exacting of time and care as in the Chesapeake (Labat 1742: 300–18).

From the point of view of the social and cultural history of tobacco, it is Labat's insistence on the Europeanness of the plant that is so striking.

There is, of course, a danger in becoming deterministic about this point: tobacco equals white labour, sugar equals black labour. The relationship among staples, labour systems and ethnic cultures is complex and is itself historically specific. What can be argued, however, is that, for all the reasons that have already been elaborated, tobacco was an eminently suitable pioneer or yeoman crop; and, as at least since the sixteenth century pioneers and yeomen have been overwhelmingly European, it is not surprising that links between culture and plants have been made. The following chapter will pursue this theme further, particularly in its elaboration in the nineteenth and twentieth centuries, in the wake of the abolition of slavery and the enormous expansion of tobacco cultivation across the globe. For now, however, we need to conclude this cultural history of tobacco by looking at Brazil.

The Chesapeake had William Tatham; the French Caribbean had Jean-Baptiste Labat; and Brazil had André João Antonil, who wrote an important description of tobacco culture in Brazil in the early part of the eighteenth century (Antonil 1965). Brazilian tobacco cultivation was concentrated in the heartland of Bahia in the north-east of the country. The production cycle in Brazil was considerably shorter than in the Chesapeake, requiring only six months from seed to harvest as opposed to nine. Seeds were sown in specially prepared seedbeds by early May, transplanted several weeks later into the fields that were prepared beginning in February, weeded, topped and suckered in July and August and harvested beginning in September (Antonil 1965: 295–9; Lugar 1977: 32). Curing normally lasted three days and then the tobacco was ready to be twisted into long cords about three inches in diameter. Over the next months these cords hung in sheds and were frequently retwisted, the moisture collected, mixed with a cocktail of anis, basil, pork fat and molasses, and the cords brushed with this mixture: this was done as much to preserve the tobacco as to give it a particular taste and aroma (Antonil 1965: 303–5). The final stage of preparation involved the re-rolling of the twists on to sticks and then three of these were packed into a roll weighing about 480 pounds, wrapped in leaves and bound in leather ready for shipment (Antonil 1965: 307–9; Lugar 1977: 32–3).

According to Antonil's description, the most intense use of labour came in the preparation of the twists and the rolls, and it was brute force, more than skill, that was needed, though someone with an intimate knowledge of the process was in charge (Antonil 1965: 313–15). The other stages of cultivation seem to have absorbed labour time in much the same way as in the Chesapeake, and also in the French Caribbean, judging from Labat's description. Cutting the tobacco plant was accorded the same level of skill in Brazil as in the Chesapeake (Antonil 1965: 311). However, unlike the Chesapeake, the production cycle did not end with the packaging of the rolls. In the Bahian Recôncavo, as apparently was also the case in the

187

Caribbean, the plant was not uprooted but rather cut at a height of one or two inches above ground level. A second growth normally occurred and, if the soil qualities were good enough, the same procedure was followed to allow for a third growth (Antonil 1965: 313). As already discussed in Chapter 6, the tobaccos from these three successive growths were destined for different markets: the first to Europe, and the second and third typically to the African coast.

Whereas Chesapeake husbandry entailed a long-fallow system, in the Bahian Recôncavo fallowing was normally combined with a routine application of animal manure. Though tobacco was the cash crop in this region, tobacco culture itself was embedded within a system of mixed farming, especially cattle-raising. Thus in Brazil the huge land areas so typical of the Chesapeake were not necessary. Soil fertility was not such a great problem in Brazilian tobacco culture and the same plot of land could be used for a much longer period of time than in the Chesapeake. Thus a rather different pattern of land utilization emerged in Brazil. Farms tended to be smaller, while the number of workers per acre tended to be greater. For example, while one worker in the Chesapeake tended 9,000 plants on a 3-acre site, two workers in the Bahian Recôncavo tended the same number of plants but on half the number of acres and with more than one growth and enough land left over for their own food requirements (Carr and Menard 1989: 413–14; Lugar 1977: 34).

In the Bahian Recôncavo as in the Chesapeake, there were no economies of scale in tobacco cultivation; expansion of output resulted from a proportionate increase in inputs (Lugar 1977: 35). Therefore small as well as large farmers were attracted to tobacco culture, in theory at least. Whatever the size of the holding, in the seventeenth and eighteenth centuries most tobacco growers employed slave labour to a greater or lesser extent. This was, of course, significantly different from the situation in the Chesapeake, part of the reason being that Brazil was not settled with indentured servants and partly because the Bahian region supported an enormous sugar industry whose labour force from the latter part of the sixteenth century was overwhelmingly African slaves (Schwartz 1978; Greenfield 1979). The number of slaves that any grower used was determined by several factors, including size of holding, relative land utilization (especially tobacco and cattle-raising), and, most importantly, whether the tobacco was processed into rolls on site – this part of tobacco cultivation often required from three to five slaves (Flory 1978: 179; Antonil 1965: 313). A small tobacco grower producing from 3,000 to 5,000 pounds of tobacco, tending cattle and raising his own food crops required the assistance of two or three slaves, if he did not do his own processing (Flory 1978: 180). Processing tobacco elsewhere seems to have been typical of small-growers in the Bahian Recôncavo (Flory 1978: 187). A moderate-sized farm with processing facilities had around twenty-five slaves (Flory 1978: 180). The Bahian tobacco society, in fact, comprised three groups. First came the large landowners, whose origins in the region derived from the grants of land given to them in the early years of settlement. The second, and by far most numerous – possibly typical – group consisted of the tenants of the first group plus small landowners. Finally, there was a group of landless subsistence farmers (Flory 1978: 193–4). The little evidence that does exist suggests that in the course of the eighteenth century the small landowners became more numerous and more important in the tobacco growing area of Brazil (Flory 1978: 192). This should not obscure the fact that, though small in terms of landholding, these tobacco growers, while not of the stature of sugar growers, were, however, far from poor, as evidenced by the fact that they were significant slave-owners (Flory 1978: 205).

By the end of the eighteenth century tobacco grown in colonial America on plantations in the Chesapeake, and mixed farms in the Bahian Recôncavo, both using slave labour, accounted for the bulk of world output. The ideal of the peasant pursuing tobacco culture with the help of family labour on small plots had been abandoned in the Americas from an early date. As late as the 1770s Spanish officials interested in stimulating tobacco cultivation in Louisiana realized that the success of any attempt to expand and extend tobacco production depended ultimately on access to slave labour (Coutts 1986; Clark 1970: 190–1). The ideal, though, did exist in Europe, especially in the Dutch provinces, and in Germany, where numerous families derived their livelihood from tobacco culture (Roessingh 1978). As for the Americas, it was slavery as an institution and the slave trade as a capitalist enterprise that shaped the nature of tobacco societies until the middle of the nineteenth century when profound social transformations changed the social construction of tobacco cultivation again. The age of the 'poor man's crop' was just about to begin.

Part IV

This vice brings in one hundred million francs in taxes every year. I will certainly forbid it at once – as soon as you can name a virtue that brings in as much revenue.

Napoleon III (1808–73)

... preaching the cult of the cigarette and distributing millions gratis so as to introduce a taste for tobacco in this particular form into regions where it was as yet unknown ... The streets of Foochow are brilliant with [BAT's] ingenious pictorial posters, which are so designed to readily catch the eye by their gorgeous coloring and attractive lettering, both in English and Chinese

British Consul, Foochow 1909, quoted in Cochran (1986: 162)

In tobacco the big gambler is the farmer himself. He has no guarantee of anything. He takes a chance on raising his crop and then when he gets it raised he doesn't know what it will bring. It's a ninety day crop growing but, as the farmers say, with the curing it takes thirteen months a year.

Sam Hobgood 1938, quoted in Daniel (1985: 184)

I love tobacco ... I love to fool with it and get the gum on my hands and clothes. I love to smell it and I love to chew it and smoke it. Annie dips. She tried to get modern and smoke, but she got sick on it and went back to her snuffbox ... nothing gives me pleasure like tobacco. I have never raised anything that I like to raise half so well ... Tobacco farmers are like gold miners always hoping to strike it rich.

Lee Johnson, tobacco farmer, quoted in Daniel (1985: 194)

8

A POOR MAN'S CROP?
The globalization of tobacco culture since 1800

By the end of the seventeenth century tobacco cultivation had become dependent upon slave labour on large holdings yet the absence of economies of scale allowed small-scale production to co-exist. Moreover, it was the absence of economies of scale that permitted the spread of cultivation as it tended to be small or even marginal growers who were in the forefront of expansion. For most of the nineteenth and twentieth centuries the social and economic history of tobacco cultivation has been characterized by a distinct dualism, between the small scale of growing operations and the giant scale of manufacturing and marketing. Only in the last few decades in the West, as mechanization has finally begun to make considerable inroads into traditional procedures of cultivation and harvesting, has the age of the small planter come under threat of extinction. In other parts of the world, in Africa and Asia especially, this dualism still exists. This chapter has two objectives: to examine the changing social relations of tobacco production; and to explore the global spread of tobacco culture from the nineteenth century until the present.

Tobacco cultivation moved out of tidewater Virginia in the eighteenth century and spread further west over the Appalachians to North Carolina, Kentucky, Tennessee and Georgia, as well as to scattered localities further west and north. By the turn of the nineteenth century, if not earlier, tobacco production had become common in areas previously unfamiliar with tobacco culture. The rapid spread of tobacco cultivation was clearly related to a general westward movement of population, to be sure, but the fact that pioneers took tobacco culture with them attests once again to the dynamic impact of the plant.

Moreover, pioneers tended not to migrate with extra labour, in particular enslaved labour, and therefore most of the early tobacco enterprises on the frontier were family-run. Yet within a relatively short time the racial profile of the society began to change as enterprising farmers turned to slave labour in an attempt to step up production. In the county of Pittsylvania, near the centre of the Old Bright tobacco belt (an area straddling the Virginia–North Carolina border), for example, slave labour was being considerably exploited by 1790, even though the area had been settled only some two or three decades earlier (Siegel 1987: 75, 176). As a proportion of the population of Pittsylvania, slaves accounted for a steadily increasing share, reaching 46 per cent in 1860, compared with 27 per cent in 1790; put another way, at the outbreak of the Civil War nearly two-thirds of landowners were also slave-holders (Siegel 1987: 176). By contrast, Lunenburg County, to the east of Pittsylvania and therefore settled at an earlier date, began its transformation sooner. The 1790 slave population share for Pittsylvania was reached in Lunenburg in 1750, and the 1860 figure on the eve of the Revolution (Kulikoff 1986: 154).

Though in time tobacco growers in the piedmont shared some features, especially the use of slave labour, with their counterparts in the Chesapeake, in one important respect they were quite different. The piedmont never developed a class of great planters as in the tidewater, and large plantations were not common. Some insight into the nature of the tobacco society to the west of the tidewater can be gained by looking more closely at Pittsylvania.

On the eve of the Civil War almost 50 per cent of the farms in Pittsylvania were under 200 acres in area, with a median size of 215 acres (Siegel 1987: 77, 79). These figures were not very different from those of Augusta, an adjoining county which, unlike Pittsylvania, derived its economic benefits from mixed farming, especially wheat and dairy farming. The main point is that despite the fact that Pittsylvania had a substantial and growing slave population co-opted to the cultivation of a single cash crop, there was little difference in the land structure in the two seemingly contrasting counties. The contrast in this part of Virginia was not between a yeoman county such as Augusta and a tobacco county such as Pittsylvania, but rather between eastern and western counties, in part between those settled early and those settled late. Mecklenburg and Fairfax counties, to the east of Pittsylvania and therefore closer to the tidewater, had larger slave populations than Pittsylvania (Siegel 1987: 88–9). The point is an important one because it underlines the fact that tobacco cultivation remained locked into a specific economic organization which, while embracing slavery, did not lend itself generally to the creation of a planter aristocracy. That this did occur in tidewater Virginia before the Revolution is due more to other forces, social and political, than to the attributes of tobacco culture. The contrast with cotton could not be more stark. Taking five main cotton states – Alabama, Mississippi, Georgia, South Carolina and Louisiana – in 1860, we find that the median number of slaves per holding averaged thirty-seven while in a sample of tobacco regions the corresponding figure was around twenty (Gray 1958: 130–1). Seen in another way, in Kentucky in 1860 only 2.3 per cent of all slaves in the state were employed on units of production that exceeded fifty slaves; for

Virginia the corresponding figure was 15 per cent, but in Louisiana the proportion stood at 50 per cent (Gray 1958: 530).

There is, though, a further feature, other than late settlement, that distinguished Pittsylvania's slaveholding pattern from that of the eastern Virginia counties. One clue is that in Pittsylvania between 1820 and 1860 the average number (mean and median values) of acres owned declined significantly, while total tobacco output increased (Siegel 1987: 79, 92). This change coincided with an increase in tenancy and, most importantly, a concerted shift towards the cultivation of Bright, or, as it came to be known, flue-cured tobacco. The details of Bright tobacco culture will be described later in this chapter, but for now it is important to know only that Bright tobacco was grown on poor soil and cured in enclosed barns, giving the leaf a distinct yellow colour, light aroma and flavour. It was first used in the nineteenth century as a wrapper for the chewing plug, and then increasingly for the plug itself. As a wrapper Bright tobacco offered distinct advantages over the darker tobacco varieties in that it did not change its colour when in contact with tobacco juices and flavourings used in chewing tobacco manufacture (Herndon 1969: 413–14). Prices for Bright tobacco were consistently above those of other leaf varieties; in the postbellum period Bright tobacco often fetched a price double that of dark fire-cured tobacco (Tilley 1948: 125). The increased demand for Bright tobacco in the antebellum period was probably responsible for the shift towards this culture and to an increase in tenancy; the price of what was once useless land surged ahead on the expectation of its newly discovered profitability – in Caswell County, North Carolina, land prices doubled in 1857 alone (Tilley 1948: 32). There is also some evidence that Bright tobacco culture was more intensive than that of the dark varieties (Tilley 1948: 37–88; Siegel 1987: 162–3). The development of Bright tobacco culture thus provided a possibility for replacing the plantation system with smaller family-run enterprises.

Until the Civil War tobacco cultivation in the United States expanded in both quantity and area. Production levels tended to fluctuate around their maximum pre-revolutionary quantity for some time into the nineteenth century, mostly because of dislocated and uncertain markets; but they began to move ahead once international conditions settled. By 1839 the United States tobacco crop was only 75 per cent of its 1790 level but on the eve of the Civil War the total crop was double that of 1839 (Herndon 1969: 423, 427). Part of this expansion was caused by the growth of cultivation in the older tobacco regions, but much more of it was accounted for by the expansion of cultivation in Kentucky and North Carolina, as well as in the western regions of Virginia (Herndon 1969: 427). The postbellum period accelerated this pattern but it also ushered in an entirely new history of tobacco cultivation, particularly in the social relations on the land.

One of the most important features of postbellum tobacco culture was the rapid spread of Bright tobacco, from its centre in what is commonly referred to as the Old Bright Belt, a rectangular area approximately 80 miles by 150 miles positioned equally on both sides of the Virginia–North Carolina border, about half way along the length of it (Tilley 1948: 12). From there Bright tobacco culture moved westward towards Winston-Salem, to the coastal regions of both North and South Carolina in the 1890s, when cotton prices collapsed below subsistence levels and farmers turned to the golden leaf for salvation (Tilley 1948: 141–50). During the First World War, the Bright Belt extended into Georgia (Daniel 1984: 430). By 1919 Bright tobacco accounted for 35 per cent of the entire tobacco crop of the United States (Tilley 1948: 357; Herndon 1969: 427). By contrast, Kentucky and Tennessee continued to concentrate on Burley tobacco in the postbellum period as they had done in the antebellum period. These two states alone accounted for 45 per cent of the nation's tobacco crop in 1919 (Herndon 1969: 427).

Alongside the geographical expansion of Bright tobacco culture, the postbellum period witnessed the restructuring, or, as some historians might argue, an acceleration, of social relations on the land. The most visible aspect of this was the rapid increase in tenancy which, though in existence before the Civil War, certainly grew dramatically after it (Tilley 1948: 94). In the Old Bright Belt the percentage of farm tenancy ranged from a low of 16 per cent to a high of 44 per cent in 1879, but this shifted profoundly to the corresponding figures of 32 per cent, and 69 per cent in 1929 (Tilley 1948: 94). Even higher proportions of tenancy were in evidence in the New Belt, in coastal North Carolina, where figures of over 80 per cent were not uncommon (Tilley 1948: 95). The breaking-up of the plantation system also resulted in a gradual decrease in farm size. In the Old Belt the average size of farms fell from a range of 115–204 acres in 1879 to a range of 58–102 acres in 1929: in the New Belt, along the South Carolina coastal plain, the range of average acreage over this period fell from 82–210 acres to 46–68 acres (Tilley 1948: 92).

Emancipation of slaves on tobacco plantations was, of course, part of a wider movement throughout the South. The plantation system in the cotton states also gave way to an enormous growth of tenancy, though this form of labour organization was extremely rare before the Civil War (Ransom and Sutch 1977: 87–105). The sugar plantations in Louisiana were not, however, transformed in this way, and in this agrarian sector the plantation system, albeit with free labour, continued (Shlomowitz 1984). Wherever tenancy became the norm it tended to revolve around some kind of share-cropping scheme. Three main kinds of contracts prevailed, but in each of them the landlord provided the land, wood and buildings. What distinguished one contract from another was the extent of the landlord's share of the output, ranging from one-quarter to one-half of the crop, the precise division depending on whether, and to what extent, he provided seed and fertilizer (Daniel 1985: 32).

Tobacco farmers, whether tenants or landlords, faced the latter part of the nineteenth and the beginning of the twentieth century with a high degree of confidence. Until 1900 prices remained fairly buoyant; flue-cured tobacco fetched between 8 cents and 12 cents per pound (Tilley 1948: 125). After 1900 tobacco prices began to rise, a phenomenon unknown in the flue-cured districts, at least, since before the Civil War (Robert 1938: 133). Between 1911 and 1920 tobacco prices soared to a remarkable level, reaching the dizzying height of 86 cents per pound in one locality in North Carolina in 1919; average flue-cured tobacco prices reached over 56 cents per pound in the same year (Daniel 1985: 35; Tilley 1948: 125). As the 1920s began, however, the bubble burst and prices crashed, in one year alone, to around 22 cents per pound (Daniel 1985: 35). For the next seven years tobacco prices maintained their level at around 20 cents per pound, on average, but in 1927 the price dipped to 17.3 cents and continued to drop for the next few years until in 1931 a level of 8.4 cents was reached, a price that had not been faced in the Bright tobacco region since the late 1890s (Badger 1980: 21; Tilley 1948: 125). At a price of 12 cents per pound, a farmer producing an average crop could be expected to lose $16 per acre (Daniel 1985: 38). Other crops, notably cotton, were not a viable alternative to a depressed tobacco economy; cotton prices during the 1920s fell even further than did tobacco (Badger 1980: 22). There was no escape, and, for the first time in their history, tobacco farmers, upholders of the Jeffersonian ideal of the small farmer, independently minded and suspicious of government, turned to Federal authorities for help.

The government was, therefore, invited to alleviate the plight of tobacco farmers, but the political scene was, as more often than not, very complicated. Not least of the problems was that the tobacco manufacturers benefited enormously from the fall in tobacco prices; between 1927 and 1930 manufacturers' profits rose by about one-third (Badger 1980: 23). By stark contrast, flue-cured growers received in 1932 one-third of what they had received in 1928 for their crop (Badger 1980: 22). Then there was the problem of what essential commodities should be included in the Agricultural Recovery Program in this section of the New Deal. In the end, tobacco was the only non-essential commodity supported by this part of the legislative package, a fact that reflected tobacco's powerful position within the nation's economy and within a more local political environment (Badger 1980: 39–40). What would have happened to tobacco had it not been included is open to some interesting speculation.

That there were problems within tobacco culture, some of which (such as overproduction) were perennial and some of which (such as marketing arrangements) were more recent, was well known. But before 1933 all attempts to counteract these were largely failures; especially those that operated on a voluntary basis. From the failures, however, it became clear that, as far as practical measures were concerned, some method had to be found of matching supply to demand so that farmers would not suffer from price depressions while tobacco manufacturers and dealers prospered. The only solution was to limit production, but how to do it was a problem (Badger 1980: 37).

The Agricultural Adjustment Act (AAA) was passed on 12 May 1933 and set up the requisite institutions, and machinery, necessary to provide tobacco growers with a route out of their economic plight. The tobacco programme took shape over the months and into the next year, but essentially what emerged was a many-sided solution that resulted in a stabilization of the tobacco economy. Output controls were established by acreage reduction, through what was called the allotment quota system, whereby farmers were allowed to cultivate only a portion of their land, the allotment, with tobacco. This was the quota for any particular farm. In return for reducing their acreage, tobacco growers would be offered a guaranteed price, as well as a benefit support, in the form of a direct payment financed by a processing tax on the manufacturers (Badger 1980: 38–98). Small growers were offered special consideration, as the terms of the payment agreements were adjusted to favour tenants over landlords (Daniel 1985: 120).

The effect of this legislation was immediate and positive. The average price paid to tobacco growers in 1933 stood at 15.3 cents, up one-third from the previous year's price and just about double the 1931 level; in income the 1933 crop brought in $112 million, compared to just over $56 million in the previous year (Badger 1980: 65). Though the AAA of 1933 was made unconstitutional in 1936 – many New Deal acts were similarly affected in this way – the essential points of it were resurrected just over one month later in the Soil Conservation and Domestic Allotment Act, which cleverly re-established the quota-payments system under the guise of conservation (Pugh 1981: 31). The AAA came back in 1938 and, though there were some minor modifications to the programme in the coming decades, the basic system operated in a similar fashion until the mid-1960s. Tobacco manufacturers, though opposing the tobacco programme vehemently, agreed the processing tax, but, perhaps not surprisingly, according to a confidential report by the United States Department of Agriculture, passed the tax on to consumers (Daniel 1985: 126).

What were the results of the New Deal legislation on tobacco culture? In the first place, acreage quotas stimulated tobacco growers to increase yields. Before the New Deal, and back to at least the 1860s, there was very little change in tobacco yields in the United States. In Kentucky, between 1866 and 1939, yields fluctuated around 800 pounds per acre and never exceeded 1,000 pounds per acre (Tilley 1948: 191; Axton 1975: 120). In the Bright flue-cured districts yields averaged around 450 pounds per acre until the mid-1890s and then jumped to about 650 pounds per acre, a level that was maintained until the early 1930s (Tilley 1948: 191, 193). On a national level yields increased by only 100 pounds per acre between 1860 and 1933 (Herndon 1969: 434). Though improved agricultural techniques, such as closer planting and higher topping, and increasing use of and improvements to fertilizers, were adopted in the tobacco-growing regions, little of this found its way into increasing productivity. The sharp rise in flue-cured tobacco yields after the mid-1890s may have been caused not by improvements to tobacco cultivation so much as by the expansion of flue-cured growing into the coastal plain where soil and climatic conditions alone were responsible for higher than average yields (Tilley 1948: 194). This regime of relatively constant productivity was shattered by the New Deal, and from that time on yields have been increasing rapidly. On an average national basis yields soared from 1,000 pounds per acre in 1940 to almost 2,000 pounds per acre in 1965 (Herndon 1969: 434). In the flue-cured districts yields rose from 922 pounds per acre in 1939 to 2,200 pounds per acre in 1964 (Mann 1981: 38); in Kentucky yields began their upward movement in 1935 when the level stood at around 800 pounds per acre – thirty years later farmers were producing 2,500 pounds per acre (Axton 1975: 136).

Many factors accounted for this dramatic change, including the use of heavy fertilization, pesticides and chemicals to prevent the growth of suckers, in addition to more intensive planting and efficient irrigation. Both growers and manufacturers shared a desire for increased productivity, and so turned to the land-grant universities in the tobacco-growing districts for scientific help. Both North Carolina State University and the University of Kentucky have been instrumental in applying scientific solutions, especially those concerned with chemicals, to tobacco cultivation. In 1949 one chemical, in particular, was found to be especially effective in controlling suckers and by 1958 it was being used extensively (Herndon 1969: 438). There is little doubt that chemical controls in the form of pesticides, fungicides, and especially products designed to inhibit sucker and secondary leaf growth, have been crucial in this considerable rise in productivity (Axton 1975: 123–4; Mann 1975: 126–7).

Another important change in tobacco culture resulted directly from the improvements to techniques that raised productivity so dramatically. The chemical controls were applied to the pre-harvesting cycle with the attendant consequence that labour requirements at this stage were drastically reduced. Between 1952 and 1983, for example, the labour expenditure on topping and suckering was reduced in flue-cured tobacco culture from 24 hours to 1.25 hours (Johnson 1984: 76). This change upset a time-honoured work routine that placed considerable pressures on labour demands both before, during and after harvesting. With chemical controls over suckering, labour demands shifted more to the harvesting stage and this placed new pressures on the share-cropping system (Mann 1981: 40).

At the same time the New Deal legislation clearly upheld the ideal of the small farm and the share-cropping system. Reference has already been made to the financial rewards given to small farmers by the government in the initial legislation. Further rulings in the 1930s singled out for particular help those growers whose allotment fell below 3.2 acres or whose output was below 3,200 pounds (Daniel 1984: 441). The number of farms in North Carolina, for example, rose from 117,000 to 150,000 between 1930 and 1950, while the average farm size fell from 5.8 to 4 acres (Green 1987: 231).

Until the middle of the 1950s, the small farmer continued a tradition of tobacco culture that stretched back to post-emancipation days and, even though changes on the farm were conspiring to undo the protective legislation of the New Deal, there was still little sign in the 1940s and early 1950s that these would overwhelm the legislative effort. Yet this did occur and the three postwar decades witnessed profound changes in tobacco culture that have obliterated a substantial part of the traditional culture. But the transformation in tobacco culture was not gradual; it was characterized by a series of distinct changes.

The first of these occurred in the 1950s when landowners began to dismiss their share-croppers, in response to the effect on the work cycle of tobacco cultivation, resulting from the technological improvements to the pre-harvest cycle, as well as other, legislative moves to shift power from small to large landowner (Daniel 1985: 262). Instead of relying on the traditional market for labour through share-cropping, landowners now sought labour for harvest time, and used mechanical means as best they could to prepare the fields for planting. In North Carolina, for example, the number of farms declined by over 37,000 between 1954 and 1959 (Daniel 1985: 262).

The second main attack on the small grower came from federal legislation that was passed between 1961 and 1968. The various pieces of legislation abruptly changed the allotment system, allowing for individual farms to amass acreage across a county, by a leasing arrangement, at the same time as changing the quota system from one based on acreage to one based on poundage; and one particular piece of legislation passed in 1968 allowed tobacco to be marketed in loose-leaf sheets, as opposed to being neatly tied (Mann 1981: 41). Flue-cured tobacco was the first to undergo legislative alteration, but Burley tobacco followed suit.

The legislation was revolutionary for tobacco culture because it provided a legal and structural framework for farm consolidation, and ushered in the forces necessary to dismantle the small farm. The impact on tobacco farm size was dramatic. Between 1964 and 1974 the average size of a tobacco farm in the North Carolina flue-cured district soared from 5.2

acres to 9.5 acres (Dalton 1981: 65); in one county alone the number of tenants over this same period collapsed from 1,834 to 361 (Daniel 1985: 266–7). Leasing, by contrast, increased substantially between 1964 and 1974, from about one-third of farms to over two-thirds (Dalton 1981: 70).

The new tobacco framework did not stop at legislation: it also paved the way for perhaps the most profound change to take place in tobacco culture: mechanization. Until the mid-1960s, tobacco culture had successfully resisted mechanization. It was not that the technical problems were insuperable, though there is little doubt that they were not easily overcome. Various inventions and machines capable of harvesting flue-cured tobacco appeared in the late 1950s and early 1960s, but none was taken up seriously by farmer or manufacturer, and in the Burley area of Kentucky mechanical devices designed specifically for Burley culture in the early 1960s seemed to have suffered a similar fate (Herndon 1969: 443–6). What retarded mechanization was a combination of the allotment system and the consequent small scale of enterprise, together with the predominant use of family labour. Once these vestiges were swept away by the legislative momentum of the 1960s, mechanization could proceed unabated. The tobacco harvester, first proposed in the mid-1960s by the giant tobacco company R. J. Reynolds, began to appear on tobacco fields in the early 1970s. It was profitable only for use on large farms – those exceeding 40 acres – and, because it produced as much as 15 per cent leaf loss, acreage quotas worked against its introduction (Berardi 1981: 48–9). Both of these obstacles were overcome by leasing arrangements, and the shift from acreage to poundage quotas (Berardi 1981: 49). In 1972 only 1 per cent of the flue-cured acreage was harvested mechanically (Martin and Johnson 1978: 656); eight years later 46 per cent of North Carolina's crop was harvested in this manner (Daniel 1984: 451).

Mechanization has also entered into the curing stage, where progress has been extremely rapid. The main change in curing was the introduction of the bulk-barn, where the tobacco leaves are packed loosely into crates, the crates are stacked up in a barn, and then hot air is forced into and around the leaves from an outside heating system (Herndon 1969: 441–2). Bulk-curing first appeared at the beginning of the 1960s – in 1962 there were fewer than 300 bulk-curing units (Akehurst 1981: 246–7); by 1976 the number of units had increased to 30,000 and by 1979 two-thirds of the flue-cured acreage was bulk-cured (Akehurst 1981: 247; Dalton 1981: 66). Bulk-curing not only allows a greater density of tobacco leaf to be cured at any one time but also considerably reduces both labour time and care, even more than changes in harvesting have done. Comparing labour requirements on a typical tobacco farm in 1952 with a large, 40-acre, farm in 1983 shows labour time in curing falling from 190 hours per acre to 61 hours per acre; if harvesting time is included, then the labour time falls from 355 hours per acre to 66 hours per acre (Johnson 1984: 77–9).

Burley tobacco cultivation has been far less affected by mechanization than has flue-cured tobacco. While legislation paved the way for the mechanization of flue-cured tobacco, no such legislation was targeted for Burley tobacco. Consequently the scale of Burley operations has remained much smaller than those for flue-cured tobacco; in 1970, for example, an average flue-cured allotment was 3 acres in size while that for Burley was 0.8 acres (Johnson 1984: 46). How long this situation will last is not clear, especially as Burley farmers are more exposed to world market conditions than in the past.

The demise of traditional tobacco culture in the flue-cured districts (and, in time, likely in the Burley districts) has not been without pain but, seen over the long term, what has happened to tobacco farmers since the Second World War has already occurred to other agricultural pursuits. The two other plantation crops of the South, cotton and rice, have been transformed into agribusiness over a longer period than tobacco, and the recent trend in tobacco culture has simply brought that plant into line (Daniel 1984). Yet while American tobacco culture is transformed beyond recognition, cultivation in other parts of the world, notably in Africa and Asia, continues to thrive on the combination of small scale and labour intensiveness. While Americans can afford to displace labour from the tobacco fields, this is not the case for the Third World, where tobacco cultivation is in many cases central to economic survival.

In global terms the United States has been, in the postwar era, a declining player in tobacco production. The United States share of total world tobacco output has been falling steadily from 29 per cent on average around 1950 to 9 per cent in the late 1980s (Grise 1990: 9). The lead has been taken by other countries, notably China, India, Brazil and Zimbabwe, on the one hand, and the countries of the European Community and Eastern Europe, on the other hand. One of the results of the changing geographical distribution of tobacco culture has been to allow for distinctly different modes of production, involving not only scale of operations and degree of mechanization but also differing social and economic relations between growers, on one side, and multinational tobacco corporations and the state, on the other. This diversity of tobacco culture globally has been a feature mostly of the twentieth century. Consumers of tobacco are generally unaware of this aspect of tobacco production, largely because manufacturing firms in the West have endeavoured to produce tobacco of similar quality over a large part of the world.

One strong contrast that exists in tobacco culture is the degree to which growers have leverage in their relationships with the large tobacco companies. In the United States tobacco farmers have, in the twentieth century at least, attempted to wrest some control over the details of the growing and marketing of tobacco from the large tobacco companies.

Elsewhere, but especially in Third World countries, the role of the tobacco company is very different.

Kenya is an excellent case illustrating relations between growers and companies and the state. In 1974 British American Tobacco (BAT) together with the Ministry of Agriculture embarked on a programme of import substitution by expanding tobacco cultivation. The programme began with recruiting 8,700 contract farmers; by 1983 the number engaged in tobacco cultivation had risen to 10,000 (Currie and Ray 1984: 1,133). The contracted farmers generally had no previous experience with tobacco, being mostly subsistence farmers, but with the help and advice of, as well as the guarantee of selling their tobacco to, BAT they were in a fairly secure position. The size of holdings was small, on average about 1 acre in extent (Currie and Ray 1984: 1133). Interestingly, the introduction of flue-cured tobacco culture has been a stabilizing force in peasant household economies, in contrast to fire-cured tobacco culture with its relatively low financial rewards and lesser status (Heald 1991).

A similar pattern emerged in Tanzania. Between 1955 and 1976 output of flue-cure tobacco increased more than tenfold from around 3 million to 30 million pounds, far outstripping fire-cured tobacco (Boesen and Mohele 1979: 17). A growing proportion, and by the mid-1970s the overwhelming proportion, was accounted for by peasant farmers. As in Kenya the scale of operations is very small – between 1 and 1.5 acres was the average size of a tobacco holding in the mid-1970s and indirect evidence would suggest that the number of large farms fell considerably in the late 1960s and early 1970s (Boesen and Mohele 1979: 35, 53). Parallel to the increase in peasant cultivation has been the growth of supervision over cultivation by outside agencies, from the state and tobacco companies. These have increasingly encroached upon the peasant producer to the extent that all aspects of cultivation, including the infrastructural demands, are controlled externally (Boesen and Mohele 1979: 126–45). In Malawi, too, tobacco is grown on small farms averaging just over one acre in size, though flue-cured tobacco is increasingly grown on large estates (Åberg 1980: 84; Muller 1978: 77). In 1985 55 per cent of the country's foreign exchange was earned by tobacco (Stebbins 1990: 231).

On the other side of the Atlantic, in Brazil, a similar, if not more intensive, relationship exists between grower and company. There tobacco is grown on small farms, using family labour, under contract either to tobacco companies or leaf exporters. Their experts guide all aspects of tobacco cultivation, from seedbed to curing and grading (Grise 1990: 31). There is more than a suggestion that tobacco growers, concentrated in the south of the country with no alternative cash crop at their disposal, do not receive a fair price for their output and are in debt to the companies and dealers (Muller 1978: 81; Cravo 1982).

Tobacco culture has traditionally rested on the twin features of labour intensiveness and small scale, in theory at least. Wherever tobacco has been grown it has absorbed more labour time than any other crop. As late as 1977 the United States Department of Agriculture reckoned that, on average, it took 281 working hours per acre to produce tobacco, compared with 42.6 for potatoes, 23 for cotton, 5.1 for maize and 2.9 for wheat (Berardi 1981: 57). Historically, however, the 'age of the small planter' has been very short, though it has appeared several times over the centuries, in the United States after emancipation, and in Brazil after the ending of the slave trade, for example. So short have these periods been that it would be more accurate to speak of the 'small grower' as an ideal, at best, or as a myth, at worst. Even where the small grower seems to predominate today, as in Brazil, Tanzania, Kenya and China, their power as small producers is almost non-existent, dependent as they are on tobacco companies, dealers and the state.

How and why did tobacco culture spread globally in the nineteenth and twentieth centuries? Until the end of the eighteenth century tobacco cultivation was inseparable from colonialism, and European overseas settlement in particular. The fact that the world's production of tobacco at the time was almost wholly concentrated in the United States, Brazil and Cuba attests to this powerful association. In the Chesapeake region alone output increased threefold in the eighteenth century reaching a maximum value of over 100 million pounds of tobacco. By any standard of measurement the New World expansion of tobacco cultivation and production was explosive. Almost all of the output was, of course, destined for European and, to a lesser extent, other overseas markets and largely in an unmanufactured form. The colonial ideology of the period ensured that colonies cultivated and the metropolis manufactured.

The developments in the history of tobacco production after 1800 could be predicted only dimly on the basis of tobacco's past. In the first place, tobacco cultivation expanded to every part of the world. Much of this expansion occurred during the twentieth century but a significant amount also took place in the nineteenth century, principally in Asia. A considerable amount of this expansion was accounted for by a new association between European settlement and tobacco culture. Non-colonial possessions, including the countries of Europe itself, also participated in this movement. In 1984, according to the Food and Agriculture Organization of the United Nations, tobacco was being grown in 115 separate countries, with total output ranging from as little as 100 metric tons in Samoa to as much as 1,526,000 metric tons in China (UN 1983–4: 540–1). Second, there has been a gradual increase in the developing world's share of tobacco cultivation (UN 1983–4: 540–1; Grise 1990: 9) Third, the expansion of tobacco cultivation has been extremely rapid in the twentieth century, during which period output has grown more than fivefold. Finally, the very nature of the product, and the way it has been consumed, has changed substantially over the two centuries under consideration. Lighter, brighter tobaccos using flue-curing have come to dominate tobacco cultivation in all parts of the world, eclipsing the heavier, darker varieties, using both fire- and air-curing methods, typical of the earlier period.

The remarkable expansion of tobacco cultivation since 1800 is a key feature of the history of tobacco since Europeans first encountered it. To understand what happened it is necessary to return to the United States, since that country has not only dominated tobacco culture until quite recently but has shaped global production, marketing and consumption.

On the eve of the American Revolution Britain imported just over 100 million pounds of tobacco from the American colonies, nearly all of it from the Chesapeake colonies (USBC 1975: 1,189–90). In the succeeding years of commercial dislocation after Independence the tobacco crop of the United States remained below the level of the pre-revolutionary period. By 1820, however, cultivation began to gather pace and output surpassed previous levels. By the outbreak of the Civil War the United States was producing over 300 million pounds of tobacco (Mulhall 1892: 42). Output doubled again by the 1880s, and in 1910 it surpassed 1 billion pounds (USBC 1975: 517). From the First World War until the middle of the 1980s, output fluctuated, though on an upward trend. Since 1945 total output has averaged 2 billion pounds annually – it is only since the mid-1980s that output has, for the first time, declined progressively (USBC 1975: 517; UN 1983–4: 540).

Within the context of an enormous increase in American production since 1800 there have been two particularly important trends. The first was that cultivation in the nineteenth century began to drift away from the traditional area of the Chesapeake tidewater inland towards the piedmont, and from the states of Virginia and Maryland. This development was not particularly dramatic though it was sustained. In 1839, for example, the states of Virginia and Maryland together accounted for just under 47 per cent of the total tobacco crop of the United States; twenty years later, in 1859, that figure had fallen to just over 37 per cent (Jacobstein 1907: 40). In 1912 the Department of Agriculture reported that these two states' share had slumped to 13.4 per cent (US Dept of Agriculture 1913: 627). Virginia's and Maryland's loss was a gain for Kentucky, Tennessee and North Carolina. By 1870 Kentucky had become the largest producer of tobacco in the United States (Jacobstein 1907: 69). Whereas these three states accounted for 46 per cent of United States output in 1839, in 1912 the corresponding figure had risen to just under 55 per cent (Jacobstein 1907: 40; US Dept of Agriculture 1913: 627). Other states, such as Ohio, Connecticut, Pennsylvania and Wisconsin, also participated in this westward movement of tobacco culture.

The migration of tobacco cultivation out of the original heartland was caused by several factors. Partly it was due to the fact that tobacco caused soil exhaustion and erosion. There was, in both Virginia and Maryland, a natural momentum to the opening of virgin lands to tobacco cultivation. It is generally accepted that tobacco planters needed 40 to 50 acres of land for each labourer, and the land could be planted with tobacco for three or four years before it was abandoned – left to its own devices, the abandoned land would regain its natural level of fertility after twenty years (Craven 1926: 69). The westward movement of tobacco was part of a general westward movement of cash crops stimulated largely by population growth. But the most important reason for the diffusion of tobacco cultivation was the growing realization that the best soils of Virginia, and the rich dark clay soils of the piedmont, did not necessarily produce the best tobacco.

Tobacco is very sensitive to the kind of soil in which it is grown. Tidewater planters, cultivating fertile, heavy soil, produced dark tobacco as a general rule. But because soil is not homogeneous, even in a small area, these planters, though using similar seed, often cultivated a considerable range of tobacco in terms of quality, weight, colour and size. The Chesapeake varieties were often subsumed under the names of 'oronoco' and 'sweet-scented', but within these distinct categories variations occurred. In the York River area of Virginia, in particular, as early as the mid-seventeenth century one particular grade of tobacco was cultivated, known as 'E. Dees', and renowned for its mildness and aroma (Tilley 1948: 6). At the time, though the land on which the tobacco was cultivated was known to be generally less fertile than other soils in the region, the connection between low soil fertility and tobacco quality was not recognized. By the end of the eighteenth century, however, there was a growing insight into this relationship, and agricultural literature of the period often made it explicit (Tilley 1948: 8). In short, the light sandy soils west of the tidewater, by semi-starving the growing plant, also deprived it of its darkness, heaviness and high nicotine content (Tilley 1948: 11). Ironically, thin soil unfit for other purposes produced a thin, lightly-flavoured and yellow leaf that came to be known, and is generally still referred to, as Bright tobacco.

The cultivation of Bright tobacco, while expanding during the first half of the nineteenth century, was not regular. The secret of producing a consistent product turned out to lie in the combination of thin soil and a new curing method developed slowly also during the first half of the nineteenth century.

Every stage in the growing of tobacco required an enormous outlay of time and care, and each stage needed to be completed as perfectly as possible. Most authorities on tobacco, agriculturalists and historians alike, would nevertheless argue that, of all the stages in the cultivation of tobacco, harvesting and curing were the most important in determining the nature of the finished product. Curing was the first stage of cultivation to undergo radical change in the nineteenth century. The objective of curing is simple enough: to continue the process of change, growth and decay that is natural in the plant, and to fix in the leaf those characteristics that are desirable, for example nicotine content, taste and combustibility, in the case of smoking tobacco. Curing is the human intervention in the life of the tobacco plant that results in a specific commodity. In short, curing involves killing the tobacco plant by denying it both moisture and food (Tilley 1948: 57). Precisely what happened in the curing stage, that is the chemical changes attendant upon leaf starvation, was not fully understood until the twentieth century, though towards the end of the nineteenth century in the United States some experts had some insight into the chemistry of ripening tobacco (Tilley 1948: 57). Until the first decade of the nineteenth century tobacco was typically cured by air or sun drying methods (Siegel 1987: 101). Fire-curing, whereby wooden fires were lit underneath the tobacco, became more popular during the 1820s. It was at this point, while planters were beginning to appreciate the effect of wood smoke on the curing of the tobacco leaf, that a chance discovery led to what can best be described as the critical step in producing a consistently high quality product. In 1839 it was discovered by accident that charcoal fires, that is fires that were nearly smokeless, turned the curing leaf towards the desired colour of bright yellow and orange more dependably and consistently (Tilley 1948: 24). The advantages of charcoal fire-curing over wood fire-curing were strongly advocated throughout the Bright tobacco growing belt, and most curing facilities appear to have turned over to the new method fairly quickly. Yet there was still a problem, in so far as fire-curing using either form of fuel was not wholly controllable. Since curing depended on maintaining even temperatures in the curing barn, while at the same time being able to control the amount of heat, fire-curing remained problematic. From as early as the 1820s, however, planters started experimenting with flue-curing, in which the heat was transferred into the curing barn from a fire outside, by way of flues. The early flues were rudimentary and progress towards improving their use proved to be very slow, partly because of the obvious success of charcoal fire-curing and partly because flue-curing required a relatively high capital investment. Indeed, at about the same time that people were working on the flue, more progress was being made in charcoal curing, especially in finding methods of consistent curing. One of the most important advances occurred in the early 1870s when one prominent planter, Major Robert L. Ragland, correctly perceived that the curing process actually consisted of three distinct stages, each of which corresponded to particular heat levels (Tilley 1948: 60). This was, of course, an extremely important insight, as it gave planters a precise guideline to achieving as consistent curing as possible; and it also gave those who advocated or were experimenting with flue-curing a decided advantage since it was much easier to control the level of heat in the barn from outside. Flue-curing began to be adopted slowly towards the last years of the 1860s and more rapidly after 1872 (Tilley 1948: 64). But it took many more decades, indeed into the twentieth century, before flue-curing can be said to have become the generally accepted method (Tilley 1948: 64–71; Gage 1937: 47). In 1919 one-third of the American tobacco crop was Bright tobacco and most of it flue-cured (Tilley 1948: 391); in the mid-1930s flue-cured tobacco accounted for 48 per cent of total production whereas in 1978 the proportion had reached 61 per cent (Akehurst 1981: 168).

The other major change that took place in tobacco cultivation in the nineteenth century was in harvesting techniques. Before the twentieth century harvesting was normally accomplished by cutting the whole plant. Cutters, whose main responsibility was the harvest, were highly skilled at their task. Because of the real danger of bruising the leaf (which made curing extremely difficult and reduced the value of the final product), cutters occupied a position of prominence in the tobacco culture (Tilley 1948: 57–8). It must be appreciated that, in harvesting by cutting, the tobacco leaves are not all at the same stage of ripeness and therefore will cure differently. This was not a serious problem as long as chewing tobacco was the main form of the product. Cigarette manufacturers, however, were not content with an average product, and demanded a cured product that was consistent (Tilley 1948: 71). Under pressure from these manufacturers and from those who were themselves experimenting with other methods, harvesting by cutting gradually gave way, after the mid-1880s, to harvesting by priming, or removing, each leaf separately (Tilley 1948: 71). Besides satisfying manufacturers, harvesting by priming had distinct economic benefits for planters; costs of harvesting were slashed, curing times fell dramatically, fewer curing barns were needed and the relative price of primed leaf rose (Tilley 1948: 73, 80–1). By 1920 cutting and cutters were all but forgotten in the Bright growing region of the United States. Priming also reinforced the fact that tobacco cultivation was highly labour-intensive and that the family was the pivot of the labour structure of tobacco culture. Mechanization was consequently very slow in advancing and it was not really until the 1970s that mechanical harvesting became an acceptable and efficient alternative to hand labour (Martin and Johnson 1978; Daniel 1985: 264–6).

The considerable developments in curing and harvesting methods, and the successful cultivation of Bright tobacco, were certainly among the most important changes in the nature of tobacco production. Flue-cured tobacco was very much the speciality of the eastern tobacco regions centred on inland Virginia and North Carolina. Further to the west, especially in Kentucky and Tennessee but also to an extent in corners of North Carolina and Virginia as well as into Ohio and Missouri, a new variety, as opposed to a new grade, of tobacco emerged. This variety, known as Burley, appears to have been discovered in 1864, by accident, growing as a mutation, in a tobacco field in Ohio (Robert 1952: 186). Its culture spread very rapidly, especially into Kentucky and Tennessee. The output of tobacco in Kentucky, for example, more than doubled between 1860 and 1890, whereas total American output rose by a mere 10 per cent (Jacobstein 1907: 69). Unlike Bright tobacco, which required a relatively elaborate curing method and a significant investment in plant, Burley tobacco achieved its excellence by air-curing. In the long run, although the United States produced other varieties of tobacco using other curing methods, especially fire-curing, Bright and Burley tobacco gradually came to dominate the entire culture. By 1970 Bright and Burley tobacco accounted for 92 per cent of the country's output; the figure was identical to this in the late 1980s (USBC 1975: 517; Johnson 1984: 100; Grise 1990: 9, 12). In addition, over the period, other changes in cultural techniques have made both Bright and Burley tobacco milder (Robert 1952: 186).

The shift in production to Bright and Burley tobacco was not only welcome for tobacco farmers moving into areas where land was considered worthless – it was reported, for example, that land values in North Carolina rocketed from 50 cents to $50 per acre (Jacobstein 1907: 38). For manufacturers, exporters and consumers the change was also significant. Manufacturers, for example, quickly realized that both tobaccos had distinct advantages over the darker, heavier varieties. Bright tobacco rapidly became the typical wrapper for plug (chewing) tobacco while Burley became the preferred variety for the filler (Siegel 1987: 102; Robert 1952: 186). The level of output of manufactured tobacco and snuff tripled between 1870 and 1900 (Johnson 1984: 16). While the adoption of both Bright and Burley varieties in chewing tobacco was important, it was overshadowed by the use of both in what was, in the second half of the nineteenth century, the relatively new industry of cigarette manufacturing. The story of the cigarette has been covered in Chapter 5, but it is well to note that the incredible growth of cigarette manufacturing would not have occurred in the absence of the changes in tobacco cultivation. Cigarette production, even before the invention of the cigarette machine, expanded exponentially. In 1870, for example, 16 million cigarettes were produced in the United States; ten years later the figure stood at 533 million (Johnson 1984: 16). Once the Bonsack cigarette machine proved to be operational, the production of cigarettes went through the roof. In 1895 output reached 4.2 billion cigarettes (Johnson 1984: 16). Exports of unmanufactured tobacco also soared. On the eve of the Civil War exports averaged 175 million pounds annually. The figure increased steadily until around the turn of the twentieth century the level of exports stood at 325 million pounds; over the same period the value of such exports rose considerably from $16 million to just under $30 million (Jacobstein 1907: 171). Consumers benefited because the final product was milder and more attractive.

The shift to cigarettes has already been noted, but in general the change to milder tobaccos was reflected in a steady increase in per capita consumption which rose from an average of 1.8 pounds in the early 1870s to an average of 5.98 pounds in the first few years of the twentieth century (Holmes 1912: 4–5). Though the increase was considerably less than in the United States, per capita consumption in the major European countries also rose in the same period (Jacobstein 1907: 45).

As stated earlier, the United States has been the world's foremost producer of tobacco since the start of commercial production in the seventeenth century. Until the nineteenth century the United States' position in the international tobacco market was unchallenged, partly because competitors were few – Brazil and Cuba in the western hemisphere, and Holland and Germany in Europe; and partly because the colonial system ensured a relatively clear segmentation of the international market. With the breakup of this system, beginning with the American Revolution and culminating in the total dissolution of colonialism in South and Central America in the course of the nineteenth century, the United States found itself with a new situation. Before embarking on a discussion of the spread of tobacco cultivation to other parts of the world, the international context as it affected United States production and exports should be outlined.

Although reliable statistical information is not available for much of the period, it would appear that, as far as production was concerned, the immediate effect of the opening of new regions to tobacco cultivation outside the United States was to reduce the American share in overall world output. Around 1800 the United States was probably responsible for as much as 70 per cent of world production. In 1884, a year for which reasonably accurate figures exist, the proportion had fallen to under 30 per cent (Mulhall 1892: 568). Output grew at a faster rate between 1870 and 1910 in North America than elsewhere in the world. In 1910 North American share of world output had improved to about 40 per cent (US Dept of Agriculture 1913: 625–6). By the middle of the 1970s, however, the share had fallen to near 18 per cent, and in 1984 it stood at a mere 13 per cent. In the special area of Bright and Burley production, the alteration in the United States' position was particularly rapid in the postwar era. In the late 1950s, for example, the United States accounted for about half of the world's output of these tobaccos, but by 1980, the figure had shrunk to 25 per cent (Johnson 1984: 100). A similar change seems to have occurred in the United States' share of world export in unmanufactured tobacco. In 1840 the United States was almost alone in exporting tobacco leaf. Only Brazil shared in tobacco export earnings. In that year the United States accounted for 87 per cent of the world export market by value (Hanson 1980: 174). With the appearance of new producers, especially the Dutch East Indies, and the continued expansion of Brazil and Cuba, the American share of the export market shrank considerably. By 1900 the

American share was 45 per cent and falling (Hanson 1980: 174). In 1910 the United States accounted for almost 40 per cent of world tobacco trade; in 1980, by contrast, the American share of the lucrative Bright and Burley export market had been reduced to under 30 per cent (US Dept of Agriculture 1913: 630; Johnson 1984: 101).

The most significant development to occur outside the United States was the rapid expansion of tobacco cultivation in Asia during the nineteenth and twentieth centuries. Nineteenth-century developments were stimulated largely by imperialism, particularly in the Dutch East Indies and in India. In the Dutch East Indies imperial control of economic resources was formalized in the 1830s under what was called the Culture System. This system, which operated mainly in the 1830s and 1840s, was designed to organize the production of export crops, primarily by the peasants of Java and, to a lesser extent, Sumatra. Peasants were compelled to allocate part of their lands to producing crops for the government. Sugar, coffee and indigo were the first crops to be included in the Culture System, but in time many others, including tobacco, were brought in. For various reasons the Culture System did not yield significant profits, and crops were dropped from the system and allowed to be cultivated on a private basis (Ricklefs 1981: 114–18). Tobacco escaped the grip of the Culture System in 1866 (Caldwell 1964: 83). Almost overnight production began to grow, partly, it has been argued, because of the advantages of private enterprise but also because one of the main tobacco estates, in the Deli district of Sumatra, successfully developed a very exportable kind of tobacco. Exports of Sumatran tobacco soared from an average level of 17 million pounds in the late 1860s to nearly 170 million pounds in the years before the First World War (Caldwell 1964: 83). The Deli region accounted for about one-third of the total crop of Sumatra (Jacobstein 1907: 182). Before the First World War, on account of this vast expansion of tobacco cultivation, the Dutch East Indies were the second largest exporters of tobacco leaf, accounting for about 18 per cent of total world exports (US Dept of Agriculture 1913: 630). Ironically, a significant proportion of total Sumatran exports went to the United States (Jacobstein 1907: 181). In the 1880s, many planters from Deli, both Germans and Dutch, were attracted to North Borneo, and there, under the administration of the North Borneo (Chartered) Company, tobacco cultivation by Europeans expanded enormously (John and Jackson 1973). On the eve of the First World War the level of output exceeded 2 million pounds (John and Jackson 1973: 105).

During the nineteenth and for a good part of the twentieth century the Dutch East Indies were the primary exporters of tobacco in Asia, but they were not the largest producer. That distinction went to India. Surprisingly, given the amount India produced, little is actually known of the history of tobacco in that country. What is known, however, is that, in terms of output, Indian production, certainly towards the end of the nineteenth century, was not far behind that of the United States. In 1884, for example, India produced 340 million pounds of tobacco, or, in other terms, roughly 80 per cent of the United States output (Mulhall 1892: 568). Just before the First World War the level of output was up to 450 million pounds, rising to 761 million pounds on average between 1935 and 1939 and 1.1 billion pounds in 1984 (US Dept of Agriculture 1913: 626; Akehurst 1981: 7; United Nations 1983–4: 541). Most of Indian production was, and still is, dark, air-cured tobacco, used largely for the domestic consumption of *bidis* and cheroots, as well as in hookahs (US Dept of Commerce 1915: 34; Akehurst 1981: 317). According to available statistics, before 1914 exports of tobacco from India represented around 5 per cent of overall production, the most important market being Aden; imports were even smaller but significantly almost three-quarters of the import level was accounted for by cigarettes, nearly all of which came from Britain (US Dept of Commerce 1915: 34–5). This pattern has been maintained throughout this century. In global terms India has been the world's second largest producer of tobacco for a good part of the nineteenth and twentieth centuries. It was only in the 1930s that India relinquished second place to China (Akehurst 1981: 7). In the postwar period India has steadily increased the share of its tobacco production destined for the export market; in 1980 17 per cent of output was exported, though the figure is low in comparison with other producing countries and represents only 5 per cent of world tobacco leaf trade (UN 1983–4: 541; Tucker 1982: 182).

China is now the world's largest producer of tobacco. In 1984 it produced one-quarter of the world's output, one-half of Asia's output and twice that of the United States (UN 1983–4: 540–1). Less than 5 per cent of output is exported (Muller 1978: 19). China has expanded its tobacco production in the twentieth century to a greater extent, and much faster, than any other country. In 1911 total production stood at 18 million pounds; seventy years later the corresponding figure was 3,400 million pounds, and during the late 1980s output averaged 4,700 million pounds annually (US Dept of Commerce 1915: 8; Grise 1990: 9).

With the single exception of the Dutch East Indies, Asian tobacco production in the nineteenth century expanded solely on the basis of meeting domestic demand. And during that period, and for a while into the twentieth century, the production concentrated almost entirely on dark, air- and sun-cured tobaccos. In China, for example, the first harvest of Bright tobacco dates from 1913, when the British American Tobacco Company purchased its first supply from Chinese farmers; the Chinese had been introduced to Bright tobacco from as early as 1906, when James Duke, the head of BAT, sent tobacco experts from North Carolina to China to experiment with American seed (Cochran 1986: 163). In 1915 BAT purchased over 2 million pounds of Bright tobacco from Chinese farmers, but only about one-quarter of it was flue-cured, the rest being sun-cured (Cochran 1986: 163). China was probably the first country in Asia to grow Bright tobacco and have it flue-cured. Certainly in 1915 neither the tobacco nor the curing process existed in India or the Dutch East Indies. Flue-cured tobacco, once in demand by BAT, came to account for an increasing share of the Chinese tobacco output. Output of flue-cured Bright tobacco quadrupled between 1920 and 1937, while imports of the same stagnated (Cochran 1980: 233). Not surprisingly, the rising output of flue-cured tobacco was paralleled by a huge increase in cigarette consumption which saw levels rise from 7.5 billion cigarettes in 1910 to 87 billion in 1928 (Cochran 1980: 234). In 1959 38 per cent of China's total tobacco production was flue-cured, at a time when India, for example, was only just embarking on the cultivation of this variety (Akehurst 1981: 170, 173). At the end of the 1970s flue-cured tobacco accounted for 60 per cent of China's production and it was, at the time, the largest producer of this variety in the world; since the late 1980s, flue-cured tobacco has accounted for almost 90 per cent of the country's tobacco crop (Akehurst 1981: 170). Flue-cured tobacco production in India, while lagging behind that of China, nevertheless accounted for 30 per cent of its total tobacco output in 1978 (Akehurst 1981: 175; UN 1983–4: 541). Even Indonesia, though a substantial producer of dark tobacco, entered the flue-cured tobacco sector; in 1975 the flue-cured crop accounted for as much as 19 per cent of total output (Akehurst 1981: 183; UN 1983–4: 541).

The shift towards flue-cured tobacco in Asia, especially in the postwar era, is part of a much more general transformation of patterns in the global production and consumption of tobacco. In Africa, especially, the development of flue-cured tobacco has had a profound effect not only on tobacco cultivation but on many aspects of the political economy of the continent.

Tobacco has been consumed and/or cultivated in Africa since the end of the sixteenth century. It was not, however, until the nineteenth century that commercial cultivation began. North Africa, especially Algeria, was the main producer in the nineteenth and a good part of the twentieth century, and almost all of the output was exported to France. The Cape Colony in South Africa cultivated tobacco from as early as 1657, but production was meagre; in 1875 the Colony boasted an output of only 3 million pounds (Akehurst 1981: 11). Elsewhere in Central and southern Africa tobacco cultivation did not begin until the end of the nineteenth century, and started in the British colonies. In 1912 total African production stood at just under 44 million pounds, 56 per cent of which originated in Algeria (US Dept of Commerce 1915: 8). In the twentieth century, generally speaking, production has been growing, though the share of Algerian production in overall output has been falling. In 1980 the African continent accounted for only 6 per cent of world output, the leading producers, in order, being Zimbabwe, Malawi and South Africa,

Zimbabwe alone producing 41 per cent of the continent's total (UN 1983–4: 540).

What is interesting about African tobacco production in general (and particularly that of the former British colonies) is the extent to which tobacco played a similar role in settlement to that which it had in the New World. In Zimbabwe, for example, tobacco was a critical component of what has been termed the 'white agricultural policy', whose origins can be dated to 1908. The British South African Company worked hard to stimulate production and, by association, European farming and settlement. In what would become a common practice, the Company appointed a tobacco expert to introduce the cultivation of Oriental tobacco (Palmer 1977: 233). Output soared, increasing ten-fold between 1909 and 1913 (Palmer 1977: 233). Though European tobacco cultivation in Zimbabwe dates from 1893, very little was produced until the encouragement of European settlement from above. Output continued to grow, despite a few setbacks, until 1925–6 when output reached 5.7 million pounds, all of it Oriental tobacco (Palmer 1977: 236). Zambia's tobacco cultivation, by contrast, was of two varieties: European farmers cultivated flue-cured tobacco, while Africans cultivated their own, indigenous, variety. Once again it was the British South African Company that encouraged the cultivation of flue-cured tobacco expressly for the South African market (Kanduza 1983: 204). Production expanded swiftly from 500 pounds in 1912–13 to 800,000 pounds in 1918–19 and reached a maximum level of just over 3 million pounds in 1927 (Kanduza 1983: 207, 216). Until 1938 the flue-cured sector was effectively closed to Africans, except, that is, as labourers (Kanduza 1983: 202). The control over tobacco production by Europeans was also reflected in Zimbabwe where the indigenous tobacco industry, situated in the Inyoka country, was allowed to wither away (Kosmin 1977). In Malawi, however, the picture was more complicated, as both flue-cured and fire-cured tobacco production was encouraged: the former was the responsibility of European estates, while the latter was produced by African tenant farmers (McCracken 1983; Chanock 1972). Since the Second World War, there has been a swing away from Oriental and Burley tobacco, primarily in Zimbabwe. In 1980, for example, Zimbabwe's share of world flue-cured tobacco production stood at 5.6 per cent while the share of world exports was 10.5 per cent, ranking third in the world (Tucker 1982: 179). In the latter years of the 1980s, 97 per cent of Zimbabwe's total tobacco output was flue-cured (Grise 1990: 9, 12).

We turn finally to South America. Throughout the nineteenth and twentieth centuries Brazil has been the chief producer and, unlike many countries, its production has been rising continuously, but especially in the postwar period. Between 1950 and 1980 output increased fourfold to over 800 million pounds (Nardi 1985: 32). As in other parts of the world, the shift to flue-cured tobacco has been particularly marked, reaching 67 per cent of total output in 1970; and in 1980 77 per cent of Brazilian tobacco exports were of this type (Nardi 1985: 33, 35); in the period 1985–8 flue-cured and Burley accounted for almost 80 per cent of Brazil's total tobacco crop (Grise 1990: 9, 12). Concurrent with the shift towards flue-cured tobacco has been the relative decline of the traditional culture based around Bahia, and the expansion of production in Brazil's southern states, predominantly from Rio Grande do Sul and Santa Catarina (Akehurst 1981: 173). Cuba stands out, by contrast, as the most important example of a country which has not gone over to flue-cured tobacco. Its production of over 100 million pounds of tobacco in 1981 was almost entirely dark air-cured tobacco (UN 1983–4; Stubbs 1985).

The shift in global tobacco culture away from dark to light tobacco, and from air- and sun-curing to flue-curing, has had a profound impact on the history of tobacco production – and consumption – in the second half of the nineteenth and the twentieth century. The shift itself occurred progressively, as far as one can tell, in the twentieth century. By 1935–9 the production level of light tobaccos was equivalent to that of dark tobacco, and growing (Akehurst 1981: 30). In 1980 88 per cent of the world's tobacco crop was light, and in the same year, flue-cured tobacco itself accounted for 43 per cent of the world's output (Tucker 1982: 178). And the proportion continues to rise – flue-cured tobacco accounted for 54 per cent of world output annually between 1985 and 1988 (Grise 1990: 9, 12). Unlike other methods of curing tobacco, flue-curing is both capital- and resource-intensive, particularly in its use of wood. Because of both capital and resource costs, flue-cured tobacco engenders an organization and management of labour strikingly different from that found in air- and sun-cured tobacco culture. It has also been the foundation of an equally remarkable transformation in the manufacture of the final product, the subject of the following chapter.

9

'TO LIVE BY SMOKE'
Tobacco is big business

Despite some early objections to the consumption of tobacco, the state seized on tobacco as a revenue generator: throughout the fiscal history of tobacco the only question asked by the state was how great a tax burden could tobacco carry before it would become self-defeating. Tobacco's tax burden has varied both over time and space but it has always played an important role in the finances of the state which, in turn, have been enlarged by the increasing consumption of tobacco.

The central role of the state in the history of tobacco needs to be understood in two ways: the level of revenue it has extracted, and the means by which it has done so. Both aspects are, of course, interrelated since the actual amount of revenue flowing into the state purse was determined to a large extent by the efficiency of collecting it. The efficiency was in itself a function of the mechanisms that were used to ensure that as few loopholes as possible existed.

Looking across a wide range of mechanisms for collecting revenue from tobacco, the most striking difference was between those states that taxed the importer, by customs and duties, and those that taxed the consumer, through excise. Both presented their own problems and brought forth some interesting solutions.

In the early modern period most European states established tobacco monopolies, both purchasing and frequently processing tobacco for consumption. The pressure to collect as much revenue as possible in the most thoroughgoing manner was increasing throughout Europe as the costs of maintaining bureaucracies, armies and navies were rising steadily; the days when governments turned to extraordinary taxation as an emergency measure were fast disappearing (Bean 1973; Parker 1974; Parker 1976). The coincidence of the increasing availability of tobacco after 1600, initially from Spanish-American sources and then from Virginia and Brazil, together with escalating governmental costs and the perception of tobacco as a luxury, encouraged governments to turn to this commodity to augment fiscal resources.

Tobacco monopolies appeared in Europe from the 1620s: the one in

Mantua, established in 1627, was probably the first of its kind (Rogoziński 1990: 65). Over the succeeding forty years most Italian states established some form of tobacco monopoly. In Spain the tobacco monopoly grew out of the monopoly which organized the American trades. In 1614 the Spanish crown authorized Seville to be the sole importer of Spanish-American tobacco, and this move led naturally to the city becoming the only manufacturing point for tobacco; indeed in 1684 the tobacco monopoly extended to Seville the sole right to manufacture tobacco (Rogoziński 1990: 68). The royal tobacco monopoly was extended to all of Spain's colonies in Latin America and the Philippines in a series of decrees beginning in 1764, thus bringing cultivation within the monopoly's jurisdiction (Hanson 1982: 150; de Jesus 1980; Deans 1984). Portugal, too, established a tobacco monopoly along the Spanish model. The first contract for the monopoly was granted in 1633, but, as this happened during the period of the Iberian union, the revenue flowed elsewhere. It was only after the end of the union, in 1640, that a proper tobacco monopoly was created in Portugal, though for two crucial years, between 1642 and 1644, Portugal operated a free trade policy for tobacco (Hanson 1982: 152). The ending of this policy and the re-establishment of the monopoly in 1644 reflected the realization that collecting revenue from tobacco was very difficult unless it was well-supervised. In the Habsburg lands monopolies farmed out to private individuals appeared to be the rule in the seventeenth and first few decades of the eighteenth century. Though the status of the tobacco monopoly fluctuated for most of the eighteenth century, in 1784 it officially became an administration of the state and, as the Tabakregie, has remained so until the present day (Rogoziński 1990: 68–9; Hitz and Huber 1975). By contrast, the states in Germany tended to enforce taxation rather than creating a monopoly; for a short time only, monopolies were established in the two largest states, Bavaria and Prussia, but essentially the German lands were a free trade area for tobacco (Rogoziński 1990: 71).

Tobacco monopolies were generally farmed out to individual entrepreneurs, though there were some important exceptions to this. The French tobacco monopoly was farmed until 1791, and turned into a state monopoly after 1810; the Spanish monopoly remained farmed until after the Second World War as did the Portuguese, though its monopoly was, since 1674, much more restricted by state action than was the case in either Spain or France (Hanson 1982: 153; Nardi 1986). The Austrian Tabakregie has been one of the longest surviving state tobacco monopolies of its kind. No monopolies have existed in Britain, the Netherlands and the United States.

Whether tobacco was to be administered through a monopoly, private or state-controlled, or left to the free market, the state was under enormous pressure to contain powerful countervailing forces for contraband. All of the evidence available suggests that in the early modern period, at least, smuggling was a perennial problem and was so highly organized in many parts of Europe as to constitute another, and very important, trading system, in a sense a parallel economy. The image of the lone smuggler operating on a small volume could not be further from the truth; most of the regulatory schemes that governments followed were designed to combat smuggling.

The problem of smuggling could be approached from two main directions. Those states with tobacco monopolies tended to prohibit domestic tobacco production in an effort to close as many channels for illicit trade as possible. The reason for this was quite simple: no monopoly, either private or state-run, could monitor and police agricultural production, and there was therefore no way of telling whether, and to what extent, a tobacco farmer was withholding some of his output for sale in the black market. Spain, Portugal, Austria and, to a lesser extent, France and the Italian states prohibited domestic cultivation by decrees authorized in the seventeenth and eighteenth centuries. England was the only country without a monopoly to follow suit but, as we have seen in Chapter 6, this decision was bound up as much with state finances as it was with colonial politics. All of these countries preferred to import their tobacco primarily from overseas since they believed there was less risk of losing revenue from maritime than from overland traffic. Because the costs of entry into the tobacco trade were rising during the seventeenth century, most European states were confident in the belief that none of the production of colonial tobacco would find its way to Europe illicitly.

Most of the European states who prohibited domestic cultivation were probably successful in this area of combating smuggling. Certainly this was true for England and for France. In both countries tobacco farmers were sacrificed at the altar of state revenues. The brutality of the measures taken by the English Crown in the West Country was no less draconian than that executed by the armed police of the French tobacco monopoly (Price 1973: 482). In Portugal the Junta da Administraçaõ do Tabaco, which had responsibility for the entire tobacco trade, including manufacturing, also tried and sentenced contrabandists (Hanson 1982: 154).

Bearing down on tobacco farmers who contravened regulations was one thing; even though it was difficult to police these regulations, it was equally difficult to cover up a field of tobacco. By the very nature of the culture, tobacco growing was typically a small-scale activity and those farmers who tried to evade the system usually lacked political influence. Contraventions of this kind were probably more of a nuisance than a threat to the state's authority. In the case of France the desire of the tobacco monopoly to eradicate domestic production entirely, contradicted the political strategy of the central authorities intent upon keeping peace in sensitive provinces, and the resulting policies were compromises.

The real threat to the revenue came not from the peasantry but from the contrabandists who handled the imported commodity. The success of the state, or its tobacco administration, in containing domestic cultivation contrasted with its problems in dealing with smuggling. Though by its very nature there are no precise measures of the volume of the contraband trade, one recent study of the English and Scottish tobacco trade in the eighteenth century suggests that as much as one-third of tobacco consumed in England in the 1730s was contraband (Nash 1982: 365). What the proportion was in other countries is not known but it is very unlikely to have been any less.

While it is one thing to profile the merchants who handled the legitimate trade in colonial tobacco, it is quite another thing to do the same for the illicit trade. Yet, as pointed out previously, the smuggling trade was lively and highly organized. Though we do not know much about it there is enough evidence to give us some indication of its pervasiveness if not its exact extent.

One of the most important obligations of the Portuguese Junta do Tabaco was to ensure that tobacco filled the state coffers, and that this should be arranged, initially, through the delivery of Brazilian tobacco into the customs warehouses, located principally in Lisbon but also in other major ports of international commerce in the country. There were, according to the Junta, several routes by which tobacco escaped the revenue net. Ships returning to Portugal from Brazil often held hidden stores of tobacco, not registered on the ship's manifest, and these would be unloaded before the inspectors arrived; some consignments were stolen instead of being warehoused; some tobacco was purchased directly from the ships on the high seas before the shipments reached shore; and some amount of tobacco that was re-exported to Spain, for example, would be smuggled back into Portugal in a manufactured state (Hanson 1982: 157–8). This illicit tobacco would then be put straight on to the black market. In Portugal itself it appears that the clergy were responsible for some large, but imprecise, part of the distribution network of black market tobacco. The preferred form of black market tobacco was snuff, and several searches by the superintendents of the Junta yielded illegal operations of snuff production in monasteries and convents. In 1700, for example, the Junta learned that the sisters of one convent were selling as much as 250 pounds of tobacco per day (Hanson 1982: 155). The Junta believed that there was not a single monastery or convent in Oporto that did not engage in contraband traffic (Hanson 1982: 155). In response to such flagrant defiance the Junta retaliated with a series of measures aimed at both ends of the tobacco network, in Bahia and Portugal, as well as by increasing penalties for offenders: for example, the standard penalty for engaging in contraband trade on the Brazilian side – two years' banishment in Angola – was increased to five, together with a stiff fine (Hanson 1982: 159). None of the measures appears to have done much to stem the tide, partly because of the impossibility of checking each single shipment, port and border crossing, not to mention each religious house, in the country, and partly because of the sheer volume of tobacco that was crossing the Atlantic. In desperation the Junta awarded a contract to a strong man from Spain, a wealthy merchant, for the distribution of tobacco in Portugal, but within two weeks of its signing, and the supposed strangling of the black market, the sisters of the Esperança convent were at it again (Hanson 1982: 161). The royal tobacco monopoly continued to be plagued by what Carl Hanson has called 'a parallel economy . . . a network of clandestine entrepreneurs' while tobacco revenues continued to be the leading source of royal income (Hanson 1982: 155, 161).

In England the customs and the state believed they were being deprived of a substantial portion of their rightful revenue by smuggling; one contemporary writer put the proportion at more than 50 per cent (Linebaugh 1991: 159). Before 1700 the smuggling trade seems to have been unorganized and based on the movement of small volumes of illicit tobacco, usually by theft, helped by judicious doses of bribes to the customs service (Rive 1929: 554–7; Linebaugh 1991: 158–9, 161). This trade was facilitated by the nature of the tobacco trade from the Chesapeake, in which tobacco was transported in unencased parcels. After much pressure from the mercantile community, Parliament, in 1699, legislated that tobacco could be exported from the Chesapeake only in hogsheads, which increased in average weight over the course of the eighteenth century, eventually reaching over 1,000 pounds (Nash 1982: 357; Shepherd and Walton 1972: 65–7).

Once the packing of tobacco was transformed in this way, the smuggling trade changed its structure, and became much more centrally organized. According to a recent study on contraband tobacco operations in the port of London, the parallel economy that unloaded Chesapeake tobacco hogsheads operated alongside and within legitimate commerce, and as much as the latter incorporated colonial tobacco cultivation, packing and shipping, the former did the same (Linebaugh 1991: 163–70). The object of all who were involved, in one way or another, in the contraband trade was the hogshead, and it was its fragility, or its liability to break up, either on the transatlantic journey, or in London, as it was being moved, that provided the contrabandist with the goods. Taking, or socking, tobacco became a customary right among London's river-working population (Linebaugh 1991: 172–3). In addition to straight theft, the customs service itself was involved in illicit tobacco trade, defrauding the revenue by underweighing hogsheads in the accounts. Under-weighing in the outports could be as high as 15 per cent of the total, true, weight, as evidenced in Bristol around 1730, but in London a figure of around 5 per cent was more likely (Nash 1982: 358–60).

The drain on royal revenue was so serious that the customs service was purged, informers placed on the waterfront, and demands for convictions raised. Indeed, the problem was so serious as to bring forth an attempt by the then Prime Minister, Sir Robert Walpole, to revolutionize the entire means by which tobacco revenue was collected; that is, by substituting an excise tax for customs duty, to shift the burden directly from the importer to the consumer (Hausman and Neufeld 1981; Price 1983; Hemphill 1985: 190–302). The excise tax scheme had a covert political purpose but overtly it was proposed as a means of undoing both socking and under-weighing (Linebaugh 1991: 178). The scheme did not go any further than that because of the enormous opposition raised from many different quarters, and socking and under-weighing continued for the time being.

Solid quantitative, as opposed to qualitative, evidence is lacking on the extent of this kind of port fraud. An estimate of it has been made which shows a rise from around 600,000 pounds at the turn of the eighteenth century to a maximum level of 2,700,000 pounds around 1730 (Nash 1982: 366). Following these estimates suggests that port frauds decreased substantially thereafter and, according to the author of these estimates, virtually disappeared by mid-century (Nash 1982: 366). They were not, however, the only form of fraud, and while their importance declined after 1730, that of other types of fraud grew to replace it.

The re-landing of re-exported tobacco, that is after the payment of the drawback, and the smuggling of tobacco into England from Scotland were the principal routes of contraband operations. Re-landing operations were carried on from the Channel Islands, Dunkirk and Ostend and, to a lesser extent, the Isle of Man. The amount of tobacco that passed into the domestic market in this way has been estimated at around one-quarter of that officially documented as being retained for home consumption (Nash 1982: 356, 362). Fraud in Scottish ports was a serious problem until the Scottish customs service was united to the English service (Barker 1954; Price 1984a; Nash 1982: 366–7).

Despite attempts by the British government to thwart the operations of the smuggler, it seems that no amount of tinkering within an unchanged taxation structure had much effect. As Nash's work shows, as soon as one form of fraud declined another form rose. Legislation bore down on the tobacco trade in the last few decades of the eighteenth century, culminating in the excising of tobacco in 1789 (Rive 1929: 566–8).

The problem of smuggling was widespread in Europe, and though we have far less information about it than we would like, there is no reason to suppose that the European experiences with smuggling and smugglers varied significantly. Certainly the problems that the French tobacco monopoly encountered with illicit tobacco trade were not very different from its British, or Portuguese, counterparts; the pattern of smuggling where certain kinds of fraud appeared under different conditions, was broadly similar in Britain, Portugal and France (Price 1973: 127–33, 446–54, 796–7;

Hanson 1982; Vigié 1989). Yet while smuggling was undoubtedly a drain on resources, both in depriving the state of revenues and in raising the costs of enforcement, its impact was not simply financial. As Peter Linebaugh has shown in his study of the criminal proletariat in eighteenth-century London, tobacco fraud was a venture that linked the slave and the planter in the Chesapeake to the river men, porters and customs officers on the Thames. A similar case can be made in the Bahia–Lisbon tobacco circuit (Hanson 1982; Lugar 1977). The driving force of this parallel economy was, of course, the state. Ironically, while smuggling deprived the state of funds, it reinforced the state's insistence on deriving revenue from taxing tobacco as it became increasingly committed to its production, distribution and consumption.

Part of the problem of smuggling stemmed from aspects of the organization not of trade but of production. Before the end of the nineteenth century tobacco manufacturing was on a very small scale. Manufacturing enterprises were limited not only in size but also in the extent of their operations. The consumer, and not the manufacturer, still did a lot of the work transforming the tobacco leaf, whether in a raw or semi-processed state, into the final product. One of the central changes in the history of tobacco was its transformation into a commodity of industrial, as opposed to commercial, capitalism.

Since its earliest introduction into Europe, consumers have taken tobacco principally by smoking, chewing or as snuff. Across Europe not only were there differences in the ways tobacco was consumed, as Chapter 4 described, but there were significant differences in the way tobacco was manufactured for consumption. The differences arose, in the first instance, between tobacco manufactured from Chesapeake and European tobacco, on the one hand, and Brazilian tobacco, on the other.

Chesapeake tobacco arrived in Europe cured but unprocessed. Once out of the hogsheads, its moisture content needed to become fixed before it could be processed any further. The aim of fixing the moisture content was to preserve the tobacco and though this was done by a variety of methods it essentially involved adding a liquor of additives (sugar and water plus other ingredients) that fermented the tobacco (Alford 1973: 8; Price 1973: 423). Once it was sufficiently moistened, the tobacco went for processing. Some of the tobacco, that destined primarily for smoking, was stripped before fermentation and then cut, ready for the consumer, as was the case in England, or spun into a roll and then cut for the consumer (Alford 1973: 8; Price 1973: 423; Rogoziński 1990: 44). Tobacco that was not intended solely for smoking passed through several other stages of manufacture, involving further bathing in vats containing a wide variety of flavourings and other additives, spinning, rolling and pressing. The most common form the tobacco took after these procedures was as a roll weighing over 100 pounds. In this form the tobacco could take on any one of several uses: it could be cut off and used as a smoking mixture, or, more commonly, as chewing tobacco; or it could be processed further, by pressing and rolling, into a shape, known as a *carotte*, destined principally to be ground into snuff (Alford 1973: 9; Price 1973: 423). European, that is Dutch and German, tobacco passed through a similar, though not identical, process of manufacture. Much of the Dutch tobacco used by the French tobacco monopoly came already processed in rolls, at least around the turn of the eighteenth century; later purchases were primarily of leaf, and the manufacturing done in France (Price 1973: 193).

Brazilian tobacco underwent a degree of processing in Bahia. Brazilian tobacco imported into Europe had already been fermented and spun into rolls. Part of the Brazilian tobacco import went straight into use as chewing tobacco, its chief use in north-western Europe, and part of it into the manufacture of snuff (Price 1973: 182).

Compared to other industrial enterprises, until the middle of the nineteenth century tobacco manufacturing was one of the least capitalized activities in both the United States and Europe. In England, for example, while tobacco was one of the country's most important imports, and certainly the main commodity from the American colonies, tobacco manufacturing was among the smallest industries (Alford 1973: 15). In Holland, too, despite the concentration of tobacco-processing facilities in Amsterdam, these were of far less importance to both the city's and the country's industrial structure, though, in the size of its labour force, tobacco manufacturing was in the top ten activities (Israel 1989: 356).

For the most part, with some important exceptions, tobacco was processed, and not manufactured, before the middle of the nineteenth century. Processing was, however, a specialized activity. The principal locations of processing, aside from that done under control of a tobacco monopoly, were, on the European Continent, in Dunkirk, Strasbourg and Amsterdam. In Britain processing took place at the port; there were facilities, therefore, in London, Bristol, Liverpool and Glasgow, but, because at the time the distinction between a tobacconist (i.e. a retailer of tobacco) and a tobacco manufacturer was unclear, most British towns could boast some degree of tobacco processing (Alford 1973: 13–14). Available evidence suggests that, on average, the processing workshops in Strasbourg, Amsterdam and Dunkirk employed one hundred persons each around the turn of the eighteenth century, a figure which, in comparison to other industrial activities of the time, suggests a fair degree of concentration (Price 1973: 487, 504; Israel 1989: 266, 356).

The largest concentration of processing activity was not, however, in those countries where tobacco importation, processing and distribution were unregulated but rather in those countries where all three were organized and controlled by a monopoly. The French tobacco monopoly was the single most important producer of processed tobacco in Europe until at least the nineteenth century. Though it processed an enormous quantity of tobacco – 20 million pounds annually, on average, in the eighteenth century – it had only a handful of manufactories; there were in the eighteenth century no more than ten such establishments at any one time (Price 1973: 411–12). Most of them were located in the north of the country – 41 per cent of total leaf tobacco was consumed by the manufactories in the Seine area on the eve of the Revolution – and, on average, they employed around a thousand workers each (Price 1973: 411–22). The Portuguese royal tobacco monopoly had only two manufactories in the country, the main one in Lisbon and the other in Oporto. Though smaller than their French counterparts, both the Lisbon and the Oporto manufactories were probably larger than the average establishments in Dunkirk, Strasbourg and Amsterdam; the Lisbon manufactory employed around 350 workers in the second half of the eighteenth century (Nardi 1986: 17). The Spanish royal tobacco monopoly concentrated its efforts on the manufactory in Seville (Perez Vidal 1959: 228–37).

There was a limit to the degree to which tobacco could be processed in a workshop. Once the leaves had been moistened and fermented and had been spun or rolled, the tobacco was ready for smoking or chewing – no further processing was necessary. For snuff, however, there was a choice of further processing stages, and this depended entirely on whether the consumer or the manufacturer prepared the snuff.

The earliest preparations of snuff were done by the consumers themselves. Starting around the middle of the seventeenth century, and then growing in popularity over the following century, recipe books, detailing myriad concoctions for the flavouring and colouring of snuff, appeared across Europe, beginning in France. It is impossible to tell how much snuff was prepared by the consumer and how much by the manufacturer, but there is little doubt that over the eighteenth century the trend was towards the centralized production of snuff, either in the manufactories of a tobacco monopoly or in the workshop of a tobacco manufacturer or tobacconist (Price 1973: 423–5). The records of the French tobacco monopoly show clearly that by 1789 more than half of the monopoly's sales consisted of manufactured snuff; *tabac ficelé*, processed tobacco that the consumer ground into snuff, accounted for about one-quarter of the total sales (Price 1973: 426). The Portuguese royal monopoly produced powdered tobacco in its manufactories from 1680, but it was not until the latter part of the eighteenth century that it began producing snuff for immediate consumption (Nardi 1986: 16, 19). The Austrian state tobacco monopoly also had a significant investment in snuff manufacture; in the first year of its operation, 1784, prepared snuff accounted for over 30 per cent of the monopoly's sales (Rogoziński 1990: 88). In England, too, the only sizeable investment in plant and machinery was in the manufacture of snuff (Alford 1973).

At the turn of the nineteenth century tobacco manufacturing was not a particularly important branch of industry and, except for those countries where tobacco monopolies controlled both distribution and manufacture, activities were not very concentrated. The only sizeable investment was in the equipment and plant for producing prepared snuff. The extent to which consumers interposed between production and consumption was still very great, and there was, at the time, little sign that much would change. Judging from information available on some tobacco firms in England as well as the records of the Austrian tobacco monopoly, it would appear that most of the changes that occurred in the sector of manufactured tobacco products for a good part of the nineteenth century were organizational in character; technical changes principally involved an increase in the mechanization of a process that remained fundamentally unaltered (Alford 1973; Hitz and Huber 1975).

The situation in the United States was not very different. Until the American Revolution there is no evidence of any tobacco manufacturing in the Chesapeake, though some snuff was made in the northern colonies (Price 1956: 14–15). Compared to the European tobacco industry, the American tobacco industry hardly existed in 1800 and even the little manufacturing that did exist was based on the production of European-style prepared snuffs. It was not until the 1820s that manufacturing began in earnest in Virginia and North Carolina, and this resulted as much from the general industrial development of the United States as from an important shift in the way tobacco was consumed. This has been described in Chapter 4 but certain aspects of it need to be restated. Americans appear to have taken particularly to chewing tobacco, shunning snuff. Some have argued that this was a manifestation of the willingness of Americans to see themselves as separate from Europe, but it is hard to prove the point. The fragmentary evidence that does exist on consumption patterns in the United States does, however, point to chewing as the major form of tobacco use without giving any clue as to why it superseded snuff taking (Gottsegen 1940: 2–10).

Processing, or manufacturing, chewing tobacco was relatively straightforward and, as this form of tobacco use was of less importance in Europe than it was in the United States, Americans developed a technology and an organization of production that differed from that in Europe. During the 1820s and 1830s tobacco manufacturing grew very quickly first in the main centres of the tobacco export trade in Virginia and North Carolina, but soon it expanded into the countryside where planters became small manufacturers (Siegel 1987: 123). Planters, and others, were undoubtedly attracted to producing chewing tobacco because of the ease of its manufacture, and the small demands it made on capital investment. In many ways the manufacture of chewing tobacco was a natural complement to tobacco culture since, in economic terms at least, the two activities were intensive of labour rather than capital. The slack time after curing, in the winter months, was perfectly suited to manufacturing. Furthermore, each small producer would, in fact, be making a unique product, partly because of the differences that existed in the nature of the leaf and partly because of the flavourings used. Liquorice and sugar were the basic ingredients but many others were used: 'sweet' spices such as vanilla and nutmeg and pungent or bitter ones including coriander and valerian (Tilley 1948: 511–14, 690–1).

The number of manufacturers in the Virginia–North Carolina tobacco belt increased considerably until the Civil War while, at the same time, tobacco manufacturing grew in importance in the regional economy; in 1860, for example, around one-third of the total value of Virginia's manufacturing wealth came from the tobacco industry (Siegel 1987: 120). Towns such as Danville, Lynchburg and Richmond were largely supported by tobacco manufacturing; in Danville this activity employed half of the labour force (Siegel 1987: 129).

Competition among the manufacturers was intense but as soon as one went out of business, another came to fill the gap, so low were the costs of entry into the trade. In addition manufacturers used a variety of means of selling their products including peddlers, commission merchants and wholesale tobacconists (Tilley 1948: 521–40). The proliferation of manufacturers and marketing methods was matched by the proliferation of brands; one manufacturer in Winston advertised forty name brands (Tilley 1948: 522). In a foreshadowing of things to come, manufacturers named their brands after people, or events, or images, or places that they believed were attractive to a particular customer. Considering the enormous number of brands available, selecting a small sector of the market by brand naming was obviously a good move (Tilley 1948: 522–4).

The concentration by manufacturers in the Virginia–North Carolina tobacco belt on chewing tobacco was part of the general preference in American society for this form of tobacco consumption. But it was also a reflection of the fact that the local tobacco was especially suited for chewing. Bright tobacco, which increased in importance in this area, especially in North Carolina, around mid-century, also found favour as a chewing tobacco, partly because of its pleasing appearance and partly because Bright, flue-cured tobacco was capable of absorbing more moisture and additives than was the dark air-cured tobacco of the tidewater. The production of cigars and snuff, for which Bright tobacco was unsuitable, was located elsewhere in the country, using imported or northern-grown tobacco (Cooper 1988; Cox 1933).

Even as late as 1860, despite the prodigious industrialization of the American economy, tobacco manufacturing remained technically backward; there were many establishments, but the average size of the factories, in terms of their labour force was small; in Richmond, one of the most important industrial centres, an average factory on the eve of the Civil War employed fewer than seventy workers (O'Brien 1978: 512). Technological and organizational changes did occur, but only on a limited scale. The most important improvements were in economies of waste and of time (Tilley 1948: 493). Other improvements were made in mechanizing the manufacture of the chewing plug, but again these were very limited.

It is not immediately clear why chewing tobacco manufacture did not follow the trajectory taken by other American industrial concerns, at least until the 1870s. But, unknown to most of those who manufactured chewing tobacco, workers and owners alike, a revolutionary change was about to transform the industry beyond recognition. In the matter of a few decades after the Civil War not only was the tobacco industry internally revolutionized, but the firms that dominated the industry emerged as some of the largest in the country.

Before one can understand this phenomenon it is necessary to discuss another form of tobacco consumption, smoking. As pointed out in Chapter 4, Americans were unique among consumers in preferring to take tobacco by chewing it in plugs and twists, as the forms were termed. The manufacture of chewing tobacco was concentrated in the Virginia–North Carolina tobacco belt; northern tobacco manufacturers concentrated on cigar production. Though tobacco factories in Virginia and North Carolina were small when compared to other contemporary industrial establishments, especially in the northern states, they were much larger than the cigar-making concerns in New York and Pennsylvania, and there were fewer of them. In 1860, for example, there were around 350 tobacco factories in the two southern states; in 1912, even after a degree of consolidation of facilities, there were still 20,000 firms making cigars in the United States; Richmond, the most important tobacco manufacturing centre in the country, produced more tobacco, by value, than did the entire state of New York (Cooper 1988: 784; Robert 1938: 187–8).

Smoking tobacco factories, on the eve of the Civil War, were in a clear minority in the southern tobacco industry; only 2 per cent of the total establishments produced smoking tobacco (Robert 1938: 170). The largest smoking tobacco factory in the region, several times larger than its nearest rival, employed only twelve workers, yet it produced over 500,000 pounds of manufactured tobacco in 1860 (Robert 1938: 183). That there was little demand for the product in the country is attested to by several sources, both statistical and literary. Yet, unforeseen by anyone it seems, a combination of factors in the years after the end of the Civil War combined to create an upsurge in the demand for, as well as in the means of production of, smoking tobacco.

There were, in the first instance, the important developments in tobacco culture itself, especially in the flue-curing of Bright tobacco and the growing popularity of Burley tobacco, both of which produced a distinctly milder, more mellow and cooler-burning smoke. This, in itself, undoubtedly led to an increasing interest in smoking tobacco, but whether it explains the phenomenal rise of smoking is another matter. Among other factors that have been suggested was the invention and availability of the friction match, and then the safety match, from the mid-nineteenth century, making smoking a mobile pastime; the replacement of the clay pipe by the briar, again after mid-century; and the possible risk of contagion of spittle (Alford 1973: 110–11; Dunhill 1924). The briar pipe was especially well-suited to the flue-cured and Burley tobaccos, while the clay and the meerschaum pipes were better suited to dark air- and fire-cured varieties (Alford 1973: 111). There is also the possibility that the Civil War heavily disrupted trade in flavourings and sweeteners, the most important of which were imported, reducing the supply of these essential ingredients and thereby raising the price of the final product: plugs that could not sell at the price, those that spoiled, and tobacco that was unflavoured could readily be turned into smoking tobacco (Tilley 1948: 497–8).

Changes in the ingredients, and the means of consuming tobacco through smoking, certainly had a bearing on the growing popularity of both the pipe and the cigar; in per capita terms, the consumption of cigars tripled between 1850 and 1870, from ten to thirty cigars per annum (Gottsegen 1940: 10, 13). But there were also important changes occurring on the supply side that may have been more significant for the transformations in tobacco production.

In the area of technical innovation, and indeed, in the extent of technological inputs, the smoking tobacco industry was far more advanced than that of chewing tobacco. Whereas chewing tobacco was intensive of capital, with a low level of embedded technology, such as vats and presses, smoking tobacco could be produced with a minimum of capital equipment, no more than some simple tools to shred the tobacco. The investment in chewing tobacco manufacture before the Civil War mostly affected the final stages of pressing the tobacco into conveniently sized shapes, but evidence suggests that these changes actually led to an increase in the costs of production (Siegel 1987: 134). By contrast, most of the technical innovations in smoking tobacco manufacture increased physical productivity through mechanization. Between 1865 and 1885 significant advances were made in the cutting or shredding of tobacco, but there was also a growing tendency to use centralized motive power by way of steam engines. Even before the invention, and diffusion, of mechanical shredding machines, smoking tobacco factories were more concentrated than their counterparts in the manufacture of chewing tobacco; two tobacco factories, equipped with steam power in 1860, produced as much smoking tobacco as all the others put together (Robert 1938: 212). Steam power, together with an improved version of the shredding machine (first patented in 1866 by a manufacturer of agricultural machinery), launched the smoking tobacco industry on a path of industrialization. The prodigious volume of smoking tobacco that the new machinery produced put severe strain on traditional methods of packing the output and, when that problem was solved in 1885, manufacturing could be said to have been mechanized (Tilley 1948: 500–1).

Smoking tobacco could be produced in different degrees of fineness, from an almost granulated to a thickly shredded form. There is no indication, unfortunately, as to the relative value of the different grades of smoking tobacco during the early years of the expansion of this manufacture. What is clear, however, is that finely shredded tobacco was being consumed before the Civil War (Tilley 1948: 508). The significance of this needs to be stressed, for, while the industrialization of tobacco manufacture was located in the smoking tobacco sector, it was the manufacture of fine smoking tobacco, packaged in a cigarette, that integrated mechanization of production with mass consumption.

The history of the cigarette has been told in Chapter 5, where it was argued that, in the United States at least, cigarettes first gained popularity during the 1870s. Though the first cigarettes were made of Turkish tobacco it soon became clear to manufacturers that flue-cured tobacco could be used most profitably in this manner (Brooks 1937: 172). The first cigarettes were manufactured in New York in 1869, by F. S. Kinney and Company, using flue-cured tobacco, and employing a largely female labour force, instructed by East European cigarette rollers, hand-rolling for the market (Tilley 1948: 508). Once Kinney entered the market, others followed suit; William S. Kimball and Company of Rochester began manufacturing cigarettes in 1876, and, at about the same time, Allen and Ginter of Richmond, and Goodwin and Company of New York (Porter 1969: 61–2; Tilley 1948: 508; Tennant 1950: 19). In 1880 these four firms are estimated to have accounted for 80 per cent of the entire cigaratte output of the country (Tennant 1950: 19). In several significant ways the cigarette industry differed from other branches of the tobacco industry. There was, in the first place, a considerable degree of concentration of production in the early enterprises. Unlike the chewing, cigar and, to a lesser extent, smoking tobacco sectors, all of which were characterized by a proliferation of small-scale enterprises, entry into cigarette production was restrictive. Partly this was because of the scarcity, and hence the high price, of labour; and partly because those firms that entered the market did so as an extension of an existing smoking tobacco business (Tennant 1950: 17; Porter 1969: 61). Second, cigarette manufacturing was located in two regions of the country, in New York which had expertise in cigar manufacturing and in tobacco marketing, and Virginia and North Carolina where there was, of course, an abundance of tobacco knowledge (Robert 1938: 223–5). Unlike the other branches of the tobacco industry, the manufacture of cigarettes, as we will see, integrated the substantial and different skills available in the north and south, and this factor, more than anything else, was crucial in the industry's incredible development. Finally, the industry settled on a specific raw material, Bright flue-cured tobacco, as its characteristic ingredient, a tobacco that was American and distinctive.

In the first year of its existence the New York firms dominated the cigarette industry; in 1881, with a total output of over 380 million cigarettes, they accounted for 72 per cent of the country's total production (Tilley 1948: 510). The firm of Allen and Ginter was no match for a giant of the industry, such as the Kinney Company who had a branch plant in Richmond, but very quickly, possibly because of its insistence on using only Bright tobacco, the Richmond firm moved into the fast track: by 1883 they had a branch plant in London and were selling throughout Europe and Australia (Tilley 1948: 509).

There are no comparative production figures for the cigarette companies in the early years of the 1880s, but during this decade their performance was eclipsed by an extremely aggressive newcomer from Durham, North Carolina. James Buchanan Duke, the youngest son of Washington Duke, who had begun manufacturing smoking tobacco outside Durham after the close of the Civil War, quickly rose within the firm to a position of power. When in 1878 the family invited two outsiders to join the partnership, it was James Duke who was clearly at the helm (Tennant 1950: 22). The partnership ended in 1885, and the firm was incorporated as W. Duke, Sons and Co. (Porter 1969: 63). Four years later Duke was the largest cigarette company in the world, and one year later, in 1890, the five principal manufacturers of cigarettes – who together produced over 90 per cent of total American cigarette production – joined to form the American Tobacco Company, one of the largest American corporations, with James Duke as its president (Tennant 1950: 22).

The story of the rise of James Duke within his own firm, and then to the very top of the American cigarette, and finally American tobacco, industry has been told elsewhere and need not be repeated (Jenkins 1927; Winkler 1942; Porter 1969; Durden 1975). What is of concern here, however, is the effect that the changes in the organization of the industry had on the nature of tobacco as a commodity.

It is generally agreed that Duke's decision to manufacture cigarettes in 1881 was taken because he believed that he could not compete effectively against W. T. Blackwell and Company, manufacturers of the leading smoking mixture of the time, Bull Durham (Porter and Livesay 1971: 201). Though the other partners were not convinced by the wisdom of the move – Durham lacked skilled rollers, for example – the cigarette industry was booming: between 1870 and 1880 total cigarette consumption in the United States had increased from 14 to 409 million units (Gottsegen 1940: 28). This prodigious growth in output was accomplished by factory girls rolling several thousand cigarettes by hand each day. Because of a lack of economies of scale, output could increase within an unchanging production technology only by adding labour. The search for a mechanical means of mass production began to move up the managerial agenda in the cigarette industry, as it had across the broad spectrum of American industry (Chandler 1990; Bruchey 1989; Roberts and Knapp 1992). And it was precisely during this decade that a large number of machines designed to make cigarettes appeared, and/or were patented in the United States, Britain and France; but none was successful, and their absorption into production was minimal (Tennant 1950: 17–18; Porter 1969: 67). In 1881, ironically in the same year that Duke began to manufacture cigarettes, James Bonsack, himself from Virginia, patented a cigarette-making machine; two years later the machine was put on the market, on a rental basis only, by the newly-formed Bonsack Machine Company (Porter 1969: 67–8; Roberts and Knapp 1992). While other manufacturers, such as the rival Richmond firm of Allen and Ginter, declined to use the Bonsack machine or had their own machines – another patented device was used by Goodwin and Company of New York – Duke took immediately to the mechanical cigarette maker and ordered two, which were installed in 1884 (Tennant 1950: 21). In that year a single Bonsack machine produced between 100,000 and 120,000 cigarettes per day, equivalent to the labour of thirty to forty hand workers (Porter 1969: 690).

Because Duke provided the Bonsack Company with the first solid order in the American cigarette industry, he was able to negotiate extremely favourable terms for himself, particularly in reducing the licence fee, and finally in obtaining exclusive rights to the machine itself (Roberts and Knapp 1992: 277–8). The control over the Bonsack machine, combined with its physical productivity, not only secured Duke a leading position in the industry but also lowered production costs; they fell by more than 50 per cent, from 80 cents to 30 cents per thousand, and, as the machine was further improved, the costs fell even more, reaching, according to United States official calculations, no more than 8 cents per thousand in 1895 (Tennant 1950: 69).

Meanwhile, the Bonsack Company renewed its efforts to corner the cigarette technology market by purchasing competing patents – a business strategy typical of the period in highly capital-intensive industries – and by 1889 this was accomplished. The convergence of technology within the industry, together with a broadly similar raw material, resulted in a manufactured product that was, in essentials, undifferentiated from one manufacturer to the other. Prices were already at rock bottom, demand was levelling off, and though the initial response to this situation was a vigorous and, in retrospect, highly significant campaign to gain market share through advertising, eventually the forces for mergers grew to dominate the competitive environment. An agreement by the major manufacturers on leaf purchases in 1889 was the final step to integration which, as we have seen, was initiated and headed by Duke himself in 1890 (Porter 1969: 69–72).

For the first few years of its existence the American Tobacco Company concentrated on the manufacture of the cigarette on which the whole combination was founded. But Duke's vision extended well beyond the cigarette, and, as the name of the firm suggests, his strategy was to control as much as possible of American manufactured tobacco. In a series of battles with producers of other products, beginning with chewing tobacco between 1894 and 1898, and then moving on to snuff in 1899 and 1900 and finally to the cigar in 1901 and 1902, Duke attempted to buy out, or ruin, as many manufacturers as possible (Porter 1969: 74–5; Burns 1982). With the exception of the cigar industry, which proved extremely difficult to capture because of its fractured structure, Duke's strategy was very successful; in 1910 the tobacco trust constructed around the American Tobacco Company accounted for no less than 75 per cent of the country's manufactured tobacco output – cigarettes, snuff, chewing tobacco, smoking tobacco and cigarillos (Tennant 1950: 27).

What happened to tobacco manufacturing in the last two decades of the nineteenth century in the United States was part of a wider movement towards the creation of enormous corporations with monopolistic or oligopolistic control over production, technology and marketing. In one extremely significant area, the tobacco trust experience differed from that of most other industrial combinations. In the restructuring of the American industrial economy in the second half of the nineteenth century it became normal for firms to integrate both forwards, into distribution, and backwards, into the acquisition of raw material supplies (Chandler 1977). Though the tobacco trust, or the American Tobacco Company, in particular, purchased subsidiary companies that made boxes, foil and pipes, it did not involve itself in owning tobacco fields (Porter and Livesay 1971: 207). Why the trust chose not to enter tobacco cultivation is not entirely clear. There is the argument that the American Tobacco Company was not threatened by the suppliers of tobacco, and therefore saw no need to control the supply directly (Porter and Livesay 1971: 297). That there was no threat is certainly true, but that does not explain the company's reluctance to control supply directly; not all of the company's strategic actions were defensive. Most likely there was no need to control supply directly because the company already had all of the bargaining power it needed because it acted virtually as a single buyer, while most farmers, as we have seen, produced too little to affect the market price (Tennant 1950: 316–41). Whatever the reason, Duke's decision not to produce his own tobacco reinforced the enormously inequitable division of power between cultivators, on the one hand, and distributors and manufacturers, on the other, that characterized tobacco culture from the beginning, and continues to do so even now.

When the American Tobacco Company was formed, all of the signatories to the agreement, James Duke excepted, expressed misgivings as to the legality of the decision. Despite their worries, the combination was formed but the question of its legal basis did not disappear, though the concern shifted from the partners to the United States Department of Justice. After several important rulings against other combinations, notably Standard Oil, the Department of Justice ruled, in 1911, that the American Tobacco Company had infringed the Sherman Act – designed to safeguard industrial competition – and the trust was broken up into constituent parts. Assets and plants were redistributed into four newly created companies – the American Tobacco Company, Liggett and Myers, R. J. Reynolds, P. Lorillard – and a handful of much smaller firms (Tennant 1950: 60–1). Until the period immediately following the Second World War the three largest companies, American Tobacco, Reynolds, and Liggett, accounted for as much as 80 per cent of cigarette output, which itself was growing in importance when compared to other manufactured products (Tennant 1950: 94). It has been only in the last four or five decades that this position has been altered, partly because of the rise in importance of some of the smaller producers, such as Philip Morris, and Brown and Williamson, but also because of the development of multibranding (Tennant 1971: 227). In 1979, Philip Morris accounted for 29 per cent of domestic American sales, and Brown and Williamson a further 15 per cent; in 1930 the corresponding figures were 0.4 per cent and 0.2 per cent respectively (Johnson 1984: 23). R. J. Reynolds and Philip Morris are now the largest cigarette producers in the United States.

James Duke's insistence on mechanizing cigarette production; his dedication to, and dependence upon, advertising as a means of increasing market share and overall demand; and his corporate strategy – these all revolutionized the American tobacco industry. Duke's innovations did not end there, however. Duke was crucial to the history of tobacco in other ways. First, his corporate strategy in controlling markets was not limited to the United States. In 1901 American Tobacco bought Ogden Ltd, one of Britain's leading tobacco manufacturers (Alford 1973: 250). By comparison with the United States the British tobacco industry was not only much smaller but far less concentrated. Small firms made up the majority of the industry and there was no movement towards concentration, as in the United States. Nevertheless, six firms, including Wills in Bristol, Cope in Liverpool, Lambert and Butler in London, John Player in Nottingham and Mitchell in Glasgow, accounted for about 20 per cent of total tobacco sales in Britain near the end of the nineteenth century (Alford 1973: 161). Ogden was not in this group, but its strength was increasing. Duke's purchase signalled his intentions not only of extending his operations beyond the American shore but also of capturing the market in Britain as well as Europe (Alford 1973: 251). The response of British manufacturers to Duke's incursion was swift and powerful. Within a few months, the thirteen largest tobacco manufacturers combined to form a new company, the Imperial Tobacco Company headed by W. D. and H. O. Wills Ltd. (Alford 1973: 263). A year later three more firms joined Imperial Tobacco (Alford 1973: 267–8). Imperial attempted to withstand competition from American Tobacco by increasing advertising, relying upon the brand strength of its chief products and by negotiating to purchase an American tobacco manufacturing firm of its own in retaliation (Alford 1973: 267–8). The strategy paid off, and in the same year, 1902, as Imperial Tobacco constituent companies were enlarged, American Tobacco settled on a negotiated truce, the results of which had a far-reaching impact on the history of tobacco.

Likened to the division of the world drawn up by Pope Alexander VI in the Treaty of Tordesillas, and signed by Spain and Portugal, the agreement between American Tobacco and Imperial Tobacco affected a similar organization of the world's tobacco market (Alford 1973: 269). American Tobacco withdrew from the British market, as Imperial did from the American market, though they agreed to retain trading rights in each other's brands: tobacco demand in the rest of the world was to be supplied by a new company, two-thirds of which was owned by American Tobacco, and the other third by Imperial. It was registered in Britain and took as its name the British American Tobacco Company Ltd (Alford 1973: 269; Cox 1989: 45–6).

Duke's second main innovative action was to tap markets for cigarettes outside the United States. He had already embarked on an export drive in 1883, when he sent one of his salesmen, R. H. Wright, on a nineteen-month world trip (Jenkins 1927: 72). The world tobacco market was not wide open, however. It was, for example, almost impossible to break into those European markets controlled by monopolies. Markets in areas of European settlement, especially Canada, Australia and South Africa, presented obstacles, especially the presence of British firms, particularly Wills (Cox 1989: 49–50; Alford 1973: 217–19). The most promising markets, therefore, appeared to be in the Far East, particularly those countries not under colonial control, and given to preferential trading structures. Indeed there is a story that Duke had already targeted China as his company's main export market upon hearing of the Bonsack machine's capabilities (Cochran 1986: 152). In 1888 Duke entrusted James Thomas with the task of opening markets in the Far East, and in 1890 sold his first cigarettes there; in 1902 the Chinese market was absorbing 1.25 billion cigarettes (Cochran 1986: 152; Thomas 1928). That exports became increasingly important to the firm is borne out by the fact that in 1898, according to one estimate, one-third of Duke's production of cigarettes was exported (Alford 1973: 217). Around the turn of the twentieth century the Asian market, principally China, accounted for 54 per cent of all cigarettes exported from the United States (Jacobstein 1907: 172 [434]). A small factory was established in 1891, but the real assault on the Chinese market did not occur until 1903 when, under the new British American Tobacco Company, the factory in Shanghai began producing cigarettes from imported American tobacco leaf (Cochran 1986: 155–6). After several years of expansion, including the opening up of factories in Hankow and Manchuria, British American Tobacco was selling 12 billion cigarettes in 1916, of which between one-half and one-third was manufactured in China, some part of it being manufactured from Bright tobacco grown in the country (Cochran 1986: 158, 163–4).

Once British American Tobacco had made successful inroads into the Chinese market, it began to exploit other opportunities, first in India, where Wills had already begun operations before the turn of the century. The strategy of gaining a foothold in the Indian market was based on the experience in China: first, the company marketed its imported products and then began to manufacture its own cigarettes (Cox 1989: 52). BAT also moved into British Malaya and the Dutch East Indies. In Africa the company first operated in Egypt – by the late 1920s four factories were manufacturing cigarettes in the country – but the rest of Africa was also opened up within a short period of time, though manufacturing facilities were not established until the 1930s (Cox 1989: 53). Ironically, Japan was the only country where Duke was forced to give up, and this after it had, in 1899, acquired a controlling interest in one of the most important Japanese tobacco firms (Durden 1976). After describing Duke as a 'capitalist . . . intending to monopolize the whole world', the Japanese government nationalized the tobacco industry, forcing Duke out in 1904 (Cochran 1986: 183–5).

The late nineteenth and the twentieth century have witnessed the total transformation of the tobacco industry from one characterized by small-scale and labour-intensive production, with either local or national distribution and marketing systems, to one characterized by multinational enterprise in all sectors. The change from one state to the other was very rapid, occurring in a span of less than two decades. The present shape of the industry has resulted clearly from the early actions by Duke, the formation, and final dissolution, of the American Tobacco Company, but there have been other factors as well.

In the first place, in terms of the total production of manufactured tobacco products, cigarettes have continued to account for an increasing proportion of the total production. In the United States, at the turn of the twentieth century, cigarettes accounted for only 3.4 per cent of leaf consumed; the figure reached 50 per cent in 1937 and continued to grow unabated to 77 per cent in 1967 (Tennant 1971: 217). Today the figure would be approaching 90 per cent, as it is in most of the developed world.

On a global level, the corresponding figure for 1980 can be estimated at around 70 per cent (Tucker 1982: 35, 173; Johnson 1984: 65).

A second important change was the explosion in the number of cigarettes that were produced, first in the United States and then throughout the world. The increase is staggering. In the United States in 1870 there were 16 million cigarettes made which, though it appears to be a lot, should be compared to the 1.2 billion cigars made in the same year (Johnson 1984: 16). The billion figure was reached in 1885, after the successful use of the Bonsack machine: in 1981 the comparable figure was 734 billion (Johnson 1984: 19). In 1980 the world consumed 4,373 billion cigarettes; in 1988 5,270 billion (Tucker 1982: 190; Grise 1990: 22).

Possible explanations of the phenomenal success of the cigarette, in both relative and absolute terms, have been offered in Chapter 5, but one factor has not been mentioned: the change in technology that made the growth in production levels possible. The Bonsack machine, at the turn of the twentieth century, was able to produce 500 cigarettes per minute. Improvements to that machine, as well as other machines, raised physical productivity considerably (Hall 1978). In 1976 one of the most widely used cigarette machines, the Molins machine (partly owned by BAT and Imperial), was rated at 5,000 cigarettes per minute (Clairmonte 1979: 265). There are few cigarette technologies available, and they are marketed globally; cigarette companies have financial interests in them (UNCTAD 1978: 18).

In the western world the date at which the cigarette has dominated consumption – 50 per cent or more, by weight of total consumption – has varied considerably. Several countries had already gone over to cigarettes in the 1920s, notably Britain, Turkey and Greece; but most switched over either during or immediately after the Second World War (Rogoziński 1990: 113). The Netherlands was later than most, and a few countries, Norway and India in particular, are still wedded to cut tobacco, as in the case of Norway, or traditional forms such as *bidis* and hookahs in India (Rogoziński 1990: 113). Japan had changed to cigarettes as early as 1923, and China probably by the same time (Rogoziński 1990: 113).

This remarkable expansion of the cigarette habit has been satisfied by a very small number of enterprises. In 1980 multinational tobacco companies furnished 35 per cent of the total output, but monopolies accounted for 55 per cent (Tucker 1982: 69). The principal monopolies are located in France, Italy, Spain and Russia in Europe; China and Japan in Asia. The largest monopolies in the world are those in Japan and China. In 1980, the former manufactured 307 billion cigarettes in thirty-six factories, had a turnover of $11 billion and employed 40,000 people directly and 460,000 indirectly (Tucker 1982: 106–7). The Japanese tobacco monopoly is a nationalized industry, but in China it is part of the Ministry of Light Industry. In 1980 the latter manufactured 750 billion cigarettes in 1,000 different brands in eighty-three factories – the Shanghai complex alone produced 41 billion cigarettes in the same year (Tucker 1982: 113–14). BAT, as the largest foreign company in China, was expelled in the 1949 Revolution, but in 1980 Chinese cigarette factories, under agreement with R. J. Reynolds, started manufacturing Camel cigarettes (Tucker 1982: 115). The largest monopoly in Europe is that of the former Soviet Union, where in 1980 350 billion cigarettes were produced: all of the Eastern European countries have their own monopolies (Tucker 1982: 116–17). SEITA, the French monopoly, MONTAL, the Italian monopoly and Tabacalera, the Spanish monopoly are large manufacturers – SEITA, for example, produced 86 billion cigarettes in 1980 – and, though they are technically no longer monopolies under the laws of the EC, they are still responsible for the production of 'domestic' brands (Tucker 1982: 109).

The monopolies may be the world's largest producers of manufactured tobacco, but all of their manufacturing facilities are located in the country of jurisdiction and, with few exceptions, their products are sold nationally: only SEITA has made any inroads into the global marketplace. It is precisely on this point that the multinationals differ, for, while there are many fewer multinational tobacco companies than monopolies, they are truly global in their operations, in both production and marketing, and, one might add, in purchasing leaf. In 1980 five tobacco multinational companies manufactured just over 1,500 billion cigarettes. Table 9.1 provides a profile of their sales, employees and cigarette production in that year.

Table 9.1 Multinational company cigarette output 1980

Company	Sales ($ billion)	Employees (number) (000)	Output (billions)
BAT	8.3	280	550
Philip Morris	5.3	72	400
R. J. Reynolds	5.0	83	280
Rothmans International	4.1	19	190
American Brands	4.3	52	110

Source: Tucker 1982: 70

With the exception of Rothmans each of the tobacco multinationals had its origin in the dissolution of the American Tobacco Company in 1911. They have all developed globally, but through a series of interlocking (and changing) agreements, they have certain areas of the world market where they predominate. BAT is the most international of the companies, operating on each continent, in manufacturing as well as sales. In 1980 its largest market was Europe, as it was for Rothmans and American Brands (Tucker 1982: 70). The United States market, the single most important market outside the monopoly system, was served almost entirely by Reynolds, Philip Morris, BAT (through Brown and Williamson) and American

Brands, the first two companies accounting for over 60 per cent of total cigarette consumption in the country (Johnson 1984: 23). None of the companies relies exclusively on tobacco for their turnover: they are all well diversified, though important differences exist in the nature and degree of their diversification (Tucker 1982: 71–103).

The cigarettes of multinational companies have been available, throughout the world, for a large part of the twentieth-century, and their availability, and penetration of domestic markets, even those protected by monopolies, continues to grow. Well behind the multinationals, as a group at least, are the national producers, those with significantly little international production and marketing. These companies include Imperial – in 1980 the company produced 70 billion cigarettes – in the United Kingdom, and Reemtsma in Germany (Tucker 1982: 104). Lorillard and Liggett are the main national companies in the United States.

Conclusion

Wayne McLaren, who portrayed the rugged 'Marlboro Man' in cigarette ads but became an anti-smoking crusader after developing lung
cancer, has died, aged 51 . . . His mother said: 'Some of his last words
were: "Take care of the children. Tobacco will kill you, and I am
living proof of it".' . . . Mr McLaren, a rodeo rider, actor and Holly-
wood stuntman . . . was a pack-and-a-half-a-day smoker for about
25 years. In an interview last week, Mr McLaren said his habit had
'caught up with me. I've spent the last month of my life in an
incubator and I'm telling you, it's just not worth it.'

Guardian, 25 July 1992

10

TO DIE BY SMOKE
Whither tobacco?

One of this book's principal objectives is to explore the processes and reasons for tobacco's pervasive global entrenchment. That entrenchment has many aspects to it, and there have been many books written about tobacco that have explored this phenomenon. Most, however, have adopted a contemporary perspective. The argument of this book is that the process of tobacco's entrenchment has a profound historical dimension that is much more than simply a background to the contemporary issues. Indeed the usual explanations offered by contemporary commentators on the 'tobacco problem', such as the addictive properties of nicotine, central government demand for revenue through taxation, underdevelopment and underemployment in the Third World and the corporate strategies of tobacco manufacturers, have all been determined historically.

Tobacco and smoking pose massive problems of health, welfare and ecology. Though there have always been dissenting voices against tobacco, not until this century, and especially since the 1950s, has tobacco been condemned from so many quarters. There is, in the West at least, a growing sense of the smoker as a social outcast, of smoking as a vice. The rights of non-smokers have been publicly acknowledged while those of the smoker have been sacrificed. Tobacco smoking has become the subject of intense ethical debate (Goodin 1989a; Goodin 1989b).

Why? There are many reasons for the changed meaning of smoking, but before looking at these it might be well to preface the discussion by emphasizing that the meaning has changed only for a small proportion of the world's population, and that these people are to be found largely in the industrialized West, predominantly among the middle classes. The meaning has not altered for most people in the Third World, and for a large proportion of the industrialized population; neither has it changed for most governments; and certainly not for multinational tobacco companies and state tobacco monopolies. Perhaps, rather than speaking about the changed meaning of smoking, as some have done, it is more accurate to describe what is happening now, as part of the historical process of definition in which several key arguments are involved. What will finally emerge remains to be seen.

Those who seek to define smoking as dangerous and the cigarette as a noxious artefact have powerful arguments on their side. First and foremost, tobacco kills. The latest reports on what is termed 'smoking-attributable mortality' make for depressing reading: the number of deaths is appalling, and has been rising. On the situation in the developed world alone, I quote:

> annual deaths from smoking number about 0.9 million in 1965, 1.3 million in 1975, 1.7 million in 1985, and 2.1 million in 1995 (and hence about 21 million in the decade 1990–99: 5–6 million European Community, 5–6 million USA, 5 million former USSR, 3 million Eastern and other Europe, and 2 million elsewhere [i.e. Australia, Canada, Japan and New Zealand]) . . . at present just under 20% of all deaths in developed countries are attributed to tobacco, but this percentage is still rising, suggesting that on current smoking patterns just over 20% of those now living will eventually be killed by tobacco . . .
>
> (Peto et al. 1992: 1,268)

Smoking-attributable mortality worldwide in 1991 was estimated at 3 million by the World Health Organization, roughly two-thirds in the developed, and one-third in the developing, world (USDHHS 1992: 91). But this pattern is certain to change as the cigarette smoking epidemic spreads through the Third World, as all commentators predict it will (Peto and Lopez 1990: 66–8; Chapman and Wong 1990).

In 1990 the United States Surgeon General stated that 'smoking represents the most extensively documented cause of disease ever investigated in the history of biomedical research' (USDHHS 1990). This outpouring of research has resulted in a second argument against tobacco use: the enormous dangers of passive smoking (USDHHS 1986). Evidence from various studies suggests that passive smoking increases the likelihood of getting lung cancer by more than 30 per cent (Goodin 1989a: 598); a recent large-scale study of non-smoking wives of smoking husbands in Japan showed that two-thirds of those women who were non-smokers and died from lung cancer were married to husbands who smoked (Hirayama 1990: 37). The decision by many governments, both central and local, to ban smoking from public places reflects the enormous impact that the statements on the health hazards of passive smoking have had on public opinion. Further prohibitions on smoking in public places are expected to come soon.

The third argument in the anti-smoking, and anti-tobacco, arsenal is the social cost of smoking. How much it costs is a matter of debate but the figures that have been proposed are devastating. In a comprehensive study of the costs of smoking, which included medical costs and costs of productivity losses because of mortality and morbidity, the United States Office of Technology Assessment calculated that smoking cost the United States anywhere between $39 billion and $96 billion annually (USDHHS 1992: 110–11). Comparable figures for other countries are relatively scarce (Markandya and Pierce 1989; FAO 1989: 22–3).

Finally, the identification of nicotine as an addictive drug has led to a change in the image of the smoker and the appropriation of a lexicon typically reserved for descriptions of addicts of hard drugs. Moreover, it has also led to a restoration of the concept of tobacco use as a disease in itself, partly by a change of language, introducing, for example, words such as 'nicotinism' and 'tobacconism'. In other words nicotine addicts are seen as needing help to quit (as most smokers attempt to do more than once in their smoking life); and the act of smoking is no longer portrayed as 'a private-regarding vice' but rather as a serious addiction: '. . . once you have become addicted to nicotine, your subsequent smoking cannot be taken as indicating your consent to the risks' (Goodin 1989a: 574, 587).

Cigarette companies have continued to deny that cigarette smoking causes disease: governments continue to contradict themselves, giving health advice and legislating for the rights of non-smokers, on the one hand, and collecting excise and duty taxes, and supporting tobacco growers with subsidies, on the other hand (Warner 1991; Joosens and Raw 1991; Grise 1990). On an international level, contradictions abound. The World Bank, for example, has been actively involved in supporting tobacco production in the developing world through loans: between 1974 and 1988 over $1.5 billion of World Bank money was placed at the disposal of tobacco projects supposedly to help the developing world (Chapman and Wong 1990: 30–1). Moreover, governments who publicly pronounce on environmental damage caused by modern agriculture, and the depletion of the world's rain forests, turn a blind eye to tobacco even when there is ample evidence that tobacco growing has profound ecological implications. Soil nutrient depletion is one of the problems, but another, and less publicized, problem is the use of wood in curing tobacco (Chapman and Wong 1990: 32). This is a very controversial subject and different studies give different results on the impact on the forests of tobacco processing. Some reports present the following statistics: one tree is felled for every 300 cigarettes; 1 hectare of tobacco requires between 0.5 and 1 hectare of woodland for curing; one in twelve trees felled worldwide is used in curing tobacco; or, each kilo of tobacco demands about 160 kilos of wood (USDHHS 1992: 125). Another report, based on detailed data for seven developing countries, presented a rather different picture, downgrading the environmental impact of tobacco processing and publishing relatively low figures for specific fuel consumption – the ratio of wood usage to tobacco. Rather than the figure of 160, these reports offered an average of 7.8, but conceded that on some individual estates the level ran as high as 40 (FAO

1989: 13). Nevertheless, while concluding that tobacco processing did not pose an ecological threat in general, the report did stress the point that many parts of the world with serious deforestation problems also grow tobacco (FAO 1989: 14; Chapman and Wong 1990: 57–63).

Whither tobacco? In recent years tobacco companies have been put in the position of having to defend their product. Decisions in court have raised issues of liability and responsibility that have far-reaching implications. Perhaps we are getting to the position, well described by Robert Goodin, of undoing a remarkable bit of cultural hypocrisy. As he so aptly states: 'Cigarettes kill 25% of their users, even when used as their manufacturers intended they be used. Suppose a toaster or lawnmower had a similar record. It would be whipped off the market forthwith' (Goodin 1989a: 588). However, it is hard to envisage this happening even in the West – and the battle in the developing world has hardly begun.

One of the real ironies of tobacco is that those who consume it are addicted to a drug, nicotine, which has powerful, but not generally harmful, results. There is nothing inherently wrong in this. Every society uses mind-altering substances to a greater or lesser extent, and both history and archaeology confirm that the history of substance use, and abuse, is as old as human kind itself. The unbelievably rapid, and permanent, absorption of tobacco in so many different cultures across the world is testimony to the overwhelming attraction that mind-altering substances have had, and continue to have. The tragedy of tobacco is that the drug is consumed in a deadly form. Nicotine, caffeine and ethanol form a triad of acceptable drugs in many of the world's cultures, where there are no specific proscriptions against them (such as occur in Islam with regards to alcohol). According to Professor M. A. H. Russell, a leading expert on nicotine and smoking, it is tobacco, not nicotine, that should be expunged. Here is his version of a possible future:

> Some time in the 21st century we may see the demise of tobacco use, but it is doubtful whether nicotine use will be abandoned. It is the impurities in tobacco and its smoke which kill, while nicotine provides most of the pleasure, stimulation and relief from stress. It is not so much the potential of purer forms of nicotine as temporary aids to smoking cessation, but their potential use for long-term self-administration which merits the most serious consideration. Conventional tobacco products may in future be as archaic as the unrefined use of alkaloids in folk medicine appears now in comparison with the modern products of the pharmaceutical industry. The principle for all drugs has been to purify them as much as possible. If the tobacco industry does not do this with its drug, the pharmaceutical industry will. It is beginning to do so already.
>
> (Russell 1987: 47)

244

However desirable this outcome, it fails to address the cigarette as a cultural artefact. As an exercise in commodity history, this book has argued that tobacco is best understood in historical terms. The cigarette is the result of a complex process of cultural accretion of which changes in cultivation, production and marketing are an essential part. Any attempt to eradicate tobacco from our lives, however well meant, will founder unless the complexity of its cultural significance is recognized.

GLOSSARY

Tobacco is categorized according to type and method of curing, as well as to whether it is light, dark or Oriental. The latter distinction refers to colour as well as quality. The table below shows the main methods of curing and types referred to in the text.

Method of curing	Type
Air	Burley
	Maryland
	Cigar
Sun	Oriental/Turkish
Fire	Virginian
Flue	Bright/Virginia

Source: Akehurst 1981: 31

Burley (mostly), Maryland and Virginian flue-cured tobacco is light; cigar tobacco and Virginian fire-cured tobacco is dark.

Types

Burley. Also referred to as White Burley, this tobacco type first appeared as a mutant in 1864, in a tobacco field in Ohio. It has a very light nature, blends well with other tobacco types and has very high absorbent characteristics. First used in chewing tobacco, it is now the essential filler of the American blend cigarette, accounting for one-third of its composition.

Bright. This type, also referred to as Virginia outside the United States, was developed in the nineteenth century, in North Carolina, together with the process of flue-curing. It is light-bodied, bright yellow in colour and relatively low in nicotine. Almost all of this tobacco is destined for cigarette production. The terms 'Bright' and 'flue-cured' are synonymous. The 'Virginia' cigarette is composed of 100 per cent flue-cured tobacco.

Oriental. This type, formerly known as Turkish, produces a very aromatic product, used primarily in American blend cigarettes, accounting for about 15 per cent of the tobacco mixture. It is mostly grown in the eastern Mediterranean and Black Sea coastal areas, its traditional home. The distinctiveness of Oriental tobacco is the result of specific environmental conditions and the chemical changes caused by sun-curing the leaf.

BIBLIOGRAPHY

MANUSCRIPT SOURCES

Arquivo Histórico Ultramarino, Lisbon: Baia, Timor, Macau
British Museum, London: Additional manuscripts

PRINTED SOURCES

Abel, E. L. (1980) *Marihuana: The First Twelve Hundred Years*, New York: Plenum.

Åberg, E. (1980) 'Agricultural aspects of growing tobacco and alternative crops', in L. M. Ramström (ed.) *The Smoking Epidemic, a Matter of Worldwide Concern*, Stockholm: Almqvist & Wiksell International.

Acuña Ortega, V. H. (1978) 'Historia economica del tabaco en Costa Rica: epoca colonial', *Anuario de Estudios Centroamericanos* 4: 279–392

Adams, K. R. (1990) 'Prehistoric reedgrass (Phragmites) "cigarettes" with tobacco (Nicotiana) contents: a case study from Red Bow Cliff Dwelling, Arizona', *Journal of Ethnobiology* 10: 123–39.

Adshead, S. A. M. (1992) *Salt and Civilization*, London: Macmillan.

Akehurst, B. C. (1981) *Tobacco* second edition, London: Longman.

Akernecht, E. H. (1964) 'Panaceas', *Documenta Geigy* n.v.: 8.

Albrecht, P. (1988) 'Coffee-drinking as a symbol of social change in continental Euorpe in the seventeenth and eighteenth centuries', *Studies in Eighteenth-Century Culture* 18: 91–103.

Alford, B. W. E. (1973) *W. D. & H. O. Wills and the Development of the U. K. Tobacco Industry, 1786–1965*, London: Methuen.

Anderson, F. J. (1977) *An Illustrated History of the Herbals*, New York: Columbia University Press.

Andrews, J. R. and Hassig, R. (1984) *Treatise on the Heathen Superstitions that Today Live among the Indians Native to this New Spain 1629 by Ruiz de Alcarón*, Norman, Oklahoma: University of Oklahoma Press.

Andrews, K. R. (1984) *Trade, Plunder and Settlement*, Cambridge: Cambridge University Press.

Anon. (1712) *The Virtues and Excellency of the American Tobacco Plant for Cure of the Diseases and Preservation of Health*, London.

Antonil, A. J. (1965) *Cultura e Opulencia do Brasil por suas Drogas e Minas* 1711, translated by Andrée Mansuy, Paris: Institut des Hautes Etudes de l'Amérique Latine.

Apperson, G. L. (1914) *The Social History of Smoking*, London: Martin Secker.

Appleby, J. C. (1987) 'An association for the West Indies? English plans for a

West Indies Company 1621–1629', *Journal of Imperial and Commonwealth History* 15: 213–41.

Arcila Farias, E. (1946) *Economia Colonial de Venezuela*, Mexico City: Fondo de cultura económico.

Arents, G. (1938) *Books, Manuscripts and Drawings Relating to Tobacco from the Collection of George Arents, Jr.*, Washington, DC: Library of Congress.

Ashton, H. and Stepney, R. (1982) *Smoking: Psychology and Pharmacology*, London: Tavistock.

Atkinson, D. and Oswald, A. (1969) 'London clay tobacco pipes', *Journal of the British Archaeological Association* 3rd series 32: 171–227.

Austen, R. A. and Smith, W. D. (1990) 'Private tooth decay as public economic virtue: the slave–sugar triangle, consumerism and European industrialization', *Social Science History*, 14: 95–115.

Axton, W. F. (1975) *Tobacco and Kentucky*, Lexington, Kentucky: The University Press of Kentucky.

Ayton, E. G. (1984) *Clay Tobacco Pipes*, Aylesbury: Shire.

Badger, A. J. (1980) *Prosperity Road: The New Deal, Tobacco and North Carolina*, Chapel Hill, North Carolina: University of North Carolina Press.

Baer, W. (1933) *The Economic Development of the Cigar Manufacturing Industry in the United States*, Lancaster, Pennsylvania: The Art Printing Company.

Baillard, E. (1668) *Discours du tabac ou il est traité particulièrement du tabac en poudre*, Paris.

Bakewell, P. (1987) 'Mining', in L. Bethell (ed.) *Colonial Spanish America*, Cambridge: Cambridge University Press.

Barbachano, A. E. (1982) 'Datos prehispaniquos del tabaquismo en America', *Boletin de la sociedad mexicana de historia y filosofia de la medicina* 6: 29–42.

Barbour, V. (1963) *Capitalism in Amsterdam in the 17th Century*, Ann Arbor, Michigan: University of Michigan Press.

Barker, T. C. (1954) 'Smuggling in the eighteenth century: the evidence of the Scottish tobacco trade', *The Virginia Magazine of History and Biography* 62: 387–99.

Baroni, W. (1970) ' "La tabaccheide" di Girolamo Baruffaldi', *Medicina nei Secoli* 7: 3–6.

Basu, C. (1988) *Challenge and Change. The ITC Story: 1910–1985*, London: Sangam Books.

BAT Industries (1990) *Annual Report and Accounts, 1990*.

Batie, R. C. (1976) 'Why sugar? Economic cycles and the changing of staples on the English and French Antilles, 1624–54', *Journal of Caribbean History* 8: 1–41.

Baudry, J. (1988) *Jean Nicot: à l'origine du tabac en France*, Lyon: La Manufacture.

Bean, L. J. and Vane, S. B. (1978) 'Cults and their transformations', in R. F. Heizer (ed.) *Handbook of North American Indians* vol. 8, California, Washington, DC: Smithsonian Institution.

Bean, R. (1973) 'War and the birth of the nation state', *Journal of Economic History* 33: 203–21.

Beckles, H. M. (1981) 'Sugar and white servitude: an analysis of indentured labour during the sugar revolution of Barbados 1643–1655', *Journal of the Barbados Museum and Historical Society* 36: 236–46.

—— (1985) 'Plantation production and white "proto-slavery": white indentured servants and the colonisation of the English West Indies, 1624–1645', *The Americas* 41: 21–45.

Beckles, H. M. and Downes, A. (1987) 'The economics of transition to the black labor system in Barbados, 1630–1680', *Journal of Interdisciplinary History* 18: 225–47.

Beer, G. L. (1959 reprinted) *The Origins of the British Colonial System 1578–1660*, Gloucester, Massachusetts: Peter Smith.

Beese, D. H. (ed.) (1968) *Tobacco Consumption in Various Countries*, London: Tobacco Research Council Research paper 6, second edition.

Benet, S. (1975) 'Early diffusion and folk uses of hemp', in V. Rubin (ed.) *Cannabis and Culture*, The Hague: Mouton.

Bennett, J. H. (1965) 'Peter Hay, proprietary agent in Barbados, 1636–1641, *Jamaican Historical Review* 5: 9–29.

Benzoni, G. (1857) *History of the New World*, Oxford: Hakluyt Society.

Berardi, G. (1981) 'Can tobacco farmers adjust to mechanization? A look at allotment holders in two North Carolina counties', in W. R. Finger (ed.) *The Tobacco Industry in Transition*, Lexington, Massachusetts: Lexington Books.

Bergstrom, P. V. (1985) *Markets and Merchants: Economic Diversification in Colonial Virginia*, New York: Garland.

Bernhard, V. (1977) 'Poverty and the social order in seventeenth-century Virginia', *The Virginia Magazine of History and Biography* 85: 141–55.

—— (1985) 'Bermuda and Virginia in the seventeenth century: a comparative view', *Journal of Social History* 19: 57–70.

—— (1988) 'Beyond the Chesapeake: the contrasting status of blacks in Bermuda, 1616–1663', *Journal of Southern History* 54: 545–64.

Best, J. (1979) 'Economic interests and the vindication of deviance: tobacco in seventeenth century Europe', *The Sociological Quarterly* 20: 171–82.

Birnbaum, E. (1957) 'Vice triumphant: the spread of coffee and tobacco in Turkey', *Durham University Journal* 49: 21–7.

Bizière, J. M. (1979) 'Hot beverages and the enterprising spirit in eighteenth century Europe', *Journal of Psychohistory* 7: 135–45.

Black, P. W. (1984) 'The anthropology of tobacco use: Tobian data and theoretical issues', *Journal of Anthropological Research* 40: 475–503.

Blakeslee, D. J. (1981) 'The origin and spread of the calumet ceremony', *American Antiquity* 46: 759–68.

Blanchard, P. (1806) *Manuel du commerce des Indes Orientales et de la Chine*, Paris.

Blegny, N. de (1687) *Le bon usage du thé, du caffé et du chocolat*, Paris.

Bödecker, H. E. (1990) 'Le café allemand au XVIIIe siècle: une forme de sociabilité éclairée', *Revue d'histoire moderne et contemporaine* 37: 570–88.

Boesen, J. and Mohele, A. T. (1979) *The 'Success Story' of Peasant Tobacco Production in Tanzania*, Uppsala: Scandinavian Institute of African Studies.

Bolton, H. E. (1987) *The Hasinai*, Norman, Oklahoma: University of Oklahoma Press.

Bowen, W. H. (1938) 'The earliest treatise on tobacco: Jacques Gohory's "Instruction sur l'herbe Petun" ', *Isis* 28: 349–63.

Boxer, C. R. (1959) *The Great Ship from Amaçon: Annals of the Macao and the Old Japan Trade, 1555–1640*, Lisbon: Centro de Estudos Históricos Ultramarinos.

Boyd-Bowman, P. (1976) 'Patterns of Spanish emigration to the Indies until 1600', *Hispanic American Historical Review* 56: 580–604.

Brandt, A. M. (1990) 'The cigarette, risk, and American culture', *Daedalus* 119, Fall: 155–76.

Braudel, F. (1981) *The Structures of Everyday Life*, London: Collins.

Brecher, E. M. *et al.* (1972) *Licit and Illicit Drugs*, Boston: Little, Brown & Company.

Breen, T. H. (1984) 'The culture of agriculture: the symbolic world of the Tidewater planter, 1760–1790', in D. D. Hall, J. M. Murrin and T. W. Tate (eds) *Saints and Revolutionaries*, New York: Norton.

—— (1985) *Tobacco Culture*, Princeton, New Jersey: Princeton University Press.

—— (1986) 'An empire of goods: the anglicization of colonial America, 1690–1776', *Journal of British Studies* 25: 467–99.

—— (1988) ' "Baubles of Britain": the American and consumer revolutions of the eighteenth century', *Past and Present* 119: 73–104.

Brennan, T. (1991) 'Social drinking in Old Régime Paris', in S. Barrows and R. Room (eds) *Drinking: Behavior and Belief in Modern History*, Berkeley, California: University of California Press.

Brenner, R. (1972) 'The social basis of English commercial expansion, 1550–1650', *Journal of Economic History* 34: 361–84.

Brockbank, W. and Corbett, O. R. (1954) 'DeGraaf's Tractatus de Clysteribus', *Journal of the History of Medicine* 9: 174–90.

Brongers, G. A. (1964) *Nicotiana Tabacum: The History of Tobacco and Tobacco Smoking in the Netherlands*, Groningen: Niemeyer.

Brooks, J. E. (1937–52) *Tobacco, its History Illustrated by the Books, Manuscripts and Engravings in the Collection of George Arents, Jr.*, 5 vols, New York: The Rosenbach Company.

Brown A. (1964 reprint) *The Genesis of the United States* volume II, New York: Russell & Russell.

Brown, I. W. (1989) 'The calumet ceremony in the southeast and its archaeological manifestations', *American Antiquity* 54: 311–31.

Bruchey, S. (1989) *Enterprise*, Cambridge, Massachusetts: Harvard University Press.

Buker, G. (1970) 'The seven cities: the role of a myth in the exploration of the Atlantic', *American Neptune* 30: 249–59.

Burnham, J. C. (1984) 'Change in the popularization of health in the United States', *Bulletin of the History of Medicine* 58: 183–97.

—— (1989) 'American physicians and tobacco use: two surgeons general, 1929 and 1964', *Bulletin of the History of Medicine* 63: 1–31.

Burns, M. R. (1982) 'Outside intervention in monopolistic price warfare: the case of the "plug war" and the Union Tobacco Company', *Business History Review* 56: 32–53.

Busbey, R. L. (1936) *A Bibliography of Nicotine*, Washington, DC: US Department of Agriculture, Bureau of Entymology and Plant Quarantine.

Butel, P. (1989) *Histoire du thé*, Paris: Editions Desjonquères.

C. T. (1615) *An Advice How to Plant Tobacco in England* facsimile edition 1978, Amsterdam: Theatrum Orbis Terrarum.

Caldwell, J. A. M. (1964) 'Indonesian export and production from the decline of the Culture System to the First World War', in C. D. Cowan (ed.) *The Economic Development of South-East Asia: Studies in Economic History and Political Economy*, London: George Allen & Unwin.

Calendar of State Papers Venetian 1607–1610, London: HMSO.

Calendar of State Papers Venetian 1617–1619, London: HMSO.

Callender, C. (1978) 'Fox', in B. G. Trigger (ed.) *Handbook of North American Indians* vol. 15, Northeast, Washington, DC: Smithsonian Institution.

Calnan, M. W. (1984) 'The politics of health: the case of smoking control', *Journal of Social Policy* 13: 279–96.

Camporesi, L. (1989) *Bread of Dreams*, Oxford: Polity Press.

—— (1990) *The Fear of Hell*, Oxford: Polity Press.

Carr, L. G. and Menard, R. R. (1989) 'Land, labor and economies of scale in early Maryland: some limits to growth in the Chesapeake system of husbandry', *Journal of Economic History* 49: 407–18.

Carr, L. G. and Walsh, L. S. (1988) 'Economic diversification and labor organization in the Chesapeake, 1650–1820', in S. Innes (ed.) *Work and Labor in Early America*, Chapel Hill, North Carolina: University of North Carolina Press.

Carr, L. G., Menard, R. R. and Walsh, L. S. (1991) *Robert Cole's World: Agriculture and Society in Early Maryland*, Chapel Hill, North Carolina: University of North Carolina Press.

Carucci, L. M. (1987) '*Kijen Emaan ilo baat*: methods and meanings of smoking in Marshallese society', in L. Lindstrom (ed.) *Drugs in Western Pacific Societies*, Lanham, Maryland: University Press of America.

Chagnon, N. A. (1983) *Yanomamö: The Fierce People* 3rd edition, New York: Holt, Rinehart & Winston.

Chandler, A. D. (1977) *The Visible Hand*, Cambridge, Massachusetts: Harvard University Press.

—— (1990) *Scale and Scope*, Cambridge, Massachusetts: Harvard University Press.

Chanock, M. L. (1972) 'The political economy of independent agriculture in colonial Malawi: the Great War to the Great Depression', *Journal of Social Science* 1: 113–29.

Chapman, S. and Wong, W. L. (1990) *Tobacco Control in the Third World: A Resource Atlas*, Penang, Malaysia: International Organization of Consumers Unions.

Chaunu, H. and P. (1955–60) *Séville et l'Atlantique (1504–1650)*, Paris: SEVPEN.

Chen, T. T. L. and Winder, A. E. (1990) 'The Opium Wars revisited as US forces tobacco exports in Asia', *American Journal of Public Health* 80: 659–62.

Christen, A. G., Swanson, B. Z., Glover, E. D. and Henderson, A. H. (1982) 'Smokeless tobacco: the folklore and social history of snuffing, sneezing, dipping and chewing', *Journal of the American Dental Association* 105: 821–9.

Clairmonte, F. F. (1979) 'World tobacco: the dynamics of oligopolistic annexationism', *Journal of Contemporary Asia* 9: 255–73.

Clark, J. G. (1970) *New Orleans , 1718–1812: An Economic History*, Baton Rouge, Louisiana: Louisiana State University Press.

Clark, P. (1988) 'The "Mother Gin" controversy in the early eighteenth century', *Transactions of the Royal Historical Society* 5th series 38: 63–84.

Clarkson, L. A. (1971) *The Pre-Industrial Economy in England 1500–1750*, London: Batsford.

Clemens, P. G. E. (1975) 'The operation of an eighteenth-century Chesapeake tobacco plantation', *Agricultural History* 49: 517–31.

Cochran, S. (1980) *Big Business in China: Sino–Foreign Rivalry in the Cigarette Industry, 1890–1930*, Cambridge, Massachusetts: Harvard University Press.

—— (1986) 'Commercial penetration and economic imperialism in China: an American cigarette company's entrance into the market', in E. R. May and J. K. Fairbank (eds) *America's China Trade in Historical Perspective*, Cambridge, Massachusetts: Harvard University Press.

Columbus, C. (1990) *Journal of the First Voyage* edited and translated by B. W. Ife, Warminster: Aris & Phillips.

Comes, O. (1893) 'Il tabacco: II introduzione del tabacco in Europa', *Atti del Reale Istituto d'incoraggimento di Napoli* series 4, 10.

—— (1900) *Histoire, géographie, statistique du tabac*, Naples: Typographie Coopérative.

Connolly, G. N. (1990) 'Political and promotional thrusts worldwide by the transnational tobacco companies', in B. Durston and K. Jamrozik (eds) *Tobacco and Health 1990: The Global War. Proceedings of the Seventh World Conference on Tobacco and Health, Perth, Western Australia*, Perth, Western Australia: Health Department of Western Australia.

Cooper, J. M. (1963) 'Stimulants and narcotics', in J. H. Steward (ed.) *Handbook of South American Indians*, volume 5 The Comparative Ethnology of South American Indians, New York: Cooper Square Publishers.

Cooper, P. A. (1983) 'Traveling fraternity: union cigar makers and geographic mobility 1900–1919', *Journal of Social History* 17: 127–38.

—— (1987) *Once a Cigar Maker: Men, Women, and Work Culture in American Cigar Factories, 1900–1919*, Urbana, Illinois: University of Illinois Press.

—— (1988) ' "What this country needs is a good five-cent cigar" ', *Technology and Culture* 29: 779–807.

Corti, E. C. (1931) *A History of Smoking*, London: G. G. Harrap.

Courtwright, D. T. (1982) *Dark Paradise: Opiate Addiction in America before 1940*, Cambridge, Massachusetts: Harvard University Press.

—— (1991) 'The rise and fall and rise and fall of cocaine in the United States', paper given at Past and Presnt Conference, London, July.

Coutts, B. E. (1986) 'Boom and bust: the rise and fall of the tobacco industry in Spanish Louisiana, 1770–1790', *The Americas* 42: 289–309.

Cox, H. (1989) 'Growth and ownership in the international tobacco industry: BAT 1902–27', *Business History* 31: 44–67.

Cox, R. (1933) *Competition in the American Tobacco Industry 1911–32*, New York: Columbia University Press.

Craven, A. O. (1926) *Soil Exhaustion as a Factor in the Agricultural History of Virginia and Maryland, 1660–1860*, University of Illinois Studies in the Social Sciences, XIII, no. 1., Urbana, Illinois: University of Illinois Press.

Craven, W. F. (1932) *Dissolution of the Virginia Company*, New York: Oxford University Press.

—— (1937) 'An introduction to the history of Bermuda', *William and Mary Quarterly* 17: 176–215, 317–61, 437–65.

Cravo, V. Z. (1982) *A lavoura de fumo em Irati*, Curitiba: Instituto Histórico, Geográfico e Etnográfico Paranaense.

Currie, K. and Ray, L. (1984) 'Going up in smoke: the case of British American Tobacco in Kenya', *Social Science and Medicine* 19: 1,131–9.

Curtin, P. D. (1969) *The Atlantic Slave Trade: A Census*, Madison, Wisconsin: University of Wisconsin Press.

Cuthbertson, D. (1968) 'Historical note on the origin of the association between lung cancer and smoking', *Journal of the Royal College of Physicians of London* 2: 191–6.

Dalton, R. (1981) 'Changes in the structure of the flue-cured tobacco farm: a compilation of available data sources', in W. R. Finger (ed.) *The Tobacco Industry in Transition*, Lexington, Massachusetts: Lexington Books.

Daniel, P. (1984) 'The crossroads of change: cotton, tobacco and rice cultures in the twentieth-century South', *Journal of Southern History* 50: 429–56.

—— (1985) *Breaking the Land*, Urbana, Illinois: University of Illinois Press.

Davey, P. (1985) 'Clay pipes from Norton Priory', in P. Davey (ed.) *The Archaeology of the Clay Tobacco Pipe: More Pipes from the Midlands and Southern England*, British Archaeological Report, British Series 146, 9.

—— (1988) *The Seventeenth and Eighteenth Century Tyneside Tobacco Pipe Makers and Tobacconists*, British Archaeological Report 192, 11.

253

Davies, K. G. (1952) 'The origins of the commission system in the West India trade', *Transactions of the Royal Historical Society* 5th series, 2: 89–107.

—— (1974) *The North Atlantic World in the Seventeenth Century*, Minneapolis, Minnesota: University of Minnesota Press.

Davis, D. L. (1987) 'Tobacco use and associated health risks', in W. R. Martin, G. R. Van Loon, E. T. Iwamoto and L. Davis (eds) *Tobacco Smoking and Nicotine: A Neurobiological Approach*, New York: Plenum Press.

Davis, R. (1969) 'English foreign trade, 1700–1774', in W. R. Minchinton (ed.) *The Growth of English Overseas Trade in the Seventeenth and Eighteenth Centuries*, London: Methuen.

—— (1973) *English Overseas Trade 1500–1700*, London: Macmillan.

—— (1979) *The Industrial Revolution and British Overseas Trade*, Leicester: Leicester University Press.

Deans, S. (1984) 'The tobacco monopoly in Bourbon Mexico', unpublished Ph. D. thesis, University of Cambridge.

de Graaf, R. (1668) *De virorum organis generationi inservientibus, de cysteribus et de usu siphonis in anatomia*, Leiden.

de Jesus, E. C. (1980) *The Tobacco Monopoly in the Philippines*, Quezon City: Ateneo de Manila University Press.

de Laguna, F. (1990) 'Tlingit', in W. Suttles (ed.) *Handbook of North American Indians*, Northwest Coast, Washington, DC: Smithsonian Institution.

de Reneaulme, P. (1611) *Specimen Historiae Plantarum*, Paris.

Dermigny, L. (1964) *La Chine et l'Occident: le commerce à Canton au XVIIIe siècle, 1789–1833*, Paris: SEVPEN.

de Smet, P. A. G. M. (1983) 'A multidisciplinary overview of intoxicating enema rituals in the western hemisphere', *Journal of Ethnopharmacology* 9: 129–66.

Devine, T. M. (1975) *The Tobacco Lords*, Edinburgh: John Donald.

—— (1976) 'A Glasgow tobacco merhcant during the American War of Independence: Alexander Speirs of Eldersie, 1775 to 1781', *William and Mary Quarterly* 33: 501–13.

—— (1984) *A Scottish Firm in Virginia 1767–1777: W. Cunninghame and Co.*, Edinburgh: Clark Constable.

de Vries, J. (1978) 'Barges and capitalism: passenger transportation in the Dutch economy, 1632–1839', *AAG Bijdragen* 21: 33–398.

Diamond, S. (1957–8) 'From organization to society: Virginia in the seventeenth century', *American Journal of Sociology* 63: 457–75.

Dickson, S. A. (1954) *Panacea or Precious Bane: Tobacco in Sixteenth Century Literature*, New York: The New York Public Library.

—— (1959) *Tobacco. A Catalogue of the Books, Manuscripts and Engravings acquired since 1942* Part I 1507–1571, New York: New York Public Library.

—— (1960) *Tobacco. A Catalogue of the Books, Manuscripts and Engravings acquired since 1942* Part IV 1610–1619, New York: New York Public Library.

Dobkin de Rios, M. (1984a) *Hallucinogens: Cross-Cultural Perspectives*, Albuquerque, New Mexico: University of New Mexico Press.

—— (1984b) *Visionary Vine*, Prospect Heights, Illinois: Waveland Press.

Dole, G. E. (1964) 'Shamanism and political control among the Kuikuru', in *Beiträge zur Völkerkunde Südamerikas*, Völkerkundliche Abhandlungen vol. I, Hanover: Kommissionsverlag Münstermann-Druck.

Doll, R. and Hill, A. B. (1952) 'A study of the aetiology of carcinoma of the lung', *British Medical Journal* 2: 1,271–86.

—— (1954) 'The mortality of doctors in relation to their smoking habits: a preliminary report', *British Medical Journal* 1: 1,451–5.

— (1956) 'Lung cancer and other causes of death in relation to smoking: a second report on the mortality of British doctors', *British Medical Journal* 2: 1,071–81.

Douglas, M. (1979) *Purity and Danger: An Analysis of Concepts of Pollution and Taboo*, London: Routledge & Kegan Paul.

Duco, D. H. (1981) 'The clay tobacco pipe in seventeenth century Netherlands', in P. Davey (ed.) *The Archaeology of the Clay Tobacco Pipe* V, Europe part 2, BAR International Series, 106(ii).

Dunhill, A. (1924) *The Pipe Book*, London: A. & C. Black

Dunn, R. S. (1973) *Sugar and Slaves*, New York: Norton.

Dunphy, E. B. (1969) 'Alcohol and tobacco amblyopia: a historical survey', *American Journal of Ophthalmology* 68: 569–78.

Dupuy, A. (1985) 'French merchant capital and slavery in Saint-Domingue', *Latin American Perspectives* 12: 77–102.

Durden, R. F. (1975) *The Dukes of Durham, 1865–1929*, Durham, North Carolina: Duke University Press.

— (1976) 'Tar heel tobacconist in Tokyo, 1899–1904', *The North Carolina Historical Review* 53: 347–63.

Earle, C. V. and Hoffman, R. (1976) 'Staple crops and urban development in the eighteenth-century South', *Perspectives in American History* 10: 7–78.

Earle, P. (1989) *The Making of the English Middle Class*, London: Methuen.

Eiden, F. (1976) 'Zur Geschichte der Tabakalkaloide', *Pharmazie in unserer Zeit* 5: 1–18.

Elferink, J. G. R. (1983) 'The narcotic and hallucinogenic use of tobacco in pre-Columbian Central America', *Journal of Ethnopharmacology* 7: 111–22.

— (1984) 'Pharmacy and the pharmaceutical profession in the Aztec culture', *Janus* 71: 41–62.

Eliade, M. (1989) *Shamanism*, London: Arkana.

Elias, N. (1978) *The History of Manners*, Oxford: Blackwell.

Elliott, Sir H. M. and Dowson, J. (eds) (1875) *The History of India as Told by its own Historians* volume VI, London: Trübner

Elliott, J. H. (1972) *The Old World and the New World*, Cambridge: Cambridge University Press.

— (1985) 'The Iberian achievment', *Princeton University Library Chronicle* 47: 25–47.

— (1987) 'Spain and America before 1700', in L. Bethell (ed.) *Colonial Spanish America*, Cambridge, Cambridge University Press.

Eltis, D. (1992) 'European slaves and plantation labour in the 17th century', paper given at Anglo-American Conference of Historians, Institute of Historical Research, July 1992.

Ernster, V. L. (1985) 'Mixed messages for women: a social history of cigarette smoking and advertising', *New York State Journal of Medicine* 85: 335–40.

Falgairolle, E. (1897) *Jean Nicot, Ambassadeur de France en Portugal au XVIe siècle: sa correspondance inédite*, Paris: Challamel.

Farmer, C. J. (1988) 'Persistence of country trade: the failure of towns to develop in Southside Virginia during the eighteenth century', *Journal of Historical Geography* 14: 331–41.

Fass, P. (1977) *The Damned and the Beautiful: American Youth in the 1920s*, New York: Oxford University Press.

Feinhandler, S. J. , Fleming, H. C. and Monahon, J. M. (1979) 'Pre-Columbian tobaccos in the Pacific', *Economic Botany* 33: 213–26.

Fernández-Carrión, M. and Valverde, J. L. (1988) 'Research notes on Spanish-American drug trade', *Pharmacy in History* 30: 27–32.

Ferrant, L. (1655) *Traicté sur tabac*, Bourges.

Ferry, R. J. (1981) 'Encomienda, African slavery and agriculture in seventeenth-century Caracas', *Hispanic American Historical Review* 61: 609–35.

Filene, P. G. (1975) *Him/Her/Self: Sex Roles in Modern America*, New York: Harcourt Brace Jovanovich.

Finau, S. A. , Stanhope, J. M. and Prior, I. A. M. (1982) 'Kava, alcohol and tobacco consumption among Tongans with urbanization', *Social Science and Medicine* 16: 35–41.

Fisher, F. J. (1976) 'London as an "engine of economic growth" ', in P. Clark (ed.) *The Early Modern Town*, London: Longman.

Fisher, J. (1985) 'The imperial response to "free trade": Spanish imports from Spanish America, 1778–1796', *Journal of Latin American Studies* 17: 35–78.

Flink, J. (1988) *The Automobile Age*, Cambridge, Massachusetts: MIT Press.

Flory, R. J. (1978) 'Bahian society in the mid-colonial period: the sugar planters, tobacco growers, merchants and artisans of Salvador and Recôncavo 1680–1725', unpublished Ph. D. dissertation, University of Texas.

Food and Agriculture Organization of the United Nations (1989) *The Economic Significance of Tobacco* FAO Economic and Social Development Paper, 85, Rome.

—— (1990) *Tobacco: Supply, Demand and Trade Projections, 1995 and 2000* FAO Economic and Social Development Paper, 86, Rome.

Foster, G. M. (1987) 'On the origins of humoral medicine in Latin America', *Medical Anthropology Quarterly* 1: 355–93.

—— (1988) 'The validating role of humoral theory in traditional Spanish-American therapeutics', *American Ethnologist* 15: 120–35.

Foust, C. M. (1992) *Rhubarb: The Wondrous Drug*, New Haven, Connecticut: Yale University Press.

Friedenwald, J. and Morrison, S. (1940a) 'The history of the enema with some notes on related procedures Part I', *Bulletin of the History of Medicine* 8: 68–114.

—— (1940b) 'The history of the enema with some notes on related procedures Part II', *Bulletin of the History of Medicine* 8: 239–76.

Fryer, J. (1899) *John Fryer's East India and Persia* volume I, London: Hakluyt Society.

—— (1912) *John Fryer's East India and Persia* volume II, London: Hakluyt Society.

—— (1915) *John Fryer's East India and Persia* volume III, London: Hakluyt Society.

Furst, P. T. (1976) *Hallucinogens and Culture*, San Francisco, California: Chandler & Sharp.

Gage, C. E. (1937) 'Historical factors affecting American tobacco types and uses and the evolution of the auction market', *Agricultural History* 11: 43–57.

Gagliano, J. A. (1963) 'The coca debate in colonial Peru', *The Americas* 20: 43–63.

—— (1965) 'The popularization of Peruvian coca', *Revista de historia de America* 59: 164–79.

—— (1979) 'Coca and popular medicine in Peru: an historical analysis of attitudes', in D. L. Browman and R. A. Schwarz (eds) *Spirits, Shamans and Stars*, The Hague: Mouton.

Galenson, D. W. (1981a) 'White servitude and the growth of black slavery in colonial America', *Journal of Economic History* 61: 39–47.

—— (1981b) *White Servitude in Colonial America*, Cambridge: Cambridge University Press.

—— (1984) 'The rise and fall of indentured servitude in the Americas: an economic analysis', *Journal of Economic History* 44: 1–26.

BIBLIOGRAPHY

Garcia-Baquero Gonzalez, A. (1976) *Cadiz y el Atlantico (1717–1778)* volume II, Seville: Escuela de Estudios Hispano-Americanos de Sevilla.

Garcia Fuentes, L. (1980) *El comercio español con America, 1650–1700*, Seville: Escuela de Estudios Hispano-Americanos de Sevilla.

Gemery, H. A. (1980) 'Emigration from the British Isles to the New World, 1630–1700: inferences from colonial populations', *Research in Economic History* 5: 179–231.

—— (1984) 'European emigration to North America, 1700–1820: numbers and quasi-numbers', *Perspectives in American History* new series 1: 283–342.

Ginzburg, C. (1990) *Ecstasies: Deciphering the Witches' Sabbath*, London: Hutchinson Radius.

Ginzel, K. H. (1990) 'A quantitative estimate of exposure of active and passive smokers to chemicals in cigarette smoke', in B. Durston and K. Jamrozik (eds) *Tobacco and Health 1990: The Global War. Proceedings of the Seventh World Conference on Tobacco and Health, Perth, Western Australia*, Perth, Western Australia: Health Department of Western Australia.

Gokhale, B. G. (1974) 'Tobacco in seventeenth-century India', *Agricultural History* 48: 484–92.

Goodin, R. E. (1989a) 'The ethics of smoking', *Ethics* 99: 574–624.

—— (1989b) *No Smoking: The Ethical Issues*, Chicago: University of Chicago Press.

Goodrich, L. C. (1938) 'Early prohibitions of tobacco in China and Manchuria', *Journal of the American Oriental Society* 58: 648–57.

Goodspeed, T. H. (1954) *The Genus Nicotiana*, Waltham, Massachusetts: Chronica Botanica.

Goslinga, C. C. (1971) *The Dutch in the Caribbean and the Wild Coast 1580–1680*, Assen: Van Gorcum.

Gottsegen, J. J. (1940) *Tobacco: A Study of its Consumption in the United States*, New York: Pitman

Gray, L. C. (1958) *History of Agriculture in the Southern United States to 1860*, Gloucester, Massachusetts: Peter Smith.

Gray, S. and Wyckoff, V. J. (1940) 'The international tobacco trade in the seventeenth century', *Southern Economic Journal* 7: 1–26.

Green, G. P. (1987) 'The political economy of flue-cured tobacco production', *Rural Sociology* 52: 221–41.

Green, W. A. (1988) 'Supply versus demand in the Barbadian sugar revolution', *Journal of Interdisciplinary History* 18: 403–18.

Greenblatt, S. (ed.) (1991) 'The New World', *Representations* 12.

Greenfield, S. M. (1979) 'Plantations, sugar cane and slavery', *Historical Reflections/ Réflexions historiques* 6: 85–119.

Grise, V. (1990) *The World Tobacco Market: Government Intervention and Multilateral Policy Reform*, Washington, DC: Department of Agriculture, Economic Research Service, Commodity Economic Division.

Guerra, F. (1961) *Nicolas Bautista Monardes: su vida y su obra*, Mexico City: Compañia fundidora de fiero y acero de Monterrey, SA.

—— (1964) 'Maya medicine', *Medical History* 8: 31–43.

—— (1966a) 'Aztec medicine', *Medical History* 10: 315–38.

—— (1966b) 'Drugs from the Indies and the political economy of the sixteenth century', *Analecta medico historica* 1: 29–54.

—— (1967) 'Mexican phantastica – a study of the early ethnobotanical sources on hallucinogenic drugs', *British Journal of Addiction* 62: 171–87.

Haberman, T. W. (1984) 'Evidence for aboriginal tobaccos in eastern North America', *American Antiquity* 49: 269–87.

Hall, R. (1978) *The Making of Molins*, London: Molins.

Hambly, W. D. (1930) 'Use of tobacco in Africa', in *Tobacco and its Use in Africa*, Field Museum of Natural History, Leaflet 29, Chicago.

Hamell, G. R. (1987) 'Strawberries, floating islands, and rabbit captains: mythical realities in the Northeast during the sixteenth and seventeenth centuries', *Journal of Canadian Studies* 21: 72–94.

Hamilton, E. J. (1976) 'What the New World gave the economy of the Old', in F. Chiapelli (ed.) *First Images of America: The Impact of the New World on the Old*, Berkeley, California: University of California Press.

Hammond, E. C. and Horn, D. (1954) 'The relationship between human smoking habits and death rates', *Journal of the American Medical Association* 155: 1,316–28.

—— (1958) 'Smoking and death rates – report on forty-four months of follow-up on 187,783 men', *Journal of the American Medical Association* 166: 1,294–308.

Hamor, R. (1957 reprint) *A True Discourse of the Present State of Virginia*, Richmond, Virginia: The Virginia State Library.

Hanson, C. A. (1982) 'Monopoly and contraband in the Portuguese tobacco trade 1624–1702', *Luso-Brazilian Review* 19: 149–68.

Hanson, J. R. II (1980) *Trade in Transition: Exports from the Third World, 1840–1900*, New York: Academic Press.

Harlow, V. T. (ed.) (1925) *Colonizing Expeditions to the West Indies and Guiana, 1623–67*, London: Hakluyt Society.

Harner, M. J. (1973) 'The role of hallucinogenic plants in European witchcraft', in M. J. Harner (ed.) *Hallucinogens and Shamanism*, New York: Oxford University Press.

Harrison, L. (1986) 'Tobacco Battered and the Pipes Shattered: a note on the fate of the first British campaign against tobacco smoking', *British Journal of Addiction* 81: 553–58.

Hart, J. (1633) *KANIKH or the Diet of the Diseased*, London.

Hattox, R. S. (1988) *Coffee and Coffeehouses: The Origins of a Social Beverage in the Medieval Near East*, Seattle, Washington: University of Washington Press.

Hausman, W. J. and Neufeld, J. L. (1981) 'Excise anatomized: the political economy of Walpole's 1733 tax scheme', *Journal of European Economic History* 10: 131–43.

Heald, S. (1991) 'Tobacco, time, and the household economy in two Kenyan societies: the Teso and Kuria', *Comparative Studies in Society and History* 33: 130–57.

Heidenreich, C. E. (1978) 'Huron', in B. G. Trigger (ed.) *Handbook of the North American Indians* volume 15 Northeast, Washington, DC: Smithsonian Institution.

Heimann, R. K. (1960) *Tobacco and Americans*, New York: McGraw-Hill.

Heiser, C. B. , Jr. (1969) *Nightshades: The Paradoxical Plants*, San Francisco, California: W. H. Freeman.

Hemming, J. (1978) *The Search for El Dorado*, London: Michael Joseph.

Hemphill, J. M. II (1985) *Virginia and the English Commercial System 1689–1733*, New York: Garland.

Hendee, W. R. and Kellie, S. E. (1990) 'Impact of US tobacco exports on the worldwide smoking epidemic', in B. Durston and K. Jamrozik (eds) *Tobacco and Health 1990: The Global War. Proceedings of the Seventh World Conference on Tobacco and Health, Perth, Western Australia*, Perth, Western Australia: Health Department of Western Australia.

Herndon, G. M. (1967) 'Indian agriculture in the southern colonies', *The North Carolina Historical Review* 44: 283–97.

—— (1969) *William Tatham and the Culture of Tobacco*, Coral Gables, Florida: University of Miami Press.

Hillier, S. (1971) 'The trade of the Virginia colony 1606–1660', unpublished Ph. D. thesis, University of Liverpool.

Hirayama, T. (1990) 'Cohort studies on smoking and mortality in Japan', in B. Durston and K. Jamrozik (eds) *Tobacco and Health 1990: The Global War. Proceedings of the Seventh World Conference on Tobacco and Health, Perth, Western Australia*, Perth, Western Australia: Health Department of Western Australia.

Hirschman, A. O. (1977) 'A generalized linkage approach to development, with special reference to staples', *Economic Development and Cultural Change* 25 supplement: 67–98.

Hitz, H. and Huber, H. (1975) *Geschichte des Österreichischen Tabakregie 1784–1835*, Vienna: Verlag der Österreichischen Akademie der Wissenschaften.

Hodne, F. (1978) 'New evidence on the history of tobacco consumption in Norway 1655–1970', *Economy and History* 21: 114–25.

Höllmann, T. O. (1988) *Tabak in Südostasien*, Berlin: Dietrich Reimer.

Holmes, G. K. (1912) 'Tobacco crop of the United States, 1612–1911', *United States Department of Agriculture, Bureau of Statistics* circular 33, Washington, DC: GPO.

Horn, J. (1979) 'Servant emigration to the Chesapeake in the seventeenth century', in T. W. Tate and D. L. Ammerman (eds) *The Chesapeake in the Seventeenth Century*, Chapel Hill, North Carolina: University of North Carolina Press.

Howard, J. H. (1957) 'The mescal bean cult of the central and southern Plains: an ancestor of the peyote cult', *American Anthropologist* 59: 75–87.

Hudson, C. (ed.) (1979) *Black Drink*, Athens, Georgia: University of Georgia Press.

Hugh-Jones, C. (1979) *From the Milk River*, Cambridge, Cambridge University Press.

Hultkrantz, A. (1979) *The Religions of the American Indians*, Berkeley, California: University of California Press.

—— (1985) 'The shaman and the medicine-man', *Social Science and Medicine* 20: 511–15.

—— (1989) 'Health, religion and medicine in native North American traditions', in L. E. Sullivan (ed.) *Healing and Restoring: Health and Medicine in the World's Religious Traditions*, New York: Macmillan Publishing Company.

Hurt, R. D. (1987) *Indian Agriculture in America*, Lawrence, Kansas: University Press of Kansas.

Innes, F. C. (1970) 'The pre-sugar era of European settlement in Barbados', *Journal of Caribbean History* 1: 1–22.

Israel, J. I. (1989) *Dutch Primacy in World Trade 1585–1740*, Oxford: Oxford University Press.

Ives, V. A. (1984) *The Rich Papers: Letters from Bermuda 1615–1646*, Toronto: University of Toronto Press.

Jacobson, B. (1981) *The Ladykillers – Why Smoking is a Feminist Issue*, London: Pluto Press.

—— (1988) *Beating the Ladykillers: Women and Smoking*, London: Gollancz.

Jacobstein, M. (1907) *The Tobacco Industry in the United States*, Studies in History, Economics and Public Law volume 26, Columbia University.

James I (1982) 'A counterblaste to tobacco', in J. Craigie (ed.) *Minor Prose Works of King James V and I*, Edinburgh: Scottish Text Society.

Janiger, O. and Dobkin de Rios, M. (1976) 'Nicotiana an hallucinogen?', *Economic Botany* 30: 149–51.

Jenkins, J. W. (1927) *James B. Duke: Master Builder*, New York: George H. Doran.

Johansen, H.-C. (1986) 'How to pay for Baltic goods', in W. Fischer, R. M. McInnis and J. Schneider (eds) *The Emergence of a World Economy*, papers of the IXth International Congress of Economic History volume I, Wiesbaden: Franz Steiner Verlag.

John, D. W. and Jackson, J. C. (1973) 'The tobacco industry of North Borneo: a distinctive form of plantation agriculture', *Journal of Southeast Asian Studies* 4: 88–106.

Johnson, H. B. (1987) 'Portuguese settlement, 1500–1580', in L. Bethell (ed.) *Colonial Brazil* Cambridge: Cambridge University Press.

Johnson, P. R. (1984) *The Economics of the Tobacco Industry*, New York: Praeger.

Jones, E. L. (1973) 'The fashion manipulators: consumer tastes and British industries, 1660–1800', in L. Cain and P. Uselding (eds) *Business Enterprise and Economic Change: Essays in Honor of Harold F. Williamson*, Detroit: Kent State University Press.

Joosens, L. and Raw, M. (1991) 'Tobacco and the European Common Agricultural Policy', *British Journal of Addiction* 86: 1,191–202.

Kalm, P. (1966 reprint) *The America of 1750: Travels in North America* volume II, New York: Dover.

Kanduza, A. (1983) 'The tobacco industry in Northern Rhodesia, 1912–1938', *International Journal of African Historical Studies* 16: 201–29.

Kanner, L. (1931) 'Superstitions connected with sneezing', *Medical Life* 38: 549–75.

Kaplan, M. , Carricker, L. and Waldron, I. (1990) 'Gender differences in tobacco use in Kenya', *Social Science and Medicine* 30: 305–10.

Karshner, M. (1979) 'The tobacco clay pipe making industry in Norwich', in P. Davey (ed.) *The Archaeology of the Clay Pipe. Britain: The Midlands and Eastern England*, British Archaeological Report, British series, 63.

Kell, K. T. (1965) 'Tobacco in folk cures in western societies', *Journal of American Folklore* 78: 99–114.

Kellenbenz, H. (1977) 'The organization of industrial production', in E. E. Rich and C. H. Wilson (eds) *The Cambridge Economic History of Europe* volume 5, Cambridge: Cambridge University Press.

Kendall, D. L. (1990) 'Takelma', in W. Suttles (ed.) *Handbook of North American Indians*, Northwest Coast, Washington, DC: Smithsonian Institution.

Kiernan, V. G. (1991) *Tobacco: A History*, London: Hutchinson Radius.

Kimber, C. T. (1988) *Martinique Revisited*, College Station, Texas: Texas A. & M. University Press.

Kingsbury, S. M. (1933) *The Records of the Virginia Company of London* volume III, Washington, DC: GPO.

Knapp, J. (1988) 'Elizabethan tobacco', *Representations* 21: 27–66.

Knauft, B. M. (1987) 'Managing sex and anger: tobacco and kava use among the Gebusi of Papua New Guinea', in L. Lindstrom (ed.) *Drugs in Western Pacific Societies*, Lanham, Maryland: University Press of America.

Körbler, J. (1968) 'Der Tabak in der Krebslehre zu Anfang des 19. Jahrhunderts', in *Atti del XXI Congresso Internazionale di Storia della Medicina* volume II, Siena: Società internazionale di storia della medicina.

Kosmin, B. A. (1974) 'The Inyoka tobacco industry of the Shangwe people: a case study of the displacement of a pre-colonial economy in Southern Rhodesia 1898–1938', *African Social Research* 17: 554–77.

—— (1977) 'The Inyoka tobacco industry of the Shangwe people: the displacement of a pre-colonial economy in Southern Rhodesia, 1898–1938', in R. Palmer and N. Parsons (eds) *The Roots of Poverty in Central and Southern Africa*, London: Heinemann.

Kroeber, A. L. (1941) 'Salt, dogs, tobacco', *Anthropological Record* 6, 1.

Kulikoff, A. (1977) 'A "prolifick" people: black population growth in the Chesapeake colonies, 1700–1790', *Southern Studies* 16: 391–428.

—— (1986) *Tobacco and Slaves*, Chapel Hill, North Carolina: University of North Carolina Press.

Kupp, J. (1973) 'Dutch notarial acts relating to the tobacco trade of Virginia, 1608–1653', *William and Mary Quarterly* 30: 653–5.

Kupperman, K. O. (1979) 'Apathy and death in early Jamestown', *Journal of American History* 66: 24–40.

—— (1984) *Roanoke: The Abandoned Colony*, Totowa, New Jersy: Rowman & Allanheld.

—— (1988) 'Errand to the Indies: Puritan colonization from Providence Island through the Western Design', *William and Mary Quarterly* 45: 70–99.

La Barre, W. (1938) *The Peyote Cult*, New Haven, Connecticut: Yale University Press.

—— (1957) 'Mescalism and peyotism', *American Anthropologist* 59: 708–11.

—— (1964) 'The narcotic complex of the New World', *Diogenes* 48: 125–38.

—— (1970) 'Old and New World narcotics: a statistical question and an ethnological reply', *Economic Botany* 24: 73–80.

—— (1977) 'Comment on Schultes', *Journal of Psychedelic Drugs* 9: 351.

—— (1980) *Culture in Context*, Durham, North Carolina: Duke University Press.

Labat, J. -B. (1742) *Nouveau voyage aux isles de l'Amérique* volume 6, Paris: Delespine.

Lamphere, L. (1983) 'Southwestern ceremonialism', in A. Ortiz (ed.) *Handbook of North American Indians*, Southwest, Washington, DC: Smithsonian Institution.

Lane, F. C. (1940) 'The Mediterranean spice trade: its revival in the sixteenth century', *American Historical Review* 45: 581–90.

Lapa, J. R. do Amaral (1970) 'O tabaco brasileiro no século XVIII', *Studia* 29: 57–144.

Larson, P. S. , Haagard, H. B. and Silvette, H. (1961) *Tobacco. Experimental and Clinical Studies: A Comprhensive Account of the World Literature*, Baltimore, Maryland: Williams & Wilkins.

Laufer, B. (1924a) *Tobacco and Its Use in Africa*, Field Museum of Natural History, Department of Anthropology Leaflet 18, University of Chicago.

—— (1924b) *The Introduction of Tobacco into Europe*, Field Museum of Natural History, Department of Anthropology Leaflet 19, University of Chicago.

—— (1930) *The Introduction of Tobacco into Africa*, Field Museum of Natural History, Department of Anthropology, University of Chicago.

Laugesen, M. and Meads, C. (1991) 'Tobacco advertising restrictions, price, income and tobacco consumption in OECD countries, 1960–1986', *British Journal of Addiction* 86: 1343–54.

La Vecchia, C. (1986) 'Smoking in Italy, 1949–1983', *Preventative Medicine* 15: 274–81.

Lawrence, S. (1979) 'York pipes and their makers', in P. Davey (ed.) *The Archaeology of the Clay Pipe. Britain: the Midlands and Eastern England*, British Archaeological Report, British series, 63.

Leclant, J. (1979) 'Coffee and cafés in Paris, 1644–1693', in R. Forster and O. Ranum (eds) *Food and Drink in History*, Baltimore, Maryland: The Johns Hopkins University Press.

Le Corbeiller, C. (1966) *European and American Snuff Boxes 1730–1830*, London: Batsford.

Lécuyer, B. -P. (1983) 'Les maladies professionelles dans les "Annales d'hygiène publique et de médecine légale" ou une première approche de l'usure au travail', *Le Mouvement Social* 124: 45–69.

Lee, P. N. (ed.) (1975) *Tobacco Consumption in Various Countries* Research paper 6, 4th edition, London: Tobacco Research Council.

León Pinelo, A. de (1636) *Question Moral*, Madrid.

Leonard, J. de L. (1967) 'Operation checkmate: the birth and death of a Virginia blueprint for progress, 1600–1676', *William and Mary Quarterly* 24: 44–74.

Lesch, J. E. (1981) 'Conceptual change in an empirical science: the discovery of the first alkaloids', *Historical Studies in the Physical Sciences* 11: 305–28.

Linebaugh, P. (1991) *The London Hanged: Crime and Civil Society in the Eighteenth Century*, London: Allen Lane.

Linton, R. (1924) *Use of Tobacco among North American Indians*, Field Museum of Natural History, Department of Anthropology Leaflet 15, University of Chicago.

Logan, M. H. (1977) 'Anthropological research on the Hot–Cold theory of disease: some methodological suggestions', *Medical Anthropology* 1: 87–112.

Lohof, B. A. (1969) 'The higher meaning of Marlboro cigarettes', *Journal of Popular Culture* 3: 441–50.

Lorenzo Sanz, E. (1979) *Comercio de España con América en la epoca de Felipe II*, Valladolid: Servicio de Publicaciones de la Diputacion Provincial de Valladolid.

Lorimer, J. (1973) 'English trade and exploration in Trinidad and Guiana, 1569–1648', unpublished Ph. D. thesis, University of Liverpool.

—— (1978) 'The English contraband tobacco trade in Trinidad and Guiana 1590–1617', in K. R. Andrews, N. P. Canny and P. E. H. Hair (eds) *The Westward Enterprise: English Activities in Ireland, the Atlantic and America, 1460–1650*, Liverpool: Liverpool University Press.

—— (ed.) (1989) *English and Irish Settlement on the River Amazon 1550–1646*, London: Hakluyt Society.

Lovejoy, P. E. (1982) 'The volume of the Atlantic slave trade: a synthesis', *Journal of African History* 23: 473–501.

Lowie, R. H. (1919) 'The tobacco society of the Crow Indians', *Anthroplogical Papers of the American Museum of Natural History* 21: 99–200.

Lucio d'Azevedo, J. (1947) *Epocas de Portugal Económico*, Lisbon: Teixeira.

Lugar, C. (1977) 'The Portuguese tobacco trade and tobacco growers of Bahia in the late colonial period', in D. Alden and W. Dean (eds) *Essays concerning the Socio-Economic History of Brazil and Portuguese India*, Gainsville, Florida: The University Presses of Florida.

McClary, A. (1980) 'Germs are everywhere: the germ threat as seen in magazine articles, 1890–1920', *Journal of American Culture* 3: 33–46.

McCracken, G. (1988) *Culture and Consumption*, Bloomington, Indiana: Indiana University Press.

McCracken, J. (1983) 'Planters, peasants and the colonial state: the impact of the Native Tobacco Board in the Central Province of Malawi', *Journal of Southern African Studies* 9: 172–92.

McCusker, J. J. and Menard, R. R. (1985) *The Economy of British America, 1607–1789*, Chapel Hill, North Carolina: University of North Carolina Press.

MacInnes, C. M. (1926) *The Early English Tobacco Trade*, London: Kegan Paul, Trench, Trubner.

McKendrick, N., Brewer, J. and Plumb, J. H. (1982) *The Birth of a Consumer Society: The Commercialization of Eighteenth-Century England*, Bloomington, Indiana: Indiana University Press.

Mackenzie, C. (1984 reprint) *Sublime Tobacco*, Gloucester: Alan Sutton.

Maclean, I. (1980) *The Renaissance Notion of Women*, Cambridge: Cambridge University Press.

MacLeod, M. J. (1973) *Spanish Central America: A Socioeconomic History*, Berkeley, California: University of California Press.

McMorrow, M. J. and Foxx, R. M. (1983) 'Nicotine's role in smoking: an analysis of nicotine regulation', *Psychological Bulletin* 93: 302–27.

Mann, C. K. (1981) *Tobacco: The Ants and the Elephants*, Salt Lake City, Utah: Olympus.

—— (1984) 'The tobacco franchise for whom?', in W. R. Finger (ed.) *The Tobacco Industry in Transition*, Lexington, Massachusetts: Lexington Books.

Marchand, R. (1985) *Advertising the American Dream: Making Way for Modernity, 1920–1940*, Berkeley, California: University of California Press.

Markandya, A. and Pierce, D. W. (1989) 'The social costs of tobacco smoking', *British Journal of Addiction* 84: 1,139–50.

Marshall, M. (1981) 'Tobacco use in Micronesia: a preliminary discussion', *Journal of Studies on Alcohol* 42: 885–93.

—— (1987) 'An overview of drugs in Oceania', in L. Lindstrom (ed.) *Drugs in Western Pacific Societies*, Lanham, Maryland: University Press of America.

Martin, P. L. and Johnson, S. S. (1978) 'Tobacco technology and agricultural labor', *American Journal of Agricultural Economics* 60: 655–60.

Martin, R. T. (1975) 'The role of coca in the history, religion, and medicine of South American Indians', in G. Andrews and D. Solomon (eds) *The Coca Leaf and Cocaine Papers*, New York: Harcourt Brace Jovanovich.

Martin, W. R. (1987) 'Tobacco and health overview: a neurobiologic approach', in W. R. Martin, G. R. Van Loon, E. T. Iwamoto and L. Davis (eds) *Tobacco Smoking and Nicotine: A Neurobiological Approach*, New York: Plenum Press.

Masironi, R. (1990) 'Smoking trends worldwide', in B. Durston and K. Jamrozik (eds) *Tobacco and Health 1990: The Global War. Proceedings of the Seventh World Conference on Tobacco and Health, Perth, Western Australia*, Perth, Western Australia: Health Department of Western Australia.

Masironi, R. and Rothwell, K. (1988) 'Worldwide smoking trends', in A. Aoki, S. Hisamichi and S. Tominaga (eds) *Smoking and Health 1987*, Amsterdam: Excerpta Medica.

Mason, J. A. (1924) *Use of Tobacco in Mexico and South America*, Field Museum of Natural History, Department of Anthropology Leaflet 16, University of Chicago.

Mathews, Z. P. (1976) 'Huron pipes and Iroquoian shamanism', *Man in the Northeast* 12: 15–31.

Mauro, F. (1960) *Le Portugal et l'Atlantique au XVIIe siècle*, Paris: SEVPEN.

—— (1986) 'French indentured servants for America, 1500–1800', in P. C. Emmer (ed.) *Colonialism and Migration: Indentured Labour Before and After Slavery*, Dordrecht: Martinus Nijhoff.

May, L. -P. (1930) *Histoire économique de la Martinique 1635–1763*, Paris: Marcel Rivière.

Meinig, D. W. (1986) *The Shaping of America*, New Haven, Connecticut: Yale University Press.

Menard, R. R. (1976) 'A note on Chesapeake tobacco prices, 1618–1660', *The Virginia Magazine of History and Biography* 84: 401–10.

—— (1977) 'From servants to slaves: the transformation of the Chesapeake labor system', *Southern Studies* 16: 355–90.

—— (1980) 'The tobacco industry in the Chesapeake colonies, 1617–1730: an interpretation', *Research in Economic History* 5: 109–77.

Menard, R. R. and Carr, L. G. (1982) 'The Lords Baltimore and the colonization of Maryland', in D. B. Quinn (ed.) *Early Maryland in a Wider World*, Detroit: Kent State University Press.

Menard, R. R., Carr, L. G. and Walsh, L. S. (1983) 'A small planter's profits: the Cole estate and the growth of the early Chesapeake economy', *William and Mary Quarterly* 40: 171–96.

Messer, E. (1987) 'The Hot and Cold in Mesoamerican indigenous and hispanicized thought', *Social Science and Medicine* 25: 339–46.

Métraux, A. (1949) 'Religion and shamanism', in J. H. Steward (ed.) *Handbook of South American Indians* volume 5 The Comparative Ethnology of South American Indians, Washington, DC: Smithsonian Institution.

Middleton, A. P. (1984 reprint) *Tobacco Coast*, Baltimore, Maryland: The Johns Hopkins University Press.

Mintz, S. (1985) *Sweetness and Power: The Place of Sugar in Modern History*, New York: Viking.

Mitchell, D. (1992) 'Images of erotic women in turn-of-the-century tobacco art', *Feminist Studies* 18: 327–50.

Moisés, R., Kelly, J. H. and Holden, W. C. (1971) *A Yaqui Life: The Personal Chronicle of a Yaqui Indian*, Lincoln, Nebraska: University of Nebraska Press.

Monardes, N. (1925 reprint) *Joyfull Newes Out of the Newe Founde Worlde*, translated by John Frampton, London: Constable.

Morgan, E. S. (1971) 'The first American boom: Virginia 1618 to 1630', *William and Mary Quarterly* 28: 169–98.

—— (1975) *American Slavery, American Freedom*, New York: Norton.

Morgan, P. D. (1988) 'Task and gang systems: the organization of labor on New World plantations', in S. Innes (ed.) *Work and Labor in Early America*, Chapel Hill, North Carolina: University of North Carolina Press.

Morison, S. E. (1974) *The European Discovery of America: The Southern Voyages, 1492–1616*, New York: Oxford University Press.

Mougne, C., MacLennan, R. and Atsana, S. (1982) 'Smoking, chewing and drinking in Ban Pong, Northern Thailand', *Social Science and Medicine* 16: 99–106.

Mui, H.-C. and L. H. (1984) *The Management of Monopoly*, Vancouver: University of British Columbia Press.

Mulhall, M. G. (1892) *The Dictionary of Statistics*, London: George Routledge & Sons.

Mullan, J. (1988) *Sentiment and Sociability: The Language of Feeling in the Eighteenth Century*, Oxford: Oxford University Press.

Muller, M. (1978) *Tobacco and the Third World: Tomorrow's Epidemic?*, London: War on Want.

Multhauf, R. (1954) 'Medical chemistry and the Paracelsians', *Bulletin of the History of Medicine* 28: 101–26.

Mundy, P. (1919) *The Travels of Peter Mundy in Europe and Asia 1608–1667* volume III Travels in England, India, China, etc. part II 1638, London: Hakluyt Society.

Munger, R. S. (1949) 'Guaiacum, the holy wood from the New World', *Journal of the History of Medicine* 4: 196–229.

Nardi, J.-B. (1985) *A história do fumo brasileiro*, Rio de Janeiro: Abifumo.

—— (1986) 'O estanco real do tabaco', *Historia* 94: 14–25.

Nash, R. C. (1982) 'The English and Scottish tobacco trade in the seventeenth and eighteenth centuries: legal and illegal trade', *Economic History Review* 35: 354–72.

Neville, W. (1957) 'England's tobacco trade in the reign of Charles I', *The Virginia Magazine of History and Biography* 65: 403–49.

Newman, R. (1991) 'The opium smoker in Chinese history', unpublished paper.

Newton, A. P. (1914) *The Colonizing Activities of the English Puritans*, New Haven, Connecticut: Yale University Press.

Nourrisson, D. (1988) 'Tabagisme et antitabagisme en France au XIXe siècle', *Histoire, économie et société* 7: 535–47.

O'Brien, J. T. (1978) 'Factory, church, and the community: blacks in antebellum Richmond', *Journal of Southern History* 44: 509–36.

Olson, A. G. (1983) 'The Virginia merchants of London: a study in eighteenth-century interest-group politics', *William and Mary Quarterly* 40: 363–88.

Orellana, S. L. (1987) *Indian Medicine in Highland Guatemala*, Albuquerque, New Mexico: University of New Mexico Press.

Ortiz, F. (1947) *Cuban Counterpoint*, New York: Knopf.

Ortiz de Montellano, B. R. (1989) 'Mesoamerican religious tradition and medicine', in L. E. Sullivan (ed.) *Healing and Restoring: Health and Medicine in the World's Religious Traditions*, New York: Macmillan Publishing Company.

—— (1990) *Aztec Medicine, Health, and Nutrition*, New Brunswick, New Jersey: Rutgers University Press.

Oswald, A. (1960) 'The archaeology and economic history of English clay tobacco pipes', *The Journal of the British Archaeological Association* 3rd series 23: 40–102.

—— (1978) 'New light on some 18th-century pipemakers of London', *London and Middlesex Archaeological Society Special Papers* 2.

Ozanne, P. (1969) 'The diffusion of smoking in West Africa', *Odu* 2: 29–42.

Pagan, J. R. (1979) 'Growth of the tobacco trade between London and Virginia 1614–40', *Guildhall Studies in London History* 3: 248–62.

—— (1982) 'Dutch maritime and commercial activity in mid-seventeenth-century Virginia', *The Virginia Magazine of History and Biography* 90: 485–501.

Pagel, W. (1982) *Paracelsus: An Introduction to Philosophical Medicine in the Era of the Renaissance* second edition, Basel and New York: S. Krager.

Palmer, R. (1977) 'The agricultural history of Rhodesia', in R. Palmer and N. Parsons (eds) *The Roots of Rural Poverty in Central and Southern Africa*. London: Heinemann.

Paper, J. (1988) *Offering Smoke: The Sacred Pipe and Native American Religion*, Moscow, Idaho: The University of Idaho Press.

Parker, G. (1974) 'The emergence of modern finance in Europe, 1500–1730', in C. M. Cipolla (ed.) *The Fontana Economic History of Europe* volume 2, Glasgow: Collins.

—— (1976) 'The "Military Revolution", 1560–1660 – a myth?', *Journal of Modern History* 48: 195–214.

Parkerson, P. T. (1983) 'The Inca monopoly: fact or fiction?', *Proceedings of the American Philosophical Society* 127: 107–23.

Patterson, J. T. (1987) *The Dread Disease: Cancer and Modern American Culture*, Cambridge, Massachusetts: Harvard University Press.

Paulli, S. (1746) *A Treatise on Tobacco, Tea, Coffee and Chocolate* English translation of 1665 original, London.

Perez Vidal, J. (1959) *España en la historia del tabaco*, Madrid: Consejo superior de Investigaciones Cientificas. Centro de Estudios de Etnologia Peninsular.

Perrot, P. (1982) 'Quand le tabac conquit la France', *Histoire* 46: 98–104.

Peter, J.-P. (1967) 'Malades et maladies à la fin du XVIIIe siècle', *Annales: ESC* 22: 711–51.

Petitjean Roget, J. (1980) *La Société d'habitation à la Martinique: un demi-siècle de formation*, Lille: Atelier Réproduction des thèses.

Peto, R. and Lopez, A. D. (1990) 'Worldwide mortality from current smoking patterns', in B. Durston and K. Jamrozik (eds) *Tobacco and Health 1990: The Global War. Proceedings of the Seventh World Conference on Tobacco and Health, Perth, Western Australia*, Perth, Western Australia: Health Department of Western Australia.

Peto, R., Lopez, A. D., Boreham, J., Thun, M. and Heath, C. Jr. (1992) 'Mortality from tobacco in developed countries: indirect estimation from national vital statistics', *The Lancet* 339 (23 May): 1,268–78.

Philaretes (1602) *Work for Chimny-sweepers; or, A Warning for Tobacconists* reprinted 1936 by Shakespeare Association, Oxford: Oxford University Press.

Philips, J. E. (1983) 'African smoking and pipes', *Journal of African History* 24: 303–19.

Pike, R. (1972) *Aristocrats and Traders: Sevillian Society in the Sixteenth Century*, Ithaca, New York: Cornell University Press.

PIEDA Plc (1992) *The Tobacco Industry in the European Community, 1990*, Edinburgh.

Pinto, V. N. (1979) *O ouro brasileiro e o comércio anglo-portugês*, São Paolo: Companhia editora nacional.

Platt, H. (1684) *Sundry New and Artificial Remedies against Famine* (London 1596), in *Collectanea Chymica*, London.

Porter, G. and Livesay, H. C. (1971) *Merchants and Manufacturers*, Baltimore, Maryland: The Johns Hopkins University Press.

Porter, P. G. (1969) 'Origins of the American Tobacco Company', *Business History Review* 43: 59–76.

—— (1971) 'Advertising in the early cigarette industry: W. Duke, Sons and Company of Durham', *The North Carolina Historical Review* 48: 31–43.

Price, J. M. (1954a) 'The rise of Glasgow in the Chesapeake tobacco trade, 1707–1775', *William and Mary Quarterly* 11: 179–99.

—— (1954b) 'The tobacco trade and the treasury 1685–1733: British mercantilism in its fiscal aspects', unpublished Ph. D. dissertation, Harvard University.

—— (1956) 'The beginnings of tobacco manufacture in Virginia', *The Virginia Magazine of History and Biography* 64: 3–29.

—— (1961) *The Tobacco Adventure to Russia: Enterprise, Politics, and Diplomacy in the Quest for a Northern Market for English Colonial Tobacco, 1676–1722*, Philadelphia: American Philosophical Society, *Transactions* 51.

—— (1964) 'The economic growth of the Chesapeake and the European market, 1697–1775', *Journal of Economic History* 24: 496–511.

—— (1973) *France and the Chesapeake*, Ann Arbor, Michigan: University of Michigan Press.

—— (1980) *Capital and Credit in British Overseas Trade: The View from the Chesapeake, 1700–1776*, Cambridge, Massachusetts: Harvard University Press.

—— (1983) 'The excise affair revisited: the administrative and colonial dimensions of a parliamentary crisis', in S. B. Baxter (ed.) *England's Rise to Greatness 1660–1763*, Berkeley, California: University of California Press.

—— (1984a) 'Glasgow, the tobacco trade, and the Scottish customs, 1707–1730', *Scottish Historical Review* 63: 1–36.

—— (1984b) 'The transatlantic economy', in J. R. Pole and J. P. Greene (eds) *Colonial British America*, Baltimore, Maryland: The Johns Hopkins University Press.

—— (1986a) 'The last phase of the Virginia–London consignment trade: James Buchanan and Company, 1756–1768', *William and Mary Quarterly* 43: 64–98.

—— (1986b) 'Sheffield v. Starke: institutional experimentation in the London–Maryland trade c. 1696–1706', *Business History* 28, 2: 19–39.

—— (1988) 'Reflections on the economy of Revolutionary America', in R. Hoffman *et al.* (eds) *The Economy of Early America*, Charlottesville, Virginia: University Press of Virginia.

—— (1989) 'What did merchants do? Reflections on British overseas trade, 1660–1790', *Journal of Economic History* 49: 267–84.

Price, J. M. and Clemens, P. G. E. (1987) 'A revolution of scale in overseas trade: British firms in the Chesapeake trade, 1675–1775', *Journal of Economic History* 47: 1–43.

Puckrein, G. A. (1984) *Little England: Plantation Society and Anglo-Barbadian Politics, 1627–1700*, New York: New York University Press.

Pugh, C. (1981) 'Landmarks in the tobacco program', in W. R. Finger (ed.) *The Tobacco Industry in Transition*, Lexington, Massachusetts: Lexington Books.

Purchas, S. (1906) *Purchas His Pilgrims* volume XVI, Glasgow: James MacLehouse & Sons.

Quinn, D. B. (1974) 'James I and the beginnings of empire in America', *Journal of Imperial and Commonwealth History* 2: 135–52.

—— (1979) *New American World* volume III English Plans for North America. The Roanoke Voyages. New England Ventures, London: Macmillan.

Rainbolt, J. C. (1969) 'The absence of towns in seventeenth century Virginia', *Journal of Southern History* 35: 343–60.

Ramsay, G. D. (1952) 'The smugglers' trade: a neglected aspect of English commercial development', *Transactions of the Royal Historical Society* 5th series 2: 131–57.

Ransom, R. L. and Sutch, R. (1977) *One Kind of Freedom*, Cambridge: Cambridge University Press.

Ratekin, M. (1954) 'The early sugar industry in Española', *Hispanic American Historical Review* 34: 1–19.

Redi, F. (1671) *Esperienze intorno a diverse cose naturali*, Florence.

Reid, A. (1985) 'From betel-chewing to tobacco-smoking in Indonesia', *Journal of Asian Studies* 44: 529–47.

—— (1988) *Southeast Asia in the Age of Commerce 1450–1680*, New Haven, Connecticut: Yale University Press.

Ricklefs, M. C. (1981) *A History of Modern Indonesia*, London: Macmillan.

Risse, G. B. (1984) 'Transcending cultural barriers: the European reception of medicinal plants from the Americas', in W.-H. Hein (ed.) *Botanical Drugs of the Americas in the Old and New Worlds*, Stuttgart: Wissenschaftliche Verlagsgesellschaft MBH.

—— (1987) 'Medicine in New Spain', in R. L. Numbers (ed.) *Medicine in the New World: New Spain, New France and New England*, Knoxville, Tennessee: University of Tennessee Press.

Ritzenthaler, R. E. (1978) 'Southwestern Chippewa', in B. G. Trigger (ed.) *Handbook of North American Indians* volume 15, Northeast, Washington, DC: Smithsonian Institution.

Rive, A. (1926) 'The consumption of tobacco since 1600', *Economic History* 1: 57–75.

—— (1929) 'A short history of tobacco smuggling', *Economic History* 1: 554–69.

Rivero Muñiz, J. (1964) *Tabaco: su historia en Cuba*, Havana: Instituto de Historia.

RJR (1987) *Annual Report*, 440.

Robert, J. C. (1938) *The Tobacco Kingdom*, Durham, North Carolina: Duke University Press.

—— (1952) *The Story of Tobacco in America*, New York: Knopf.

Roberts, B. W. C. and Knapp, R. F. (1992) 'Paving the way for the Tobacco Trust: from hand rolling to mechanized cigarette production by W. Duke, Sons and Company', *The North Carolina Historical Review* 69: 257–81.

Roberts, R. S. (1965) 'The early history of the import of drugs into Britain', in F. L. Poynter (ed.) *The Evolution of Pharmacy in Britain*, London: Pitman Medical.

Robicsek, F. (1978) *The Smoking Gods: Tobacco in Maya Art, History and Religion*, Norman, Oklahoma: University of Oklahoma Press.

Roessingh, H. K. (1976) *Inlandse Tabak, Expansie en Contratie van een Handelsgewas in de 17de en 18de Eeuw*, Wagenigen: AAG Bijdragen 20.

—— (1978) 'Tobacco growing in Holland in the seventeenth and eighteenth centuries: a case study of the innovative spirit of Dutch peasants', *Acta Historiae Neerlandicae* 11: 18–54.

Rogoziński, J. (1990) *Smokeless Tobacco in the Western World, 1550–1950*, New York: Praeger.

Rosenblatt, M. B. (1964) 'Lung cancer in the 19th century', *Bulletin of the History of Medicine* 38: 395–425.

Rosenblatt, S. M. (1962) 'The significance of credit in the tobacco consignment trade: a study of John Norton and Sons, 1768–1775', *William and Mary Quarterly* 19: 383–99.

—— (1968) 'Introduction', in F. N. Mason (ed.) *John Norton and Sons Merchants of London and Virginia*, Newton Abbot: David & Charles.

Rothmans International (1991) *Annual Report and Accounts, 1991*.

Royal College of Physicians (1962) *Smoking and Health*, London: Pitman Medical.

Russell, H. S. (1980) *Indian New England before the Mayflower*, Hanover, New Hampshire: University Press of New England.

Russell, M. A. H. (1987) 'Nicotine intake and its regulation by smokers', in W. R. Martin, G. R. Van Loon, E. T. Iwamoto and L. Davis (eds) *Tobacco Smoking and Nicotine: A Neurobiological Approach*, New York: Plenum Press.

Russell-Wood, A. J. R. (1987) 'The gold cycle, c. 1690–1750', in L. Bethell (ed.) *Colonial Brazil*, Cambridge: Cambridge University Press.

Rutter, J. A. and Davey, P. J. (1980) 'Clay pipes from Chester', in P. Davey (ed.), *The Archaeology of the Clay Tobacco Pipe. III: Britain: the North and West*, BAR, British Series, 78.

Ryan, M. (1981) 'Assimilating New Worlds in the sixteenth and seventeenth centuries', *Comparative Studies in Society and History* 23: 519–38.

Sahagún, Fray B. de (1950–69) *The Florentine Codex: General History of Things in New Spain*, Santa Fe, New Mexico: School of American Research and the University of Utah.

Sahlins, M. (1988) 'Cosmologies of capitalism: the trans-pacific sector of "the world-system" ', *Proceedings of the British Academy* 74: 1–51.

Saignes, T. (1988) 'Capoche, Potosí y la coca: el consumo popular de estimulantes en el siglo XVII', *Revista de Indiás* 48: 207–35.

Salaman, R. N. (1949) *The History and Social Influence of the Potato*, Cambridge: Cambridge University Press.

Samson, O. W. (1960) 'The geography of pipe smoking', *Geographical Magazine*, 33: 217–30.

Sangar, S. P. (1981) 'Intoxicants in Mughal India', *Indian Journal of History of Science* 16: 202–14.

Sapolsky, H. M. (1980) 'The political obstacles to the control of cigarette smoking in the United States', *Journal of Health Politics, Policy and Law* 5: 277–90.

Sass, L. J. (1981) 'Religion, medicine, politics and spices', *Appetite* 2: 7–13.

Satow, E. M. (1878) 'The introduction of tobacco into Japan', *Transactions of the Asiatic Society of Japan* 6: 68–86.

Sauer, J. D. (1976) 'Changing perception and exploitation of New World plants in Europe, 1492–1800', in F. Chiapelli (ed.) *First Images of America: The Impact of the New World on the Old*, Berkeley, California: University of California Press.

Schama, S. (1987) *The Embarrassment of Riches*, London: Collins.

Schleiffer, H. (1979) *Narcotic Plants of the Old World*, Monticello, New York: Lubrecht & Cramer.

Schmitz, R. (1985) 'Friedrich Wilhelm Sertürner and the discovery of morphine', *Pharmacy in History* 27: 61–74.

Schnakenbourg, C. (1968) 'Note sur les origines de l'industrie sucrière en Guadeloupe au XVIIe siècle, 1640–1670', *Revue française d'histoire d'outre-mer* 55: 267–315.

—— (1980) 'Le "terrier" de 1671 et le partage de la terre en Guadeloupe au XVIIe siècle', *Revue française d'histoire d'outre-mer* 67: 37–54.

Schudson, M. (1985) *Advertising, The Uneasy Persuasion*, New York: Basic Books.

Schultes, R. E. (1972) 'Hallucinogens in the Western Hemisphere', in P. T. Furst (ed.) *Flesh of the Gods*, London: George, Allen & Unwin.

—— (1977) 'Mexico and Colombia: two major centres of the aboriginal use of hallucinogens', *Journal of Psychedelic Drugs* 9: 173–6.

Schultes, R. C. and Hofmann, A. (1979) *Plants of the Gods*, New York: McGraw Hill.

Schumpeter, E. B. (1960) *English Overseas Trade Statistics, 1697–1808*, Oxford: Oxford University Press.

Schwartz, S. B. (1978) 'Indian labor and New World plantations: European demands and Indian responses in Northeastern Brazil', *American Historical Review* 83: 43–79.

—— (1985) *Sugar Plantations in the Foundation of Brazilian Society*, Cambridge: Cambridge University Press.

—— (1987) 'Plantations and peripheries, c. 1580–c. 1750', in L. Bethell (ed.) *Colonial Brazil*, Cambridge: Cambridge University Press.

Scott, A. and C. (1966) *Tobacco and the Collector*, London: Max Parrish.

Scrimgeour, E. M. (1985) 'Cigarette smoking in Papua New Guinea: a model for keeping a people down', *New York State Journal of Medicine* 85: 420–1.

Seaton, A. V. (1986) 'Cope's and the promotion of tobacco in Victorian England', *Journal of Advertising History* 9, 2: 5–26.

Shammas, C. (1978) 'English commercial development and American colonization 1560–1620', in K. R. Andrews, N. P. Canny and P. E. H. Hair (eds) *The Westward Enterprise: English Activities in Ireland, the Atlantic and America, 1460–1650*, Liverpool: Liverpool University Press.

—— (1990) *The Pre-Industrial Consumer in England and America*, Oxford: Oxford University Press.

Shepherd, J. F. and Walton, G. M. (1972) *Shipping, Maritime Trade and the*

Economic Development of Colonial North America, Cambridge: Cambridge University Press.

Shirley, J. W. (1983) *Thomas Harriot: A Biography*, Oxford: Oxford University Press.

Shlomowitz, R. (1984) 'Plantations and smallholdings: comparative perspectives from the world cotton and sugar cane economies, 1865–1939', *Agricultural History* 58: 1–16.

Siegel, F. F. (1987) *The Roots of Southern Distinctiveness: Tobacco and Society in Danville, Virginia, 1780–1865*, Chapel Hill, North Carolina: University of North Carolina Press.

Siegel, R. K. , Collings, P. R. and Diaz, J. L. (1977) 'On the use of *tagetes lucida* and *nicotiana rustica* as a Huichol smoking mixture: the Aztec "Yahutli" with suggestive hallucinogenic effects', *Economic Botany* 31: 16–23.

Silver, S. (1978) 'Chimariko', in R. F. Heizer (ed.) *Handbook of North American Indians* volume 8, California, Washington, DC: Smithsonian Institution.

Silvette, H. , Larson, P. S. and Haag, H. B. (1958) 'Medical uses of tobacco, past and present', *Virginia Medical Monthly* 85: 472–84.

Slack, P. (1979) 'Mirrors of health and treasures of poor men: the uses of the vernacular medical literature of Tudor England', in C. Webster (ed.) *Health, Medicine and Mortality in the Sixteenth Century*, Cambridge: Cambridge University Press.

Slicher van Bath, B. H. (1986) 'The absence of white contract labour in Spanish America during the colonial period', in P. C. Emmer (ed.) *Colonialism and Migration*, Dordrecht: Martinus Nijhoff.

Smith, C. R. (1978) 'Tubutulabal', in R. F. Heizer (ed.) *Handbook of North American Indians* volume 8, California, Washington, DC: Smithsonian Institution.

Smith, J. G. (1979) *The Origins and Early Development of the Heavy Chemical Industry in France*, Oxford: Oxford University Press.

Smith, W. D. (1984) 'The function of commercial centers in the modernization of European capitalism: Amsterdam as an information exchange in the seventeenth century', *Journal of Economic History* 44: 985–1,005.

—— (1991) 'Complications of the commonplace: respectability, imperialism, and the origins of the custom of sugaring tea', unpublished paper.

Society (1775) *Reports of the Society Instituted in the Year 1774 for the Recovery of Persons Apparently Drowned*, London.

Soltow, J. H. (1959) 'Scottish traders in Virginia, 1750–1775', *Economic History Review* 12: 85–98.

Spence, J. (1975) 'Opium smoking in Ch'ing China', in F. Wakeman Jr. and C. Grant (eds) *Conflict and Control in Late Imperial China*, Berkeley, California: University of California Press.

Springer, J. W. (1981) 'An ethnohistoric study of the smoking complex in eastern North America', *Ethnohistory* 28: 217–35.

Stannard, J. (1966) 'Dioscorides and Renaissance materia medica', *Analecta Medica Historica* 1: 1–21.

Stebbins, K. R. (1990) 'Transnational tobacco companies and health in underdeveloped countries: recommendations for avoiding a smoking epidemic', *Social Science and Medicine* 30: 227–35.

Steensgaard, N. (1974) *The Asian Trade Revolution of the Seventeenth Century*, Chicago: University of Chicago Press.

—— (1985) 'The return cargoes of the Carreira in the 16th and early 17th century', in T. R. de Souza *Indo-Portuguese History: Old Issues, New Questions*, New Delhi: Concept.

Stein, R. L. (1988) *The French Sugar Business in the Eighteenth Century*, Baton Rouge, Louisiana: Louisiana State University Press.

Steinfeld, J. L. (1985) 'Smoking and lung cancer', *Journal of the American Medical Association* 253: 2,295–7.

Stella, B. (1669) *Il tabacco*, Rome.

Stewart, G. G. (1967) 'A history of the medicinal use of tobacco 1492–1860', *Medical History* 11: 228–68.

Stewart, O. C. (1987) *Peyote Religion: A History*, Norman, Oklahoma: University of Oklahoma Press.

Strachey, W. (1953) *The Historie of Travell into Virginia Britannia*, London: Hakluyt Society.

Stubbs, J. (1985) *Tobacco on the Periphery*, Cambridge: Cambridge University Press.

Talbot, C. H. (1976) 'America and the European drug trade', in F. Chiapelli (ed.) *First Images of America: The Impact of the New World on the Old*, Berkeley, California: University of California Press.

Taylor, P. (1984) *Smoke Ring: The Politics of Tobacco*, London: Bodley Head.

Tedeschi, J. (1987) 'Literary piracy in seventeenth-century Florence: Giovanni Battista Neri's *De iudice S. inquisitionis opusculum*', *Huntington Library Quarterly* 50: 107–18.

Teigen, P. M. (1987) 'Taste and quality in 15th- and 16th-century Galenic pharmacology', *Pharmacy in History* 29: 60–8.

Tennant, R. B. (1950) *The American Cigarette Industry*, New Haven, Connecticut: Yale University Press.

—— (1971) 'The cigarette industry', in W. Adams (ed.) *The Structure of American Industry* 4th edition, New York: The Macmillan Company.

Thirsk, J. (1974) 'New crops and their diffusion: tobacco-growing in seventeenth-century England', in C. W. Chalkin and M. A. Havinden (eds) *Rural Change and Urban Growth 1500–1800*, London: Longman.

Thomas, J. A. (1928) *A Pioneer Tobacco Merchant in the Orient*, Durham, North Carolina: Duke University Press.

Thompson, J. E. (1970) *Maya History and Religion*, Norman, Oklahoma: University of Oklahoma Press.

Thornton, A. P. (1921–2) 'Lord Sackville's papers respecting Virginia 1613–31', *American Historical Review* 27: 493–538.

Tiedemann, F. (1854) *Geschichte des Tabaks und andere ähnlicher Genußmittel*, Frankfurt-A-Main: Brönner.

Tilley, N. M. (1948) *The Bright Tobacco Industry, 1860–1929*, Chapel Hill, North Carolina: University of North Carolina Press.

—— (1985) *The R. J. Reynolds Tobacco Company*, Chapel Hill, North Carolina: University of North Carolina Press.

Tooker, E. (ed.) (1979) *Native North American Spirituality of the Eastern Woodlands*, London: SPCK.

Trigger, B. G. (1986) *Natives and Newcomers*, Manchester: Manchester University Press.

—— (1991a) 'Distinguished lecture in archeology: Constraint and freedom – a new synthesis for archeological explanation', *American Anthropologist* 93: 551–69.

—— (1991b) 'Early North American responses to European contact: romantic versus rationalist interpretations', *The Journal of American History* 77: 1,195–215.

Trouillot, M.-R. (1981) 'Peripheral vibrations: the case of Saint Domingue's coffee revolution', in R. Rubinson (ed.) *Dynamics of World Development*, Beverly Hills, California: Sage.

— (1982) 'Motion in the system: coffee, color and slavery in eighteenth-century Saint-Domingue', *Review* 5: 331–88.

Troyer, R. J. (1984) 'From prohibition to regulation – comparing two anti-smoking movements', *Research in Social Movements, Conflict and Change* 7: 53–69.

Tucker, D. (1982) *Tobacco: An International Perspective*, London: Euromonitor.

Turnbaugh, W. A. (1975) 'Tobacco, pipes, smoking and rituals among the Indians of the Northeast', *Quarterly Bulletin of the Archaeological Society of Virginia* 30: 59–71.

— (1980) 'Native North American smoking pipes', *Archaeology* 33: 15–22.

Turner, N. J. and Taylor, R. L. (1972) 'A review of the Northwest Coast tobacco mystery', *Syesis* 5: 249–57.

Tyler, J. W. (1978) 'Foster Cunliffe and Sons: Liverpool merchants in the Maryland tobacco trade, 1738–1765', *Maryland Historical Magazine* 73: 246–79.

United Nations (1983–4) *Statistical Yearbook 1983/4*, New York: United Nations.

United Nations Conference on Trade and Development (1978) *Marketing and Distribution of Tobacco*, New York: United Nations.

United States Bureau of the Census (1975) *Historical Statistics of the United States, Colonial Times to 1970*, Washington, DC: GPO.

United States Department of Agriculture (1913) *Yearbook of the Department of Agriculture*, Washington, DC: GPO.

United States Department of Commerce (1915) *Tobacco Trade of the World*, Bureau of Foreign and Domestic Commerce, Special Consular Reports, 68, Washington, DC: GPO.

United States Department of Health, Education and Welfare (1964) *Smoking and Health*. Report of the Advisory Committee to the Surgeon General of the Public Health Service, Washington, DC: GPO.

United States Department of Health and Human Services (1980) *The Health Consequences of Smoking for Women*. A Report of the Surgeon General. Public Health Service. Office on Smoking and Health, Washington, DC: GPO.

— (1986) *The Health Consequences of Involuntary Smoking*. Public Health Service. Centers for Disease Control. Center for Health Promotion and Education. Office on Smoking and Health, Rockville, Maryland.

— (1988) *The Health Consequences of Smoking. Nicotine Addiction*. A Report of the Surgeon General. Public Health Service. Centers for Disease Control. Center for Health, Promotion and Education. Office on Smoking and Health, Publication no. (CDC) 88–8406.

— (1989) *Reducing the Health Consequences of Smoking. 25 Years of Progress*. A Report of the Surgeon General. Public Health Service. Centers for Disease Control. Center for Chronic Disease Prevention and Health Promotion. Office on Smoking and Health, Publication no. (CDC) 89–8411.

— (1992) *Smoking and Health in the Americas*. Public Health Service. Centers for Disease Control. National Center for Chronic Disease Prevention and Health Promotion. Office on Smoking and Health, Publication no. (CDC) 92–8419, Atlanta, Georgia.

Vallance, P. J. , Anderson, H. R. and Alpers, M. P. (1987) 'Smoking habits in a rural community in the highlands of Papua New Guinea in 1970 and 1984', *Papua New Guinea Medical Journal* 30: 277–80.

Van Dantzig, A. (1980) *Les hollandais sur la côte de Guinée à l'époque de l'essor de l'Ashanti et du Dahomey 1680–1740*, Paris: Société Française d'Histoire d'Outre-mer.

van Peima, B. (1690) *Tabacologia*, The Hague.

Verger, P. (1964) 'Rôle joué par le tabac de Bahia dans la traité des esclaves au Golfe du Bénin', *Cahiers d'études Africaines* 4: 349–69.

—— (1976) *Trade Relations between the Bight of Benin and Bahia from the 17th to the 19th Century*, Ibadan: Ibadan University Press.

Vigié, M. and M. (1989) *L'Herbe à Nicot*, Paris: Fayard.

Vogel, V. J. (1970) *American Indian Medicine*, Norman, Oklahoma: University of Oklahoma Press.

von Gernet, A. (1982) 'Interpreting intrasite spatial distribution of artifacts: the Draper Site pipe fragments', *Man in the Northeast* 23: 49–60.

—— (1988) 'The transculturation of the Amerindian pipe/tobacco/smoking complex and its impact on the intellectual boundaries between "savagery" and "civilization", 1535–1935', unpublished Ph. D. dissertation, McGill University.

—— (1992) 'Hallucinogens and the origins of the Iroquoian pipe/tobacco/smoking complex', in Hayes, C. F. III (ed.) *Proceedings of the 1989 Smoking Pipe Conference*. Research Records no. 22 of the Rochester Museum and Science Service, Rochester, New York.

von Gernet, A. and Timmins, P. (1987) 'Pipes and parakeets: constructing meaning in an Early Iroquoian context', in I. Hodder (ed.) *Archaeology as Long-Term History*, Cambridge: Cambridge University Press.

Wafer, L. (1934) *A New Voyage and Description of the Isthmus of America*. Oxford: Hakluyt Society.

Wagley, C. (1977) *Welcome of Tears: The Tapirapé Indians of Central Brazil*, Oxford: Oxford University Press.

Wagner, P. (1990) *Eros Revived: Erotica of the Enlightenment in England and America*, London: Paladin.

Wake, C. H. H. (1979) 'The changing patterns of Europe's pepper and spice imports, ca. 1400–1700', *Journal of European Economic History* 8: 361–403.

Wald, N. and Nicolaides-Bouman, A. (1991) *UK Smoking Statistics* 2nd edition, Oxford: Oxford University Press.

Waldron, I. (1991) 'Patterns and causes of gender differences in smoking', *Social Science and Medicine* 32: 989–1005.

Waldron, I., Bratelli, G., Carriker, L., Sung, W.-C., Vogeli, C. and Waldman, E. (1988) 'Gender differences in tobacco use in Africa, Asia, the Pacific, and Latin America', *Social Science and Medicine* 27: 1,269–75.

Walker, I. C. (1971) *The Bristol Clay Tobacco Pipe Industry*, Bristol: City Museum.

—— (1983) 'Nineteenth-century clay tobacco-pipes in Canada', in P. Davey (ed.) *The Archaeology of the Clay Tobacco Pipe* VIII, America, BAR International Series, 175.

Walker, R. B. (1980) 'Medical aspects of tobacco smoking and the anti-tobacco movement in Britain in the nineteenth century', *Medical History* 24: 391–402.

Walsh, L. S. (1989) 'Plantation management in the Chesapeake, 1620–1820', *Journal of Economic History* 49: 393–406.

Warner, K. (1986) *Selling Smoke*, Washington, DC: American Public Health Association.

—— (1990) 'Tobacco taxation and economic effects of declining tobacco consumption', in B. Durston and K. Jamrozik (eds) *Tobacco and Health 1990: The Global War. Proceedings of the Seventh World Conference on Tobacco and Health, Perth, Western Australia*, Perth, Western Australia: Health Department of Western Australia.

—— (1991) 'Tobacco industry scientific advisors: serving society or selling cigarettes?', *American Journal of Public Health* 81: 839–42.

Warren, J. C. (1919) 'The "pulmotor" of the eighteenth century', *Annals of Medical History* 2: 14–20.

Watkins, G. (1979) 'Hull pipes: a typology', in P. Davey (ed.) *The Archaeology of the Clay Pipe. Britain: the Midlands and eastern England*, British Archaeological Report, British series, 63.

Watts, D. (1987) *The West Indies: Patterns of Development, Culture and Environmental Change since 1492*, Cambridge: Cambridge University Press.

Webster, C. H. (1979) 'Alchemical and Paracelsian medicine', in C. H. Webster (ed.) *Health, Medicine and Mortality in the Sixteenth Century*, Cambridge: Cambridge University Press.

Wells, P. K. (1979) 'The pipemakers of Lincolnshire', in P. Davey (ed.) *The Archaeology of the Clay Pipe. Britain: the Midlands and Eastern England*, British Archaeological Report, British series, 63.

Wells, R. V. (1975) *The Population of the British Colonies in America before 1776*, Princeton, New Jersey: Princeton University Press.

West, R. and Grunberg, N. E. (1991) 'Implications of tobacco use as an addiction', *British Journal of Addiction* 86: 485–8.

Wetherell, C. (1984) ' "Boom and bust" in the colonial Chesapeake economy', *Journal of Interdisciplinary History* 50: 185–210.

Wilbert, J. (1972) 'Tobacco among the Warao of Venezuela', in P. T. Furst (ed.) *Flesh of the Gods*, London: George Allen & Unwin.

—— (1979) 'Magico-religious use of tobacco among South American Indians', in D. L. Browman and R. A. Schwarz (eds) *Spirits, Shamans and Stars*, The Hague: Mouton.

—— (1987) *Tobacco and Shamanism in South America*, New Haven, Connecticut: Yale University Press.

—— (1991) 'Does pharmacology corroborate the nicotine therapy and practices of South American shamanism?', *Journal of Ethnopharmacology* 32: 179–86.

Wilkinson, J. (1986) *Tobacco*, Harmondsworth: Penguin Books.

Williams, N. (1957) 'England's tobacco trade in the reign of Charles I', *The Virginia Magazine of History and Biography* 65: 403–49.

Winkler, J. K. (1942) *Tobacco Tycoon: The Story of James Buchanan Duke*, New York: Random House.

Wrigley, E. A. and Schofield, R. S. (1981) *The Population History of England, 1541–1871*, London: Edward Arnold.

Wroth, L. C. (1954) 'Tobacco or codfish: Lord Baltimore makes his choice', *Bulletin of the New York Public Library* 58: 523–34.

Wynder, E. L. and Graham, E. A. (1950) 'Tobacco as a possible etiologic factor in bronchigenic carcinoma. A study of six hundred and eighty-four proved cases', *Journal of the American Medical Association* 143: 329–36.

Zahedieh, N. (1986) 'Trade, plunder and economic development in early English Jamaica, 1655–89', *Economic History Review* 39: 205–22.

Zenk, H. B. (1990) 'Siuslawans and Coosans', in W. Suttles (ed.) *Handbook of North American Indians*, Northeast Coast, Washington, DC: Smithsonian Institution.

INDEX

Adshead, S. A. M. 15
advertising 103; *see also* cigarettes
Africa, tobacco cultivation in 213–14
agaric family 21
Agricultural Adjustment Act 198
Agricultural Recovery Program 197
Alarcón, Ruiz de 28–9
alkaloid research 115–16
allotment system 200
Amazon, English and Irish settlements 138
American Tobacco Company 103, 232; dissolution of 233
Amerindian cosmology 22; *see also* tobacco in Amerindian life
Amersfoort 142
Amsterdam, tobacco trade in 154
Anglo-Spanish truce 133
anti-cigarette campaign 118–19
anti-tobacco legislation, United States 119
anti-tobacco societies 117–18; in Britain 118; in France 117–19; in United States 118
Antonil, André 68, 187; on chewing tobacco 68; on snuff 84
Atlantic slave trade 162
Austria, tobacco monopoly 217; *see also* tobacco monopolies
ayahuasca 24

Bahia 162; curing of tobacco in 187; fallow system in 188; growing cycle of tobacco in 187–8; slave labour in 188–9; structure of landownership in 188–9; tobacco cultivation in 140, 163
Baillard, Edmé, on snuff 45, 74, 79
Barbados: labour in tobacco cultivation

on 180; quality of tobacco on 182; settlement of 139; tobacco cultivation on 139, 179–80; *see also* English West Indies
Bermuda: fallow system in 183–4; population of 183; settlement of 138; tobacco cultivation on 183–4; tobacco exports from 138; *see also* English West Indies
bidis 12, 95
Blackwell, W. T. and Company 100–1, 230
Bonsack, James 102, 231
Bonsack cigarette-making machine, output of 102, 209, 231, 236
Bontekoe, Cornelis 76
Brazil, tobacco cultivation in 203, 214–15
briar pipe 228
Bright tobacco 193–6, 198, 206, 209, 227–8, *see also* flue-cured tobacco; in China 212–13
British American Tobacco (BAT) 10, 234; in Africa 235; in Asia 235; in China 212–13, 235, 237
brus 96
Burley tobacco 202, 208–9, 227–8

calumet 33
Camel cigarettes, advertising of 104, 108–9; *see also* cigarettes
Cameo cigarettes 102
Camporesi, Piero 42
Cardenas, Juan de 44
Carr, Julian 100–1
Chang Chieh-pin 52–3
Chesapeake: ethnic distribution in 169,